PHYTOLITH ANALYSIS

An Archaeological and Geological Perspective

PHYTOLITH ANALYSIS

An Archaeological and Geological Perspective

Dolores R. Piperno

Department of Anthropology
Temple University
Philadelphia, Pennsylvania

ACADEMIC PRESS, INC.
Harcourt Brace Jovanovich, Publishers
San Diego New York Berkeley Boston
London Sydney Tokyo Toronto

ACADEMIC PRESS, INC.
1250 Sixth Avenue, San Diego, California 92101

United Kingdom Edition published by
ACADEMIC PRESS INC. (LONDON) LTD.
24–28 Oval Road, London NW1 7DX

Library of Congress Cataloging in Publication Data

Piperno, Dolores R.
Phytolith analysis.

Bibliography: p.
Includes index.
1. Archaeological geology. 2. Plant remains
(Archaeological) I. Title.
CC77.5.P56 1987 930.1 87-11353
ISBN 0–12–557175–5 (alk. paper)

PRINTED IN THE UNITED STATES OF AMERICA

87 88 89 90 9 8 7 6 5 4 3 2 1

Contents

3. PHYTOLITH MORPHOLOGY

4. FIELD TECHNIQUES AND RESEARCH DESIGN

5. LABORATORY TECHNIQUES

Preface

There exists a group of plant microfossils that exhibit all the attributes necessary to achieve legitimacy and prominence in paleoethnobotany and paleoecology: production in large numbers, durability in ancient sediments, and sufficient morphological specificity to allow identification of a wide range of taxa. Called phytoliths, these microfossils are particles of silica derived from the cells of living higher plants. Until recently, phytoliths have not received much attention in their own right as contributors of paleobotanical data. Their attributes and behavior in modern and fossil contexts have been insufficiently understood, and, as a result, the paleobotanical community has tended to dismiss them as unreliable indicators of past plant usage and environments. Currently, only a few archaeobotanists do phytolith analysis, and only a minority of field archaeologists collect soils and other materials for phytolith survey.

Phytolith Analysis: An Archaeological and Geological Perspective describes the significance and potential of phytolith analysis in archaeobotanical and geobotanical research, as demonstrated by studies carried out heretofore by phytolith specialists working in many regions of the world. It presents a summary of the methods and principles that have guided me in my own phytolith work. I explore the enormous value of phytolith applications in archaeology and show that aspects of phytolith science, such as production and morphological specificity, cited (mainly by those with a superficial level of understanding) as inscrutable and unyielding to refinement, can indeed be elucidated and are subject to control through careful, systematic study. In developing phytolith science into a mature discipline, we face a task no more severe than palynologists tackled when they first applied pollen studies to reconstruction of Quaternary paleoenvironments. The task will be difficult, challenging, and extremely rewarding.

The organization and content of the book reflect these goals. The first chapter gives a brief history of phytolith research. Chapter 2 provides a synthesis of phytolith production patterns in higher plants. Sometimes seen as random, undependable, and occurring mostly in monocotyledons, phytolith patterns are shown to be widespread and faithfully replicated in all kinds of higher plants and all their different structures. Chapter 3, "Phytolith Morphology," is the longest of the book because it deals with one of the most important and least developed aspects of phytolith science. I believe that readers will be as fascinated by the myriad distinctive shapes of phytoliths as I was when over a period of several years I observed different kinds of siliceous bodies virtually every time I extracted a set of modern specimens. Chapter 3 also shows how the shapes of phytoliths of a single species are often uniform, regardless of where in the world a plant may grow. Chapters 4 and 5 deal with field and laboratory techniques, respectively. They offer a current "how to" approach to research design and laboratory methods and guide the field archaeologist and phytolith specialist through the basic kinds of phytolith studies. Chapter 6 recapitulates some of the information presented in the first five chapters and presents the fundamental principles of the method and theory of paleoecological phytolith analysis. It also suggests areas in which future research may prove profitable and presents the first results of modern phytolith studies in tropical forests. Chapters 7 and 8 detail results of the work carried out thus far in archaeological and paleoecological reconstruction. The appendix shows that it is indeed possible to classify the myriad of phytolith shapes isolated from plants and to construct regional phytolith keys.

This book is intended for a number of audiences on different levels: those beginning in phytolith science, the experienced specialist, and archaeologists and geologists who are interested in applying phytolith studies to their own research. I hope that it will provoke discussion and the development of phytolith analysis as a more refined instrument of paleoecology. I also hope that it will push the archaeological profession to a new level of understanding and greater use of phytolith-analytical studies.

Acknowledgments

I would like to thank a number of people and institutions without whom this book would not have been written. From 1983 to 1985 I had the good fortune to hold a postdoctoral fellowship from the Smithsonian Tropical Research Institute (STRI) in the Republic of Panama. It was there that much of my phytolith research and writing was carried out, and the period was so full of beneficial intellectual encounters that it would be difficult to list them. The excellent staff and organization of STRI provided the utmost in good facilities and afforded me every possible courtesy. My association with STRI was an extremely profitable and enjoyable one. I would like especially to thank Richard Cooke, Olga Linares, and Martin Moynihan for much support and encouragement; Sylvia Churgin and the rest of the STRI library for filling numerous interlibrary loan requests; Stephen Hubbell, Robin Foster, and Sue Williams for their generous help with work in the Hubbell–Foster plot on Barro Colorado Island and use of their forest census data; Hugh Churchhill and Gordon McPherson, curators of the herbarium of the Missouri Botanical Garden housed at STRI, for permission to sample plants from the herbaria folders; and good and true friends Georgina de Alba and Myra Shulman, who read parts of the manuscript and remained confident during times of doubt that the book would be completed.

I thank Anthony Ranere, who offered both support and encouragement over the past seven years. Freddy Chen and William Dement of the Tropic Test Center in Panama took many of the scanning electron photographs for me and provided facilities for other microscope photography. Mireya D. Correa A. of the University of Panama gave permission to sample plants at the university herbarium. Figures 3.1, 3.2, 3.3, 3.4, 3.5, and 3.11 were kindly and ably drawn by Arcadio Romaniche, who became fascinated with phytolith morphology. The rest of the figures, including all the phytolith diagrams, were done by Doris Weiland in her usual high-quality style. At Temple University I had access to the fine facilities of the Laboratory of Anthropology, directed by Muriel Kirkpatrick, where many phytolith extractions were carried out. Major Goodman and John Doebley kindly provided many of the maize and teosinte varieties. Rita Piperno took time out of a well-earned retirement to faithfully and loyally type each version of the manuscript. She also demonstrated the vitality and intellectual vigor of senior citizens by learning in record time how to use a computer and word processor. To both of my parents, who instilled a sense of confidence and an awareness of where one has come from, I am grateful.

I thank the National Science Foundation, the Wenner-Gren Foundation for Anthropological Research, and the Smithsonian Institution for research grants that helped fund the research without which this book could not have been written. Figure 3.4 is redrawn from Miller-Rosen (1985) by permission of Arlene Miller-Rosen. Figure 6.3 is redrawn from Robinson (1979) by permission of the Center for Archaeological Research, the University of Texas at San Antonio. Information on the topographic contour intervals and coordinates of Figure 6.4 were provided by Stephen Hubbell. Table 7.4 is taken from Pearsall (1981b) by permission of Deborah Pearsall. Plate 20 was taken by and used by permission of Deborah Pearsall. Table A.1 of the Appendix is from Brown (1984) by permission of Academic Press. Table A.2 of the Appendix is from Brown (1986) by permission of Dwight Brown. Plates 94 and 95 are reproduced from Piperno (1985b) by permission of Academic Press. Plates 24, 26, 32, 34, and 35 are reproduced from Piperno (1984) by permission of the Society for American Archaeology. The following plates are reproduced from Piperno (1985a) by permission of Elsevier Science Publishers: 1, 3, 4, 7, 9, 12, 15, 16, 40, 44, 46–50, 52–63, 65–73, 82, and 83.

PHYTOLITH ANALYSIS

An Archaeological and Geological Perspective

1

History of Phytolith Research

Introduction

This volume addresses the paleoecological significance of a group of plant fossils that has been known and researched under the umbrella of various names. At one time or another called opal phytoliths, silica phytoliths, silica cells, plant opal, biogenic opal, or simply phytoliths, they are particles of hydrated silica formed in the cells of living plants that are liberated from the cells upon death and decay of the plants. Phytoliths with taxonomic significance are produced in great numbers by many families of higher plants and exhibit a remarkable durability in soils and sediments over very long periods of time. However, for many years phytoliths were not in the mainstream of any archaeological, geological, or ecological science, for reasons to be explored in this chapter.

It was during the 1970s that archaeologists first engaged in the application of a paleobotanical discipline that was in many respects fragmented, in some respects at a standstill, and thought by some botanists to be of little utility in paleoecological reconstruction. We arrive at the impression that pedologists and botanists who 20 years before had initially applied phytolith analysis to environmental research did not quite know where to go next: fundamental questions of production, taxonomy, and preservation had not been satisfactorily explored, and scientists never made problems studied with the technique sufficiently explicit, or even appropriate, in order to encourage the necessary basic research.

There was a curious lack of interest in phytoliths by botanists who were carrying out numerous pollen studies of Quaternary vegetational and climatic history, even as they sought ways to improve the accuracy of the paleobotanical record by eliminating biases introduced by limitations of pollen data. Existing projections of the taxonomic significance of phytoliths were pessimistic when few kinds of plants had been studied in any detail. Archaeologists then began to apply the technique at a time when there was a lack of appreciation of phytoliths by the paleobotanical community and many basic research issues remained unresolved.

This is the reverse of ordinary circumstances. Prehistorians are often criticized by biologists and other prehistorians for naively and incorrectly adopting approaches developed in biology without considering their basic assumptions and relevancy to anthropology. There have been complaints of uncritical borrowing from biology on the part of archaeologists for such approaches as sociobiology, and most recently, optimal foraging theory and its applications (Sih and Milton, 1985). In the case of phytolith research, it can be argued that it is the archaeobotanists who have begun to carry out the kind of research necessary to develop the technique into a sound paleoecological discipline. In so doing, they have adapted, not merely adopted, the approach to archaeological problems, enabling a better fit of the data to questions of past human behavior. The results thus far are extremely promising and are the topics of subsequent chapters. In order to understand the status of phytolith analysis as a paleoecological discipline when prehistorians first entered the modern scene in 1971, a brief summary of phytolith history and nonarchaeological applications is now presented.

The history of phytolith research can be conveniently divided into four periods:

1. The discovery and exploratory stage, from 1835 to *ca.* 1900, when phytoliths were initially observed in living plants, named, and isolated from some environmental contexts.
2. The botanical phase, from *ca.* 1900 to 1936, the center of which was Germany. Phytoliths were first widely recognized in Europe as being derived from plant tissue, and studies of production, morphology, and taxonomy began in earnest. The first archaeological applications of phytolith analysis were carried out.
3. The period of ecological research, from 1955 to *ca.* 1975, when applications by western soil scientists and botanists to several issues of paleobotanical importance were first initiated. Grass phytolith morphology was intensively examined, and studies were begun on siliceous bodies from deciduous and coniferous tree flora. Phytolith analysis was first applied to studies of North American vegetational history.
4. The modern period of archaeological phytolith research, beginning *ca.* 1971, when archaeobotanists began systematic investigations into phytolith production and morphology in modern plants and phytolith frequencies and distributions in archaeological and geological sediments.

The Discovery and Exploratory Period

In the early part of the nineteenth century pioneering researchers of the microscopic world were discovering smaller and smaller particles, and so phytoliths were first observed in living plants in 1835 by Struve, a German botanist. Paradoxically, this is one year before pollen grains were initially discovered in pre-Quaternary sediments (Faegri and Iverson, 1975). One hundred and fifty years would pass, sixty-five years longer than for palynology, before phytolith analysis achieved the status of an independent paleoecological discipline.

At first, phytoliths were studied as somewhat puzzling constituents of soils and windblown dust and not as inclusions of living plants. Shortly after the H.M.S. Beagle left England in 1831, the ship anchored off the Cape Verde Islands, several hundred miles west of the North African coast. Darwin (1909, p. 5) wrote,

> Generally, the atmosphere is hazy; and this is caused by the falling of impalpably fine dust, which was found to have slightly injured the astronomical instruments. . . . I collected a little packet of this brown-coloured fine dust, which appeared to have been filtered from the wind by the gauze of the vane at the masthead . . . Professor Ehrenberg finds that this dust consists in great part of the infusoria with siliceous shields, and of the siliceous tissue of plants. In five little packets which I sent him, he has ascertained no less than sixty-seven organic forms. The infusoria, with the exception of two marine species, are all inhabitants of fresh water.

Before phytoliths were being isolated from living plants, they were described in aerosols that had obviously been transported hundreds of miles from their source areas on land.

Ehrenberg, a German scientist, had already enjoyed an international reputation as a plant and animal microbiologist, having pioneered microbial work in Germany during the first third of the nineteenth century. He had observed similar kinds of siliceous bodies in soil samples sent to him from all over the world (Ehrenberg, 1841, 1846), called them "Phytolitharia" from the Greek meaning "plant stone," and developed the first phytolith classification system. Ehrenberg's classification (1854) recognized several dozen types of phytoliths (some of the 67 organic forms that Darwin refers to), which he placed in four "paragenera," the prefix *para* denoting an artificial, parataxonomic system used to name specimens that are isolated parts of living forms and considered unlikely to be identifiable to particular living species (Blackwelder, 1967). Three of his paragenera were assigned to the family Gramineae, and one to the Equisetaceae, thus he recognized that differences in phytolith morphology existed at least on the family level. Ehrenberg's parataxonomic system foreshadowed taxonomic classifications developed by modern researchers in which smaller units (tribes, genera, species) are determined from collections of phytoliths recovered from fossil contexts. Ehrenberg's pastel illustrations of phytoliths remain some of the most exquisite ever done. Phytolith research seems to have been carried on sporadically during the remainder of the nineteenth century, with several scientists noting their occurrence in plants and soils (e.g., Gregory, 1855; Ruprecht, 1866).[1]

The Botanical Period of Research

The onset of the twentieth century marks the beginning of the second period of phytolith research, the botanical period, which began about 1895 and lasted until *ca.* 1936. In Europe, especially Germany, from where much of the literature of the period would originate, botanists widely recognized the derivation of phytoliths from plant tissue and embarked on the first systematic explorations of phytolith production, taxonomy, and intraspecific variation. Quantitative studies of silica content in modern plants were first carried out and mechanisms of silica deposition in plant cells proposed.

Numerous accounts speak of the occurrence of *Kieselkörper* (silica bodies) in many families of the plant kingdom, including dicotyledons and ferns (e.g., Grob, 1896; Haberlandt, 1914; Werner, 1928; Netolitzky, 1929; Mobius 1908a,b; see also Prat, 1936, for an example of important work by a French scientist).[2] For example, Mobius (1908a) cited the Chrysobalanaceae, Dilleniaceae, Palmae, Orchidaceae, Urticaceae, and Hymenophyllaceae, a family of ferns (the genus *Trichomanes,* but not *Hymenophyllum*) for their prominent phytolith production. Netolitzky (1929) spoke of extensive silicification in the Podostemaceae, Chrysobalanaceae, Burseraceae, Palmae, Musaceae, Cannaceae, and Marantaceae, among others. We show in Chapter 2 how their results have been used as an important guide to research design and duplicated by modern researchers.

In some plants, such as grasses, ferns, nettles, and other dicotyledons, phytolith shapes were described and drawn by German botanists and morphology was shown to be highly significant. There is a rich body of information to be found in this literature, including tabulations of which plant families are and are not accumulators of silica, and even a cursory examination of it leads us to wonder why phytoliths from grasses and a few other nondicotyledenous plants were examined so intensively to the virtual exclusion of others during the early phase of the modern period of research.

[1]There is, however, a detailed description of phytoliths from several dicotyledon families in DeBary (1884), the original of which was first published around 1877. DeBary cites several interesting studies by German scientists of phytoliths in the Urticaceae, Moraceae, and Acanthaceae. This work was a precursor to the intensive work carried out later by German botanists and described in the next section.

[2]What is offered here is a brief survey of a large and important literature. (Those interested in further examples and discussion may consult Formanek, 1899; Netolitzky, 1900, 1914; Neubauer, 1905; Solereder, 1908; Frohmeyer, 1914; Frey-Wyssling, 1930a,b; and Bigalke, 1933.)

Since the main body of the botanical period research was published in German, phytoliths remained largely unnoticed in the English-speaking scientific world. Curiously enough, minerological studies of soils from Australia (Leeper *et al.*, 1936; Nicholls, 1939) had revealed the presence of phytoliths in large numbers, but they were thought to be sponge spicules and were described as such. These results were not given great weight for they would have had rather startling implications for the genesis of the sediments, invoking as they might have Creationist accounts of the Great Flood. The coming to power of the Nazi regime in Germany and the subsequent onset of World War II sadly brought a halt to the botanical period of research. Few publications by German scientists can be found after 1936.

The Period of Ecological Phytolith Research

It was not until the mid-1950s that concerted research began again, with considerable interest shown for the first time by English-speaking scientists. The years between 1955 and 1975 saw the inception of the period of ecological phytolith research, when botanists, soil scientists, agronomers, and geologists applied phytolith analysis as an index of environmental history. Following the lead of Smithson's recognition of phytoliths in British soils (Smithson, 1956, 1958) Baker (1959a,b) showed that many of the particles previously described as sponge spicules in Australian deposits were in fact phytoliths. Beavers and Stephen (1958) were among the first in the United States to report the occurrence of phytoliths in North American soils, and Kanno and Arimura (1958) helped to initiate the modern period of research in Japan.

Aspects of Holocene forest–prairie transitions in North America were subsequently studied by American soil scientists (Jones and Beavers, 1964a,b; R. L. Jones *et al.*, 1963; Witty and Knox, 1964; Verma and Rust, 1969; Wilding and Drees, 1968, 1971; Norgren, 1973). Annual phytolith production rates of above-ground grass parts were used to estimate how long soils had been under dominant grassland or forest cover. It was shown that soils developed under long periods of grass vegetation commonly contained 5–10 times more biogenic opal than those formed under forested environments, although soils in prairie–forest transition regions and in areas where prairies recently invaded forests yielded similar quantities.

These analyses are complicated by a number of biotic and abiotic factors, such as annual and multiannual climatic variation, soil pH, and concentrations of iron and aluminum that influence either dry matter production or the amount of silica in individual plant species. They are also complicated by the fact that grass root systems may contribute as much silica annually as above-ground parts (Geis, 1978; Sangster, 1978). Consequently, estimations of the duration of major vegetational formations existing on soils from annual inputs of plant biomass or phytoliths may not be accurate. Other studies of phytoliths in soils were undertaken by Brydon *et al.* (1963), Riquier (1960), and Pease and Anderson (1969).

Increased recognition of phytoliths led to their discovery in other contexts, including calcareous Wisconsin-age loess and till (R. L. Jones *et al.,* 1963), sedimentary rocks and other geologic materials of Cretaceous, Tertiary, and Quaternary age (Baker, 1960a; Jones, 1964; Gill, 1967; Weaver and Wise, 1974), deep sea cores (Kolbe, 1957; Parmenter and Folger, 1974) and atmospheric dusts (Baker, 1960b; Folger *et al.,* 1967; Twiss *et al.,* 1969). Baker's (1959a) original estimate that phytoliths did not survive under conditions in present-day soils for more than 1000 years was rendered doubtful by Wilding's (1967) radiocarbon determination of 13,300 years B.P. on occluded carbon from pedogenically derived phytoliths. The occurrence of phytoliths from geological formations millions of years old made it clear that biogenetic opal of plant origin was stable over very long periods of time under a range of mechanical and chemical weathering processes.

Coincident with the separation of phytoliths from soils were studies that undertook detailed descriptions of phytoliths in living plants known or assumed to have been present during pedogenesis. The necessary modern comparative basis for fossil plant identification was thus established. Grasses were most intensively studied with early work being carried out by Metcalfe (1960), Parry and Smithson (1964, 1966), Blackman and Parry (1968), and Blackman (1971), among a host of others. Twiss *et al.* (1969) developed a classification system that permitted the discrimination of three subfamilies of grasses, a scheme that with some modification is still in use today. Even with the large amount of attention devoted to phytolith morphology in grasses some important aspects, such as three-dimensional structure and size, remained poorly studied. Consequently, a consensus emerged that phytoliths within subfamilies were highly redundant and could not be used to identify genera and species of grasses. Chapter 3, which focuses on phytolith morphology, shows that this is not always the case. Studies of phytolith morphology in nongrass monocotyledons included those by Tomlinson on the Palmae (1961) and other families (1969); Mehra and Sharma (1965) and Metcalfe (1971) on the Cyperaceae (sedges); and Stant (1973) on the Commelinaceae.

Studies of nonmonocotyledenous species were fewer in number and emphasized phytoliths found in the foliage of deciduous (temperate, broadleaf) and coniferous trees and the wood of tropical trees (e.g., Amos, 1952; Bamber and Lanyon, 1960; Brydon *et al.,* 1963; Wilding and Drees, 1968, 1971; Verma and Rust, 1969; Rovner, 1971; Geis, 1973; Klein and Geis, 1978; Scurfield *et al.,* 1974; Kondo and Peason, 1981). Morphological differences between phytoliths from grasses and dicotyledenous angiosperms, and grasses and conifers were found to be readily achievable. Genus-specific phytoliths were isolated by Brydon's group from Douglas fir. Lanning (e.g., 1960, 1961, 1966) undertook a number of studies of phytoliths in various nonarboreal dicotyledons, such as strawberry, raspberry, and *Helianthus,* but efforts focused on the production and localization of silica in plants and no detailed descriptions or comparisons of phytolith shapes were made available.

Two findings emerged from these studies that promised to complicate the identification of specific plants in soil phytolith assemblages. First, a multiplicity of phytolith shapes often occurred in the same plant and second, redundancies in phytolith shape were often present between quite unrelated species. Moreover, tree phytoliths were often poorly represented in soil assemblages. However, this conclusion may well reflect an artificial underrepresentation caused by the methodological focus on the 20–50-μm (coarse silt-sized) soil fraction. It has been shown that over 75% of siliceous particles from angiosperm forest floor samples occurs in size classes less than 20 μm in diameter (Wilding and Drees, 1971; Geis, 1972), and tree phytoliths greater than 50 μm in diameter, which would be present in the sand portion of fractionated soils, may be diagnostic.

Erroneous impressions were also developed concerning the significance of phytolith morphology. Despite the discrepancy in the amount of attention devoted to different plant taxa, the few species for which detailed phytolith descriptions were made available, and the promising morphological characters observed in the few dicotyledons intensively studied, a consensus arose that phytoliths had the greatest potential in monocotyledons, especially grasses, that nongrass phytolith morphology tended to be highly redundant, and that dicotyledons were nonproductive areas of research. There is little justification for any of these premises, a subject that is fully explored in Chapter 3.

The 20-year period from 1955 to 1975 also saw investigations into the chemical and physical properties of phytoliths (e.g., Lanning *et al.*, 1958; Jones and Beavers, 1963; Jones and Handreck, 1963; Lanning, 1960), the function of silica in plants, and the timing and nature of phytolith formation and dissolution (Blackman, 1968, 1969; Blackman and Parry, 1968; Wilding, 1967). Numerous techniques were developed to isolate, identify, and quantify silica in plants and sediments.

However, during the last five years of this period, studies of phytoliths in soils and sediments decreased markedly. Despite the plethora of pollen studies in a paleoecological context, botanists concentrated on the documentation of phytoliths in a limited number of modern grasses using the scanning electron microscope (some examples include Hayward and Parry, 1975; Sangster, 1977a; and Bennett and Sangster, 1981).[3] While these studies have extended knowledge of the areas of deposition, microstructure, and surface details of phytoliths, it is obvious that a technique with considerable paleoecological potential has not been well exploited recently by either the botanical or geological community.

[3]See also Sangster and Parry (1976a–c), Parry and Kelso (1975), Kunoh and Akai (1977), and Dinsdale *et al.* (1979) for further examples of phytolith documentation using scanning electron microscopy.

Those fewer studies oriented toward paleoecological reconstruction include those of Palmer (1976), Schreve-Brinkman (1978), and Bukry (1979, 1980). Palmer used silica bodies still embedded in organic tissue to tentatively identify savanna grass genera in East African lake deposits of Pleistocene age. She demonstrated the importance of phytoliths in documenting the past climate of regions where grasses are important elements in the fossil record. Studies of such regions are usually hindered by the indistinguishability of grass pollen grains. Schreve-Brinkman (1978) recognized the occurrence of grass phytoliths in Late Pleistocene cores from Colombia that she was analyzing for pollen. Phytolith evaluations of the sediments agreed well with pollen data. The highest numbers of grass phytoliths coincided with high numbers of grass and Compositae pollen, which had indicated conditions of maximum cold and subparamo–paramo vegetation.

Bukry recognized the occurrence of grass phytoliths in deep-sea cores off Northwest Africa (1979) and the tropical Eastern Pacific ocean (1980) whose source areas were Africa and Central America, respectively. Transport of phytoliths to the coring localities was considered to be primarily aeolian, a major factor in assessing the climatic significance of phytolith distributions. In the Eastern Pacific core, phytolith occurrence was less frequent in Upper than in Lower Quaternary sediments, possibly suggesting less dry conditions in the Upper Quaternary. Also, fresh-water diatoms were far scarcer in the Eastern Pacific as compared to the African assemblages, a factor that argued for a general lack of dried up lake beds and streams, and hence, wetter conditions in Central America than in Northwest Africa during the latter stages of the Pleistocene. These are conclusions of considerable importance to archaeologists interested in New World tropical paleoenvironments and colonization during the Late Glacial period, when Paleo-Indian populations first settled the region.

Another brief but very important surge in the application of phytoliths to paleoecology was represented by the work of botanists studying the origins and taxonomic relationships of angiosperms (Thomassen, 1983, 1984; Thomassen *et al.,* 1986; Smiley and Huggins, 1981). They observed phytoliths in fossil plants preserved in Miocene-age rocks from North America. The shapes and locations of phytoliths in the fossil grasses, sedges, and deciduous trees helped determine the taxonomic assignment of these plants and suggested possible affinities to extant species. These studies also provided fossil evidence for the stability of morphogenetic patterns of silica in plants and for their change over time, two cardinal attributes of genetically controlled systems. Phytoliths were clearly not the casually and haphazardly occurring plant structures that some had thought them to be.

The most recent phytolith work by botanists has concentrated on some traditional lines of research as well as addressing new biological questions; studying depositional areas, developmental stages of silicification, ultrastructural characteristics of silicified cells, and their morphological function (Hodson *et al.,* 1984; Sangster, 1985; Perry *et al.,* 1984; Kaufman *et al.,* 1981, 1985; Hodson, 1986).

Regrettably, while such an orientation is essential to fundamental knowledge of siliceous secretions of plants, the inattention of paleobotanists to phytoliths is difficult to understand.

The Modern Period of Archaeological Phytolith Research

The use, if not systematic application of phytolith analysis in archaeology, has considerable antiquity, and again, German scientists were prominent in this research (Netolitzky, 1900, 1914; Schellenberg, 1908; Edman and Söderberg, 1929). Netolitzky (1914) carried out studies of archaeological spodograms (the inorganic residue, including phytoliths, remaining after plant tissue is incinerated) from Swiss and other European sites and was able to correct some misidentified specimens on the basis of phytolith criteria. Cereal grains from these sites had apparently often been identified as something other than *Panicum miliaceum* or *Setaria italica,* but Netolitzky noted that grain-bearing ash samples contained silicified husks characteristic of these two species of millet. He also identified wheat and barley phytoliths from other sites in Europe (Netolitzky, 1900).

Schellenberg (1908) identified silicified remains of wheat and barley in ceramic fragments from a prehistoric settlement in Turkey. These plants had apparently been incorporated into the clay during pottery making. In their identifications, Schellenberg and Netolitzky paid particular attention to the shape of silicified epidermal and short cells from floral glumes, criteria which would be used 50 years later by modern archeobotanists studying the origins of cereal agriculture in the Near and Far East (Miller, 1980; Miller-Rosen, 1985; Watanabe, 1968, 1970). Edman and Söderberg (1929) identified scrapings from the inside of a Chinese Neolithic potsherd as the siliceous remains of rice.

Much of the early modern research into agricultural origins and dispersals occurred in the Near East, and here Helbaek (1961, 1969) noted the occurrence of silica skeletons, a term also applied to pieces of silicified epidermis that have been released from plants by burning or heating, in Neolithic and later period ash heaps and pottery. Helbaek recognized differences in silicified epidermis from the husks of wheat, rice, millet, and barley, results that further indicated a basis for the identification of these domesticates in the absence of macrobotanical materials. Watanabe (1955, 1968, 1970) was able to detect the presence of silicified remains of rice and millet in prehistoric deposits from Japan. Dimbleby (1967) drew attention to the occurrence of discrete phytoliths in archaeological soils.

Phytoliths never received more than a passing interest in the early archaeobotanical literature, partly because the abundance of carbonized plant material recovered from Neolithic sites in the Near East served to deemphasize their potential value. In both of Dimbleby's comprehensive reviews of archaeobotany (1967, 1978) phytoliths were placed in the category "other plant remains." It is no accident then that the first emergence of a systematic approach to phytolith analysis in archaeology occurred in regions such as Eastern North America and the tropics, where botanical preservation was very poor and evidence bearing on important questions of agricultural origins and environmental history remained inferential (Carbone, 1977; Pearsall, 1978; Piperno, 1984, 1985a). Issues such as the origins and dispersals of seed and root cropping and the antiquity and nature of tropical forest modification had proven to be infractile to resolution with pollen or macrobotanical remains. New methods were needed to elucidate old problems and raise interpretations above the level of conjecture.

The 1971 article by Rovner evaluating the potential of phytoliths in archaeology, which appeared in *Quaternary Research,* did much to stimulate the interests of prehistorians in the technique. It was shortly after that archaeobotanists undertook the exacting comparative and quantitative efforts that mark a paleobotanical discipline. The rest of the story is not quite history because it is too recent, and it forms the purpose and body of this book.

2

The Production, Deposition, and Dissolution of Phytoliths

Introduction

Phytoliths are a result of the process by which certain living higher plants deposit solid silica in an intracellular or extracellular location after absorbing silica in a soluble state from groundwater. The term phytolith, from the Greek meaning plant stone, has sometimes been used to indicate all forms of mineralized substances secreted by higher plants, be they siliceous or calcareous in composition. "Phytoliths" reported from archaeological coprolites (for example, Bryant, 1974; Bryant and Williams-Dean, 1976) are all calcium oxalate crystals, not siliceous bodies, and are seldom diagnostic of any one plant family let alone particular genera or species.

Although it might seem logical to use the term phytolith for any kind of mineral secreted by plants, Bryant's statements concerning the lack of taxonomic specificity of phytoliths could prove misleading if taken to mean both calcareous and siliceous bodies. As we shall see in Chapter 3, the shapes of silicified cells in many plants are highly distinctive. Therefore, the term phytolith is probably less confusing if used in a more restrictive sense to indicate only silicified remains of plants. In all cases dealt with in the present book this will be its usage.

Numerous other terms have been proposed for silica bodies found in plants, such as opal phytoliths, plant opal, opaline silica, and even grass opal. Of these, the latter is the most restrictive, because it implies a specific origin, whereas the others simply imply silicon constituents derived from higher plants. Phytoliths can also fall under a more generalized nomenclatural rubric, that called biogenic silica or simply bioliths, all-inclusive terms for silicon found in lower plants and animals as well as higher plants. These terms distinguish silica derived from living systems from silica of inorganic and mainly pedalogic origin.

Silica also occurs commonly as a component of plant and animal microorganisms. It forms the well-known incrustations of cell walls in unicellular plants of the phylum Bacillariophyta, the diatoms. In the animal phylum Protozoa, the Radiolarians and Heliozoans have diagnostic siliceous secretions in one form or another, and the phylum Porifera includes the sponges, which secrete numerous skeletons of siliceous spicules up to several millimeters in length. Simpson and Volcani (1981) provide an excellent review of silicon as it is found in various biological systems. All of these siliceous secretions may eventually accumulate as residue products in soils and geological deposits. Diatoms and sponges can inhabit both marine and fresh-water environments, while heliozoans are to be found strictly in fresh water and radiolarians solely in marine situations. Their significance in reconstruction of ancient landscapes is addressed in Chapters 7 and 8.

The Development of Phytoliths in Higher Plants

The degree of development of phytoliths in a plant relates to a number of factors, including the climatic environment of growth, the nature of the soil, the amount of water in the soil, the age of the plant, and most importantly, the taxonomic affinity of the plant itself. The process starts when plants absorb soluble silica in groundwater through their roots and ends when the silica, sometimes at a very early stage of plant development, is laid down as solid infillings of cell walls, cell interiors, or intercellular spaces. In between, this cycle can be slowed down, speeded up, or modified in some way by a number of biotic and abiotic factors, which are discussed in the sections that follow.

Uptake of Soluble Silica by Plants

Except for oxygen, silicon is by far the most abundant of all the earth's elements, accounting for as much as all the other 90 elements together. It is contantly dissolving and precipitating over a large part of the earth's surface. The former of these processes provides the raw material of phytolith formation. Soluble silica is mainly derived from the weathering of silicate minerals such as quartz and feldspar. The extent of weathering depends on a number of factors including climate, topography, bedrock character, and the amount of water passing through the soil. Warm, wet climatic regimes, for example, are thought to accelerate the process (Dunne, 1978), liberating greater quantities of soluble silica. Some of this remains long enough in groundwater to be taken up by plants.

Plants absorb silica through their roots, and it is carried upward to the aerial organs in the transpiration stream via the water-conducting tissue called xylem. The form of silica in soil solution that enters plants is monosilicic acid, $Si(OH)_4$, between soil pH values of 2 and 9 (Barber and Shone, 1966; Jones and Handreck, 1963; McKeague and Cline, 1963a,b). It is recognized that plants of a single species can contain different concentrations of solid silica when grown in different soils. This is a completely different phenomenon from the differential occurrence of phytoliths in different taxa, which now looks to be primarily genetically controlled and which is dealt with later in this chapter. Variable content of silica within species is thought to depend mainly on environmental factors which regulate the concentrations of dissolved silica available to plants. A number of laboratory studies have shown that amounts of silica (measured as percent of dry weight) in some silicon accumulating species increase in direct proportion to amounts of dissolved silica in the growth medium (Jones and Handreck, 1965; Okuda and Takahashi, 1961). It is important to consider here those factors affecting levels of dissolved silica in soils.

There is a marked effect of pH on silica solubility, some of which is associated with absorption of monosilicic acid by iron and aluminum oxides. These sesquioxides, particularly the latter, commonly adsorb or bind silica onto their surfaces, effectively removing it from solution. Adsorption decreases on either side of a maximum at about pH 9.5 (Wilding *et al.,* 1977). One might think, therefore, that highly weathered, lateritic soils containing high concentrations of iron and aluminum oxides, such as are commonly found in tropical regions, could be especially poor in soluble silica, and hence in phytolith-producing vegetation.

In general, however, tropical soils have much greater concentrations of silica than temperate soils, especially those which are poorly drained (Siever, 1957). These higher concentrations in tropical soils are presumably due in part to the acceleration of the weathering process on silicate minerals described previously. Leaching is relatively rapid and the residence time of water in the soil is too short to allow extremely high concentrations of silica to build up (Dunne, 1978). However, the considerable supply of available silica coupled with the ability of tropical root systems to very efficiently absorb dissolved soil nutrients, contributes to high concentrations of silica in many tropical plants (Lovering, 1959; Riquier, 1960; Piperno, 1985a). It has been known for some time that tropical forests contribute large inputs of phytoliths into soils (Lovering, 1959; Riquier, 1960).

Levels of monosilicic acid in soils are also affected by pH in a manner that is independent of the presence of iron and aluminum sesquioxides. The capacity of soil particles to adsorb soluble silica changes as pH changes, and the concentration of monosilicic acid has been found to increase on either side of a minimum between pH 8 and 9 (McKeague and Cline, 1963b). Thus, acidic soil environments have more free silica available to enter plants.

There are studies indicating that the uptake of silica by plants increases with increasing water content and perhaps temperature of soils (reviewed by Jones and Handreck, 1967). Rice *(Oryza sativa)* is thought to be one of the highest, if not the highest accumulator of solid silica in the Gramineae (Wadham and Parry, 1981).

Jones and Handreck (1967) comment that wetland rice almost certainly contains much more silicon than dryland rice because values of 10–15% silicon dioxide are commonly reported from the straw of the former variety. Miller (1980) has used this apparent tendency of high silicon accumulation in wetter conditions to distinguish between dry farming and irrigated agriculture. She found that grasses grown in warm, moist environments, such as floodwater fields in Egypt and high rainfall areas in Belize, exhibited a more complete silicification of the epidermis as a result of increased silica uptake than grasses from regions where rainfall agriculture was practiced. As a result, archaeological phytoliths occurred in the form of large aggregates of silicified cells instead of the discrete particles that are more often isolated from midden soils (Chapter 7).

Experimental data also exist showing that concentrations of dissolved silica are higher in soil mixtures that have higher water–soil ratios, although the increase was not infinite and leveled to a slower rate of silica dissolution a day or two after applications of water (McKeague and Cline, 1963a). Conversely, drier soils have lower levels of dissolved silica.

There is evidence to suggest that soils with high contents of dissolved organic matter show increases in amounts of soluble silica, while paradoxically, the presence of nitrogen and phosphorus in significant amounts is thought to lead to decreases in the concentration of silica in the plant. High concentrations of nitrogen and phosphorus in cultivated fields caused a decrease in the concentration of silica in barley, wheat, and other crops grown in these areas (Jones and Handreck, 1967). Studies concerned with these problems have been reviewed by Jones and Handreck (1967).

Another source of increased levels of soluble silica for uptake by root systems of standing vegetation appears to be made available by a fairly rapid dissolution of some types of phytoliths in decayed litter. It appears that phytoliths formed from the incrustations of plant cell walls are very vulnerable to weathering processes because they are not present (or are drastically underrepresented) in A horizons underneath vegetation from which they are derived. For example, it was found that Panamanian tropical forests contribute considerable quantities of epidermal and hair cell phytoliths with unsilicified lumina, but these were rarely recovered from modern soils (Special Topic 2: Chapter 6). Accounts of phytolith distributions from A horizons underneath North American forests also rarely mentioned deciduous tree silica, though this may have resulted, in part, from restricted attention to the coarse silt fractions of soils (Wilding and Drees, 1968, 1971; Verma and Rust, 1969). Whether most of the dissolved silica from organic litter is added to groundwater circulation and made available for recyling by plants or is swept off by erosive agents must depend on site-specific climatic, topographic, and soil conditions (Lovering, 1959).

Some contrasting results concerning phytolith dissolution in organic litter are provided by Bertoli and Wilding (1980). From studies of leachates collected from natural soil–litter (pine and beech fir) environments, they determined that in waters at 22°C the dissolution rate was quite slow, resulting in low concentrations of silica (0.5–2 μg/mL), and concluded that phytoliths are not an important labile silica source affecting soluble silica levels in soils. However, this may have to do with the type of phytoliths produced by the standing vegetation, or with differential degrees of solubility under different environmental regimes. High levels of silica-dissolving substances in tropical soils like tannins, for example, may contribute to increased dissolution of some kinds of phytoliths.

In summary, factors thought to regulate the levels of monosilicic acid in soils, and subsequently to some extent in plants, include the presence of iron and especially aluminum oxides, pH, dissolution of phytoliths from decayed litter, temperature, water supply, organic material, and possibly dissolved nitrogen and phosphorus.

In considering the effect that levels of dissolved silica may have on the subsequent phytolith content of plants, an obvious question is, Could silica-free plants result under natural conditions simply because not enough monosilicic acid was available in groundwater to make phytolith formation possible? The experiments of Jones and Handreck (1965) are of importance here because they demonstrated a six-fold decrease in silica content, from 3.9% to 0.4% of dry plant weight, in oat leaves grown in experimental soils with depauperate amounts of silicic acid. Such a situation, if commonplace in nature, would greatly diminish the value of paleoecological phytolith analysis because one could never determine beforehand from what plants to expect significant representation in a fossil phytolith assemblage. In reality, the chances of such an occurrence are remote. Constant replenishment of soluble silica in groundwater appears to occur through rock and sediment weathering and dissolution of soil phytoliths. Plants known to be high accumulators of solid silica, such as grasses, palms, banana, sedges, nettles, and composites have been found to be high accumulators no matter from what part of the world individual plants are studied (Metcalfe and Chalk, 1950; Tomlinson, 1961, 1969; Piperno, 1983, 1985a).

Levels of silica in terrestrial sediments have never been found to be so low as to cause a cessation or near cessation of phytolith production in species that characteristically produce high amounts (which probably relates to phytolith functions in plants), although levels of production may certainly vary in different plants of the same silicon accumulating species. Furthermore, it seems that such plants have the ability to capture and accumulate monosilicic acid from sometimes dilute soil solutions. Blackman (1969) grew grasses in a silica-poor medium and still observed significant short-cell phytolith production. It has been suggested that grasses are able to solubilize silica from clay particles in a silica-free solution (McNaughton and Tarrants, 1983). These are important points that will be discussed shortly.

Mechanisms of Soluble Silica Uptake

There are two major ways by which plants are thought to absorb soluble silica, and though varying opinions exist as to which is most important, both probably have significant roles to play in the transfer of monosilicic acid from groundsoil into roots and then to aerial organs of plants. The two proposed mechanisms of silica uptake are active transport of monosilicic acid by metabolic processes, and passive, nonselective flow along with other elements through the transpiration stream. In the former pathway, the plant actually expends energy during silica absorption, and in the latter it does not.

There is a considerable amount of evidence that passive uptake and conduction of monosilicic acid occurs in certain plants (see Lewin and Reismann, 1969; and most recently, Raven, 1983, for reviews). The well-established fact that the amount of silica present in some laboratory-grown, silicon-accumulating plants is directly proportional to the amount of dissolved silica in soils alone argues for such a relationship. Jones and Handreck (1965) showed that they could closely predict silicon content in oats, a dryland grass, by knowing simply the concentration of silicic acid and the amount of water transpired. Such a relationship would be expected with a passive uptake.

On the other hand, there is convincing evidence for the active transport of soluble silica in some plants. Okuda and Takahashi (1964) found that silicic acid appeared to be entering the xylem sap of rice shoots against a concentration gradient, while Barber and Shone (1966) showed that while the initial uptake of silicic acid by barley roots conformed to the passive diffusion of a nonpolar solute, the movement of silicic acid across the root into the transpiration stream did not. Van der Worm (1980) demonstrated active uptake in sugar cane, wheat, and rice. It seems that angiosperm roots may exhibit either an active or passive absorption, depending on the species, and that in some taxa both processes are involved, each operating in different locations of the plant.

I have mentioned the significant evidence available to indicate that, independent of environmental variables, some families of plants consistently produce low to null amounts of phytoliths, while others deposit solid silica in copious quantities. An often cited example of this phenomenon is the contrast between silica content of the Gramineae and Leguminoseae; grasses commonly have 10–20 times the amount of legumes. In some families, such as the Araceae, studies of numerous species from widely divergent environmental regions have failed to reveal the presence of any phytoliths at all (Tomlinson, 1969; Dahlgren and Clifford, 1982; Piperno, 1985a). The appropriate evidence bearing on this issue will be presented shortly, but for now we shall explore some possible explanations for the tremedous variation seen between species in the deposition of solid silica.

It stands to reason that plants with inherently low concentrations of phytoliths must have some mechanism for either rejecting the entry of silicic acid at the root surface or preventing it from passing from the root into aerial organs of plants. From their studies on clover *(Trifolium incarnatum)*, Jones and Handreck (1969) concluded that the roots of nonsilica accumulators may possess a hypothetical barrier located in the epidermis, which allows monosilicic acid into the transpiration stream at a lower rate than water. In this situation, preferential exclusion of silicic acid has taken place. Such potential mechanisms have been experimentally studied by Parry and Winslow (1977). Root severance in pea seedlings *(Pisum sativum)*, normally a low silicon accumulator, resulted in a heavy accumulation of silicon (but not discrete opaline deposits) in cell walls of the leaves and tendrils, whereas plants grown in silica solution with roots intact showed no such accumulation. This indicates that a mechanism of silica exclusion exists in the roots and is probably located at the external surface of roots, since no silicon was detected within the roots of intact plants.

Parry and Winslow (1977) further pointed out that root hairs of some plants with inherently low amounts of solid silica, such as *Vicia faba* and *Ricinus communis* are enveloped by a thin layer of fatty material that is similar in behavior to cutin and suberin. Such a substance, while maintaining a permeability to water, might well constitute a barrier to monosilicic acid. It is significant that no such deposits have been found on the root surfaces of maize and barley, plants that are characteristically high phytolith producers. Studies of metabolic processes (van der Worm, 1980) have indicated active exclusion of monosilicic acid by the roots of some plants, a further indication that cellular silicification, or the lack of it, is a nonrandom process. We have, then, at least the beginning of an explanation for the differential production of phytoliths by plants.

Deposition of Solid Silica and Its Timing

Once monosilicic acid enters plant tissues a process begins whereby some of it is polymerized and forms solid deposits of silicon dioxide (SiO_2) in and around plant cells. These deposits are often referred to as opal or opaline silica in an analogy to the mineral opal deposited by geological processes, but Lewin and Reimann (1969) remind us that their behavior is rather more like that of a silica gel because they are hydrated. In plant tissues there are three loci of silica deposition: (1) cell wall deposits, often called membrane silicification, (2) infillings of the cell lumen, and (3) in the intercellular spaces of the cortex. Patterns of localization are quite similar in species and whole families of plants, regardless of environmental conditions of growth.

The mechanisms behind the polymerization of silica in plants are not well understood. They have for the most part been studied in the aerial organs of grasses, where transpiration or water loss has been implicated as a major contributing factor (Jones and Handreck, 1967; Sangster and Parry, 1971; Raven, 1983). In actively transpiring cells, a supersaturated solution of silicic acid is thought to occur, leading to the precipitation of silica. In those areas of the plant where water loss is highest, such as the flag or uppermost leaf and floral bracts of grasses, substantially more silica is deposited than in other areas.

The relationship between transpiration and deposition is by no means an inviolate one, as becomes clear when one looks at silicification patterns on a cellular level. Hayward and Parry (1973) point out that in barley, in all but the flag leaf which because of its position must transpire for longer lengths of time, the highest levels of silica are found in cells not directly associated with loss of water; namely, the idioblasts (which give rise to phytoliths such as dumbbells and cross shapes), trichomes (hair cells) and schlerenchyma. In intercostal areas (between veins), where most transpiration takes place, fewer cells are silicified. This is a pattern I have observed in many other grasses.

A similar pattern, but to a more extreme degree, occurs in some of the Leguminoseae, where schlerenchyma is silicified but other actively transpiring tissues often are not (Piperno, 1983, 1985a). In palms, mesophyll tissue is always extensively silicified, while epidermal cells are not (Tomlinson, 1961; Piperno, 1983, 1985a). There are many other cases of localized silica deposition that cannot be attributed to a passive, transpiration-mediated mechanism (Parry *et al.,* 1984). In this light, Commoner and Zucker's (1953) proposed enzyme system for deposition that is synthesized or segregated in some cells only takes on particular relevance. To account for such highly specific polymerization sites attention has been given to the possibility of phylogenetically determined silica distribution patterns (Kaufman *et al.,* 1981; Parry *et al.,* 1984).

Blackman and Parry some time ago (1968) distinguished between typical and atypical silicification in grass leaves. The former takes place in costal (over the vein) areas, invariably occurs, and results in the familiar dumbbell, saddle, and cross-shaped bodies. The latter takes place intercostally (between the veins) in cells not primarily associated with deposition, such as stomatal complexes and long or fundamental cells, and is more random and sporadic in occurence. Since atypical deposition often is found in older plants, they suggest that it may be a result of environmental factors such as an abundant supply of soluble silica. In other words, the plant has an excess supply of monosilicic acid, which fills up in secondary areas after primary deposition sites have been used up. Blackman (1969) further postulated that very young typical silica cells might have special metabolic porperties closely associated with subsequent deposition of silica, such as enymes that degrade cell content after they disintegrate to permit the cell to become filled with silica. By observing the developmental anatomy of typical silica cells Blackman noticed some striking changes, such as a rapid and complete loss of the nucleus and cytoplasm disappearance at an early stage of leaf formation. The implication is, hence, very strong that many silica deposition sites in plants are under active, metabolic control.

Other factors shown to affect silicification in grass leaves are mechanical damage and senescence, both of which resulted in the deposition of huge amounts of solid silica in some grass species (Parry and Smithson, 1963). In many plants older leaves have been shown to have substantially more silica than younger leaves (Blackman, 1968, 1969; Lanning and Eleuterius, 1985). This may have to do either with the increased availability of deposition sites as the plant grows and matures, or crucial changes in cellular environment brought on by maturation that are needed to acquire silica.

With regard to the precise timing of silicification, only grass leaves have been studied in any detail, with results indicating that deposition time is not the same for all species. It can occur quite early or fairly late in the development of the plant. Sangster (1977a) detected idioblast silica in two day old crab grass leaves *(Digitaria sanguinalis)*. In rye *(Secale cereale)*, however, idioblast silica was not detected until the final maturation process when the leaf was fully expanded (Blackman and Parry, 1968). L. H. P. Jones *et al.* (1963) observed that substantial portions of the oat leaf blade were silicified within six weeks, and impregnation of cellular tissues continued throughout the maturation of the plant. Apparently, initial silica deposits and perhaps even some laid in later stages are not rigid, since they do not prevent the extension of immature cells. In the floral bracts (glumes and lemma) of some grasses, silicification occurs quite early, before inflorescence emergence (Sangster *et al.*, 1983; Hodson *et al.*, 1985).

The Occurrence of Phytoliths in the Plant Kingdom

Although opaline silica accumulates in the cells of numerous plant taxa, phytoliths are not uniformly produced throughout the plant kingdom. As has been discussed previously the reasons behind the differential production of solid silica by higher plant taxa are far from clear, but what is essential for our purposes is to achieve some basic, yet comprehensive understanding of patterns of production in modern plants so that we can properly interpret the fossil phytolith record. Knowing which families and genera of plants are likely to leave behind tangible proof they once comprised existing vegetation is necessary to unravel the significance of unrepresented plants or "blind spots" that always form a part of paleobotanical profiles. Was the plant really not present, or can we simply not expect any silicified evidence for it?

We also need to know about possible influences of environmental variables on silica accumulation. Are phytoliths to be found in plants from all regions where archaeological research is undertaken, or are they more prone to formation only in certain latitudes, climatic zones, or soil types? Within a single species, are silicification patterns constant from region to region, or do they vary, and if so, by how much? Some of these are undoubtedly issues over which phytolith specialists will continue to grapple for some years to come, just as palynologists have struggled with questions of differential pollen production, transport and preservation for the last 65 or so years.

For some of these questions, however, we already have answers. Phytoliths are worldwide in occurrence, and though degrees of silicification may certainly vary within a species, one can often accurately predict which plants will produce substantial amounts of diagnostic or nondiagnostic phytoliths and which will form no phytoliths at all. In many species, silicification patterns in plant tissues and plant structures are remarkably constant from region to region, a phenomenon that harkens back to the recent discussion of genetic control of silicification. In sum, sufficient information exists about the production of phytoliths to make possible a comprehensive and thorough review of the subject, and form a sound basis for the interpretion of archaeological and geological phytolith assemblages.

The botanical literature produced during the 1960s and 1970s often contains statements that phytolith formation is essentially limited to monocotyledons and occurs only rarely in dicotyledons. These statements ignore so much variation in production among monocotyledons, dictoyledons, gymnosperms, and lower vascular plants, as to present a very misguided picture of phytolith production. They do not take into account the work by German scientists during the botanical period of research of the first third of this century and efforts by Amos, Lanning, Geis, Klein, Wilding, Drees, and more recently, archeobotanists who have demonstarted significant phytolith formation in all kinds of plants. What follows next is an overview of phytolith production and depositional patterns in the plant kingdom, which brings together literature dating from the early part of this century to the present. It is based on my analysis of over 1000 species of New World tropical plants, comprising over 70 different families, and studies in other regions of the world carried out by botanists, archaeobotanists, and geologists.

Solid deposits of silica can be found in measureable amounts in many plants. However, the term *measurable* often includes nondescript fragments that have no value in plant identification. In the discussion and tables that follow, the term phytolith refers only to microscopically recognizable shapes, not amorphous pieces or traces of silica detectable only by microchemical methods that would not be taxonomically useful.

Table 2.1 contains patterns of phytolith production for various groups of plants that have been divided into three main categories: pteridophytes (plants reproducing by spores), gymnosperms, and angiosperms. Angiosperms have been further subdivided into monocotyledons and dicotyledons. I have drawn on data compiled by myself and various other investigators and arbitrarily denoted three categories of production: often common to abundant, often uncommon to rare or absent, and not observed.

Table 2.1
Patterns of Phytolith Production in Higher Plants[a]

I. Angiosperms

Monocotyledon families where phytolith production has not been observed

Alismataceae, Amaryllidaceae, Araceae, Burmanniaceae, Butomaceae, Cartonemataceae, Cyclanthaceae, Dioscoreaceae, Eriocaulaceae, Hydrochariataceae, Iridaceae, Liliaceae, Mayacaceae, Pandanaceae, Xyridaceae

Monocotyledon families where phytolith production is often uncommon to rare, or absent

Commelinaceae, Juncaceae, Pontederiaceae, Smilacaceae

Monocotyledon families where phytolith production is often common to abundant

Bromeliaceae, Cannaceae, Cyperaceae, Gramineae, Heliconiaceae, Marantaceae, Musaceae, Orichidaceae, Palmae, Zingiberaceae

Dicotyledon families where phytolith production is often common to rare, or absent

Amaranthaceae, Apocynaceae, Bignoneaceae, Bixaceae, Caricaceae, Chenopodiaceae, Combretaceae, Convulvulaceae, Flacourtiaceae, Guttiferae, Labiatae, Lauraceae, Malvaceae, Malphigiaceae, Melastomataceae, Myrsinaceae, Myrtaceae, Nymphaceae, Polygonaceae, Rubiaceae, Sapindaceae, Solanaceae, Tiliaceae

Dicotyledon families where phytolith production is often common to abundant

Acanthaceae, Annonaceae, Aristolochiaceae, Burseraceae, Cannabaceae, Chloranthaceae, Chrysobalanaceae, Compositae, Cucurbitaceae, Dilleniaceae, Euphorbiaceae, Loranthaceae, Moraceae, Piperaceae, Podostemaceae, Rosaceae, Sterculiaceae, Ulmaceae, Urticaceae, Verbenaceae

II. Gymnosperms

Families where phytolith production is often uncommon to rare

Pinaceae

III. Pteridophytes

Families where phytolith production is often common to abundant

Equisetaceae, Hymenophyllaceae, Selaginellaceae

[a]In the construction of this table the taxonomies of Dahlgren and Clifford (1982), Metcalfe and Chalk (1979), and Benson (1957) were followed.

Notes: About 50% of the 28 species of Polypodiaceae thus far studied by the author have contributed high numbers of phytoliths. In the Fabaceae, the author has observed significant phytolith production to occur much more often in the Mimoisodeae than in other subfamilies. In the Hymenophyllaceae, no phytoliths were observed in the three species of *Hymenophyllum* examined by the author. In this family, production may be confined to the genus *Trichomanes*.

The category "often common to abundant" signifies that a great many species within a family, usually well over 50% of the total studied, produce significant amounts of phytoliths, that when expressed as a percentage of dry plant weight would approach, equal, or be greater than the 2–5% values commonly reported for grasses. The category "often uncommon to rare or absent" indicates that phytoliths have not been observed in many species (usually over 50%) of the total examined from a family, and when observed usually occurred in small amounts that would probably quantify as less than 0.5% of dry plant weight. The category "not observed" indicates families that are on the whole lacking in silica, allowing that more studies might reveal their presence in certain species.

In tandem with Table 2.1, Tables 2.2, 2.3, and 2.4 present specific information on the degrees and patterns of silicification for particular plant species, with an emphasis on Old and New World plant domesticates and wild species of importance for economic and vegetational reconstructions. Major plant domesticates are listed in boldface type. For some plants in Tables 2.2–2.4, quantitative evaluations of percentage silica as carried out by various workers are given. Where the same species sampled from different environmental zones has been quantitatively assessed, replicate values are provided. For many plants, estimations of silicon production are given, drawn largely from my own work, and also studies by Metcalfe (1960) and Tomlinson (1961, 1969). My estimates are broken into four categories: not present (NP), abundant (A), common (C), and rare (R). Some of my ratings of production may tend to err on the low side because I extracted somewhat less plant material than did other workers, but this should hold true mainly for the NP category, as examination of additional material from some species might show that levels denoted as not present should be changed to rare or uncommon. In Tables 2.2–2.4 patterns of production are organized by way of the particular cell or tissue type that is silicified. Detailed descriptions will be provided of each kind of cell and tissue as well as resulting derived phytoliths in Chapter 3.

Production and silicification patterns in Tables 2.1–2.4 are largely confined to results from studies of the aerial portions of herbacious plants and the foliage of woody plants, with exceptions as noted. While they may seemingly offer a biased picture of these characteristics, the tables were arranged in this fashion for two reasons. First, silica is often concentrated in the aerial parts of herbs and foliage of shrubs and trees, and second, the aforementioned parts contribute silicified structures that are discrete, easily described, and classifiable, and, therefore, amenable to recovery and identification in sediments. Inconsistencies and ambiguities resulting from such a presentation will be clarified in the following section, when the distribution of silica in various organs of plants is reviewed in some detail.

In relation to Tables 2.1–2.4 some explanatory comments and refinements can now be made, and conclusions can be drawn about the nature of phytolith production. In the first place, it is clear that numerous families of plants are consistent accumulators of significant amounts of solid silica and are high phytolith

Table 2.2
Production and Silicification Patterns in Pteridophtyes and Gymnosperms[a]

	Production	Silica (%)	Epidermis	Sub-epidermal	Endodermal	Sclereids	Tracheids	Unknown origin
Pteridophytes								
Hymenophyllaceae								
Trichomanes osmundoides	A							A
T. pinnatum	A							A
Hymenophyllum plumosum	NP							
H. nigrescens	NP							
Polypodiaceae								
Adianthum concinum	C		C^b					
A. decoratum	A		C^a,C^b					
Asplenium auritum	NP							
Cyclopeltis semicordata	NP							
Selaginellaceae								
Selaginella arthritica	A							A
S. haematodes	A							A
Gymnosperms								
Pinaceae								
Abies balsamea[1]	R-NC	0.18	*	*			*	
Larex decidua[1]	C	1.4	*	*			*	
Picea glauca[1]	C	1.1	*	*			*	
Picea mariana[1]	R-NC	0.17	*	*	*		*	
Pinus banksiana[1]	R-NC	0.18	*	*	*		*	
Pinus strobus[1]	R	0.09	*	*			*	
Podocarpus neriifolia[2]	R-NC	0.12						
Podocarpus magnifolius	NP							
Pseudotsuga menjiesii[1]	NC	0.29		*	*		*	
Tsuga canadensis[1]	R-NC	0.02, 0.19, 0.10 0.12, 0.09		*			*	
Tsuga caroliniana[1]	NC	0.29		*	*		*	

[a]Production categories: A, abundant; C, common; NC, not common; R, rare; NP, not present; Epidermis categories: [a] = anticlinal, [b] = elongate with two undulating ridges; * = observed in plants; 1., Klein and Geis (1978); 2., Lanning (1966).

Table 2.3
Production and Silicification Patterns in Monocotyledons[a]

	Production	Silica (%)	Epidermis	Hair cells	Sub-epidermal	Sclereids	Tracheids	Unknown origin
Marantaceae								
Calathea violaceae	A		NC[e]		C[h]-A[l]			
Calathea altissima	NC				NC[l]			
Leaf								
Seed	A		A					
Stromanthe lutea								
Leaf	C				C[j]-C[k]			
Stromanthe lutea								
Seed	A		A					
Maranta arundinaceae								
Arrowroot								
Leaf	A				C[g]-N[i]-R[j]-C[k]			
Seed	A	11	A					
Palmae								
Acrocomia sp.[15]	A				A[h]			
Acrocomia vinifera	A				A[h]			
Chrysophila warscewiczii	C				C[i]			
Bactris gasipaes								
Peach palm								
Leaf	A				C[h]			
Fruit	C				C[h]			
Inflorescence	C				C[h]			
Euterpe panamensis	A				A[i]			
Oenocarpus panamensis								
Leaf	A				A[i]			
Fruit	A				A[i]			
Sabal minor	A				A[i]			
Sabal minor[4]	A	2.1						
Sabal etoria[5]	A	3.1						

Serenoa repens[14]	A	5.2	
Serenoa repens[5]	A	3.7	
Scheelia sp.[15]	A		A[i]
Scheelia zonensis			
Leaf	A		A[i]
Petiole	A		A[i]
Stem	A		A[i]
Araceae			
Anthurium denudatum	NP		
Monstera sp.	NP		
Spathiphyllum phryniifolium	NP		
Colocasia esculenta[1]			
Taro			
Leaf	NP		
Tuber	NP		
Xanthosoma sagittifolium			
Otoy			
Leaf	NP		
Tuber	NP		
Cyclanthaceae			
Carludovica palmata	NP		
Cyclanthus biparticus	NP		
Ludovia integrifolia	NP		
Orchidaceae			
Epidendrum anceps	A		A[h]
Maxillaria variabilis	A		A[h]
Rodriquezia secunda	A		A[h]
Liliaceae			
Antherium macrophyllum	NP		
Smilacina paniculata	NP		
Yucca aloifolia[4]	NP (?)	0.03	
Alismataceae			
Sagittaria lancifolia	NP		
Amaryllidaceae			
Hymenocallis pedalis	NP		

(continues)

Table 2.3 (*Continued*)

	Production	Silica (%)	Epidermis	Hair cells	Sub-epidermal	Sclereids	Tracheids	Unknown origin
Bomarea alleni	NP							
Iridaceae								
Cipura paludosa	NP							
Neomarica gracilia	NP							
Butomaceae								
Limnocharis flava	NP							
Smilacaceae								
Smilax panamensis	NP							
Smilax lanceolatum	R							
Xyridaceae								
Gleichenia bifida	NP							
Xyris jupicai	NP							
Dioscoreaceae								
Dioscorea sp.								
Yam-yampi								
Leaf	NP							
Tuber	NP							
Dioscorea sp.								
Yam-ñame								
Leaf	NP							
Tuber	NP							
Dioscorea sp.[1]								
Yam								
Leaf	NP							
Tuber	NP							
Dioscorea sp.[2]								
Leaf	NP							

Musaceae							
Musa sp.							
Plantain	A						
Musa sp.[1]							
Banana, leaf	A				A[m]		
Corm	NP						
Musa sp.[6]							
Banana, leaf	A				A[m]		
Commelinaceae							
Athyrocarpus persicariaefolium	A		C[b]	C[f]-C[g]			C
Commelina diffusa	R					R	
Cyperaceae							
Cyperus chorizanthus	A		A[h]				
Eleocharis nodulosa	A		A[h]				
Fimbristylis annua	R		R[h]				
Rhynchospora barbaya	NC		NC[h]				
Cyperus polystachyas[2]	C	1.9					
Bromeliaceae							
Bromelia balanense	C		C[d]				
Tillandsia polystachya	NP						
Ananas comosus							
Pineapple							
Leaf	A		A[d]				
Fruit	A		A[d]				
Aechmea magdalenae	C		C[d]				
Heliconeaceae							
Heliconia sp.	A		C[e]		C[k]-A[m]		
Heliconia sp.	A		C[e]		A[m]		
Heliconia sp.[6]	A				A[m]		
Zingiberaceae							
Costus villosissimus	C		C[c]				
Dimerocostus uniflorus	A		A[c]				
Cannaceae							
Canna indica	A				A[k]-C[l]		

(*continues*)

Table 2.3 (*Continued*)

	Production	Silica (%)	Epidermis	Hair cells	Sub-epidermal	Sclereids	Tracheids	Unknown origin
Canna edulis[3]								
Achira	C							
Gramineae								
Oryza sativa[7]								
Rice	A		A					
Oryza sativa[8]								
Leaf	A		A					
Glume	A		A					
Hordeum distichum								
Barley								
Glume[9]	A		A					
Triticum dicoccum								
Emmer wheat[7]								
Leaf	A		A					
Glume[9]	A		A					

Triticum monococcum			
Einkorn wheat			
Glume[9]	A		A
Zizania aquatica[10]	A	6.0	
Zizanopsis miliacea[11]	A	7.4	
Zea mays			
Leaf	A	2.8,2.6*	
Husk	R-NC	.97,0.6*	
Tassel	C	1.0,2.2*	
Leaf[12]	A	7.3	
Husk[12]	R-NC	0.9	
Tassel[12]	A	4.3	
Kernel[12]	R	.01	
Root[13]	NP		

[a]Major plant domesticates are boldface.

Production categories: A, abundant; C, common; NC, not common; R, rare; NP, not present, NP(?) may be in the form of nondescript fragments. Epidermis categories: a, anticlinal; b, polyhedral; c, irregularly angled or folded surfaces; d, spherical spinulose; e, other. Hair-cell categories: f, segmented; g, non-segmented. Subepidermal categories: h, conical to hat-shaped; i, sperical spinulose; j, spherical nodular; k, spherical rugulose; l, irregularly angled or folded surfaces; m, with troughs.

*Values listed are for two separate races.

References: 1. Wilson (1982); 2. Ayensu (1972), 3. Pearsall (1979), 4. Lanning and Eleuterius (1983), 5. Kalicz and Stone (1984), 6. Tomlinson (1969), 7. Metcalfe (1960), 8. Watanabe (1968), 9. Miller-Rosen (1985), 10. Lanning and Eleuterius (1981), 11. Lanning and Eleuterius (1983), 12. Lanning *et al.* (1980), 13. Bennett and Sangster (1982) 14. Lanning and Eleuterius (1985), 15. Tomlinson (1961).

Table 2.4
Production and Silicification Patterns in Dicotyledons[a]

	Production	Silica (%)	Epidermis	Hair cells	Hair bases	Cystoliths	Mesophyll	Stomata	Sclereids	Tracheids	Rods	Unknown origin
Chloranthaceae												
Hedyosmum sp.	A		A[a]-C[b]							NC		
Piperaceae												
Piper flagellicuspe	C		C[b]	C[d]								
Piper pseudoasperi	A		C[b]	C[c]	NC					NC		
Peperomia galoides	NP											
Peperomia dodeatheotophylla	NP											
Menispermaceae												
Odontocarya tamoides	C			C[d]								C
Amaranthaceae												
Amaranthus sp.	NP											
Polygonaceae												
Polygonum sp.	NP											
Dilleniaceae												
Curatella americana	A	24.3	C[b]	C[d]	C							
Davilla aspera	A		C[a]-C[b]	C[d]	C							
Davilla rugosa	A		NC[a]-C[b]	C[d]	C							
Tetracera volubilis	A	21.3	C[a]	NC[d]	C							
Sterculiaceae												
Guazuma ulmifolia	C		C[b]	R[c]	C				NC		NC	
Waltheria indica	C		C[b]	NC[d]	C		C			NC		
Theobroma cacao (cacao)	NC							NC	NC	NC		
Chrysobalanaceae												
Hirtella racemosa												
Leaf	C		NC[b]		C				C			
Fruit	C		C									
Seed	NP											
Hirtella triandra												
Leaf	A		C[b]		A				A		C	

Fruit	A		A								
Licania hypoleuca											
Leaf	A		NC[b]					NC	A	NC	C
Fruit	NP										
Seed	NP										
Rosaceae											
Rubus praecipuus	R		R[b]								
Fragaria virginiani[5]	A	9.4									
Guttiferae											
Mammea americana											
Mammey	NP										
Myrtaceae											
Psidium guajava											
Guava	NP										
Convolvulaceae											
Ipomoea batatas											
Sweet potato											
Leaf	NP										
Tuber	NP										
Fagaceae											
Quercus macrocarpa[1]	NC	0.44	*		*		*				
Quercus rubra[1]	R-NC	0.15	*		*		*				
Cannabaceae											
Cannabis sativa[9]	C			*		*					
Malvaceae											
Malvaviscus arboreus	C		C[b]	C[d]	C		NC		NC		
Sida ciliaris	R		R[b]								
Malvastrum americanum	NP										
Gossypium barbadense											
Cotton											
Leaf	NC						NC		NC		
Pod	NP										
Fiber	NP										

(*continues*)

Table 2.4 (*Continued*)

	Production	Silica (%)	Epidermis	Hair cells	Hair bases	Cystoliths	Mesophyll	Stomata	Sclereids	Tracheids	Rods	Unknown origin
Ulmaceae												
Celtis iguanea	NP											
Celtis occidentalis[1]	A	3.4	*	*			*	*				
Trema micrantha	A		C[b]	A[d]	C							
Ulmus americana[1]	A	3.3	*	*			*	*				
Ulmus americana[2]	A	5.1	*	*			*	*				
Moraceae												
Artocarpus altilis	A		C[b]	C[d]	C							
Brosimum discolor	A		A[b]	A[d]								
Castilla elastica	C			C[d]							NC	
Cecropia peltata	A		C[b]	C[d]	C							
Chlorophora tinctoria	A		C[b]	C[d]	C			C		C		
Ficus americana	C		C[b]		C	NC		NC				
Morus rubra[1]	A	3.8	*	*	*		*	*				
Morus rubra[8]	A	3.1										
Trophis racemosa	A		NC[b]	C[d]	C		NC				NC	
Urticaceae												
Boehmeria aspera	A	7.3	C[b]	C[d]		A						
Laportea aestuans	C		NC[b]	NC[d]	NC	C						
Myriocarpa densiflora	A		NC[b]	C[d]		A						
Pauzolzia obliqua	A			C[d]		A						
Pilea acuminata	C			R[d]		C						
Urera elata	A		C[b]	C[d]	R	A						
Cucurbitaceae												
Cayaponia citrullifolia	A		C[b]	C[c]-NC[d]	C		C				NC	
Gurania makoyana	C		NC[a]	C[c]								
Luffa cyclindrica												
Sponge gourd	C			C[c]	C							
Melothria fluminensis	C		C[b]	C[c]	C							

Melothria scabra	A		C[b]	C[c]	C					
Pittiera grandiflora	A		C[b]	C[c]	C	NC				
Sicyos echinocystoides	A		C[a]	A[c]	C	C				
Cucurbita pepo										
Squash										
Leaf	A	6.6	NC[a]	C[c]	C					
Rind[3]	P,ND									P,ND
Lagenaria siceraria										
Bottle gourd										
Leaf	C		C[a]	C[c]	NC	NC				
Rind[3]	P,ND									P,ND
Annonaceae										
Guatteria dumetorum										
Leaf	A		A		C		NC		R	
Fruit	NP									
Seed	NP									
Unonopsis pittieri										
Leaf	C		C							
Fruit	NP									
Seed	NP									
Annona muricata										
Guanabana	NC		NC[a]						R	
Anaxagorea panamensis	NP									
Lauraceae										
Persea americana										
Avocado										
Leaf	C						C	C		
Rind	NP									
Sassafras albidum[1]	R	.07								
Betulaceae										
Carpinus caroliana[1]	C	0.6	*	*	*	*	*			
Ostrya virginiana[1]	NC	0.3	*	*			*			
Salicaceae										
Populus deltoides[1]	C	0.9	*			*	*			

(*continues*)

Table 2.4 (*Continued*)

	Production	Silica (%)	Epidermis	Hair cells	Hair bases	Cystoliths	Mesophyll	Stomata	Sclereids	Tracheids	Rods	Unknown origin
Aceraceae												
Acer saccharum[1]	A	2.6										
Acer negundo[1]	NC	0.3										
Chenopodiaceae												
Chenopodium	NP											
Chenopodium album[4]	NP											
Bombacaceae												
Bombacopsis quinata	NP											
Bombacopsis sessiles	C		NC[b]					C		C		
Cavanillesia platanifolia	NC		NC[b]							R		
Ceiba pentandra	NP											
Burseraceae												
Bursera simaruba												
Leaf	NP											
Seed	A		A									
Fruit	NP											
Protium costaricense												
Leaf	C		C[a]									
Fruit	C		C									
Protium panamense												
Leaf	A		A[a]					C		NC		
Fruit	C		C									
Seed	NP											
Tetragastris panamensis												
Leaf	A		A[a]									
Fruit	C		C									
Seed	NP											
Trattinickia aspera												
Leaf	A		A[b]	C[d]-NC[c]	NC		NC					
Seed	A		A									

Juglandaceae										
Juglans nigra										
Leaf[1]	NC	0.3	*		*		*			
Nut[7]	P,ND									
Carya ovata[1]	C	0.5	*	*	*		*			
Caricaceae										
Carica papaya										
Papaya	NP									
Bixaceae										
Bixa orellana										
Achiote	NP									
Malphigiaceae										
Byrosonima crassifolia	NC		NC[b]						NC	
Solanaceae										
Petrea volubilis	NP									
Solanum nigrum	NP									
Capsicum annuum										
Chile pepper	NP									
Boraginaceae										
Bourreria pulchra	A		C[b]	C[d]	NC				C	
Cordia alba	C			C[d]	C					
Cordia lutea	A		C[b]	NC[d]	C				C	
Ehretia anacua	A		C[b]	C[d]	C					
Hackelia mexicana	A		C[b]	C[c]	C					
Heliotropium angiospermum	C		C[b]		C		NC			
Heliotropium indicum	A			C[d]	C					
Verbenaceae										
Lantana camera	A		C[a]-C[b]	C[c]-C[d]					R	
Petraea sp.	NP									
Labiatae										
Hyptis suaveolens	R		R[b]	R[c]-R[d]				R		
Acanthaceae										
Aphelandra runcinata	NC		NC[b]							
Blechum brownei	NP									
Dicliptera cochabambensis	NP									
Justicia pringlei	C		C[a]-C[b]	C[c]-NC[d]		C			R	R

(*continues*)

Table 2.4 (*Continued*)

	Production	Silica (%)	Epidermis	Hair cells	Hair bases	Cystoliths	Mesophyll	Stomata	Sclereids	Tracheids	Rods	Unknown origin
Mendoncia coccinea	A			A^d		NC						
Odontonema bracteolata	C		C^a-C^b			C						
Rubiaceae												
Borreria densiflora	NC		NC^b	NC^d								
Borreria latifolia	NP											
Borreria verticulata	R		R^b	R^d								
Coffea arabica[2]												
Coffee	R-NC	0.2										
Compositae												
Bidens sp.	A		C^b	C^c-NC^d			NC					
Baccharis cassinaefolia	R		R^b									
Calea urticifolia	A		C^a-C^b	C^c-C^d								
Eclipta alba	C	C^a	C^b	C^c-NC^d	C					C		
Elephantopus mollis	A		C^a-C^b	C^d	C		C					
Erigeron maximis	NC		NC^a	NC^c								
Melanthera hastata	C		C^b	C^c	C							
Vernonia argyropoppa	NC		C^b	NC^c-NC^d	C						R	
Wulffia baccata	A		C^a-C^b	C^c	C							
Helianthus annuus												
Sunflower												
Leaf[6]	A	2.4										
Achene[7]	P,ND											
Iva frutescens[4]	C	1.8										
Bignoneaceae												
Cydista aequinoctalis	NP											
Tabebuia neochrysantha	NP											
Tabebuia palustris	R		R^b							R		
Rhizophoraceae												
Rhizophora mangle	NP											
Melastomataceae												
Clidemia taurina	NP											
Miconia argentea	R										R	
Combretaceae												

Avicennia nitida	NP											
Conocarpus erectus	NC							NC	NC			
Laguncularia racemosa	NC		NC[b]					NC				
Loranthaceae												
Phithirusa pyrifolia	C		NC[b]							NC		C
Phorodendron piperoides	R									R		
Phrygilanthus corymbosus	C									C		C
Struthanthus marginatus	R		R[b]									
Euphorbiaceae												
Acalypha diversifolia	C		C[b]	R[d]	C						NC	
Bernardia macrophylla	C		C[a]-C[b]	NC[d]	C		C					
Euphorbia eichleri	NP											
Hieronyma oblonga	NC		NC[b]						NC	C		
Hura crepitans	NP											
Tragia volubilis	R		R[b]	R[d]								
Manihot esculenta												
Manioc												
Leaf	R	0.4		R[d]								
Tuber	NP											
Wood and bark	NP											
Leguminoseae												
Acacia farnesiana	R		R[b]									
Enterolobium cyclocarpum	NP											
Pithecolobium latifolium	C		C[a]					C	C			
Prosopis juliflora	NP											
Gymnocladus dioica[1]	R-NC	0.2	*	*			*					

[a]Major plant domesticates are boldface.

Production categories: A, abundant; C, common; NC, not common; R, rare; NP, not present; P,ND, phytoliths present, production level not determined. Epidermis categories: a, anticlinical; b, polyhedral. Hair cell categories: c, segmented; d, nonsegmented.

*Observed in plants.

References: 1. Geis (1973), 2. Lanning (1966), 3. Bozarth (1986), 4. Lanning and Eleuterius (1983), 5. Lanning (1960), 6. Lanning (1972), 7. Bozarth (1985), 8. Lanning and Eleuterius (1985), 9. Dayanandan and Kaufman (1976).

producers no matter from what part of the world they are examined. Conversely, many families are consistent nonaccumulators and low phytolith producers, independent of the environmental conditions of growth. The compatible results obtained from production studies of researchers working on the same families and genera drawn from widely different environmental regions show this to be true. Examples of families where phytolith production is high include the Palmae, Urticaceae, Moraceae, Marantaceae, Dilleniaceae, Burseraceae, Cucurbitaceae, Compositae, Chrysobalanaceae, Hymenopyllaceae, and Selaginellaceae. Examples of families characterized by poor silicon accumulation include the Chenopodiaceae, Rubiaceae, Araceae, Cyclanthaceae, Melastomataceae, and Smilacaceae. Therefore, in addition to families of the monocotyledonae, many dicotyledons and plants reproducing by spores are demonstrated to be high producers of phytoliths. They possess a siliceous content equal to or more than the values of 2–5% commonly reported from grasses. Furthermore, it is seen that phytolith formation in monocotyledons is not invariably high; in fact, many families have yielded no phytoliths at all. The characterization of monocots as silicon accumulators and other plants as nonaccumulators should be unceremoniously laid to rest.

One should certainly be cognizant of nongenetically controlled (environmental) factors that may affect silica uptake and concentration, but it is clear that taxonomic affinities of plants are the prime determinant of their phytolith content. When I first started to construct a modern phytolith comparative collection, I looked to information in Metcalfe and Chalk (1950) and Tomlinson (1961, 1969) for indications of which families commonly deposit silica. Their information, in turn, had been compiled partly independently and partly on the basis of studies carried out during the botanical period of research, on plants predominantly from Old World temperate and tropical regions. Results of my own subsequent research on mainly New World tropical species corresponded almost entirely with the information of Metcalfe, Chalk, and Tomlinson. Families that had been listed as notable concentrators of solid silica were indeed high phytolith producers and those listed as being limited or nonaccumulators were invariably found to have low or nonlevels of phytoliths in tropical species. (Compare results compiled by early twentieth-century German botanists, Ch. 1, p. 4, with Tables 2.1–2.4.) My results, in turn, agreed well with those published during the last 25 years by North American researchers who worked on related plants (e.g., Wilding and Drees, 1968, 1971, 1974; Geis, 1973).

In addition to familial correlation, subfamilies and tribes also tend to have similar patterns of deposition. For example, the Astereae, a tribe of the Compositae, commonly produce low amounts of phytoliths when other subfamilies in the Compositae produce phytoliths in substantial numbers (Piperno, 1983, 1985a; Lanning and Eleuterius, 1985). The Mimosoideae subfamily of the Fabaceae (Leguminoseae) tends to concentrate silica to a much greater degree than others in the family

(Solereder, 1908; Piperno, 1985a). The Pooideae (Festucoid) subfamily of grasses characteristically concentrate lower amounts of solid silica than do other Gramineae.

It is evident also that within families and subfamilies the same kinds of cells or tissue are persistently silicified. For example, for every species in the Palmae, Orchidaceae, and Marantaceae, subepidermal (mesophyll) tissue is virtually the sole locus of silica deposition, while in the Cyperaceae and Gramineae, silicification is predominant in the epidermis. In the Moraceae and many other dicotyledon families, epidermis, including hair cells and hair bases, forms the dominant numbers of phytoliths. Cystoliths are produced in high numbers by the Urticaceae, Moraceae, and Boraginaceae but are found in few other families. It is clear that the production of phytoliths in modern species is very regular and faithfully replicated, and there is little reason to suspect that it has changed much over the period of recent plant evolution. We show in Chapter 3 how the shapes of phytoliths in plants are also faithfully produced. For example, the silicified mesophyll of the Palmae and Marantaceae invariably gives rise to discrete, family-specific shapes (Table 2.3). Again, such patterns argue strongly for a genetic control of silicification, a factor that imparts sound paleoecological validity to phytolith analysis and opens interesting possibilities for its application to modern plant taxonomy. Like other microscopic features, such as chromosome number, and gross macroscopic characteristics, phytoliths may contribute important data on plant relationships.

The quantification of the amount of silica in plants, either by absolute or estimated methods, is important not only for comparative studies of silicification in modern species, but also for appraisements of what species will be represented in a paleobotanical phytolith record. Palynologists, for example, usually have a sound idea of which types of pollen might be underrepresented in (or absent entirely from) pollen sequences because of differential production. Pollen production levels are usually directly proportional to the probability of success in fertilization. Hence, plants using insects and animals as dispersal agents make and contribute far less pollen than do wind-pollinated species. As a result, pollen from entimophilous species, such as manioc *(Manihot esculenta)* and sweet potato *(Ipomoea batatas)* is seldom recovered from archaeological deposits.

In the preceding discussion certain families of plants were described as high or low producers of phytoliths, and it was shown how deposits of solid silica have been isolated in measurable amounts, in quantities ranging from less than 0.1% to more than 25% of the dry vegetal matter, from numerous plants. At what level must a "measureable" percentage be to achieve significant representation in soils, and what kinds of percentages do high and low phytolith producers contribute? We know that plants described as silicon accumulating, such as grasses, palms, squashes, nettles, and composites, contribute silica levels of at least 1 to 2%, and they form important constitutents of soil profiles. One may wonder, however, about the archaeological and geological visibility of many plants and plant structures whose levels of production do not reach about 0.5–1% (for example, the husks of maize and leaves of manioc). This category also applies to many plants listed in Table 2.2 as having uncommon to rare production. They will probably be either significantly underrepresented or not present in soil phytolith assemblages. Chapters 6, 7, and 8 fully discuss the occurence of phytoliths in modern and fossil soils.

Patterns of Silica Deposition and Distribution in Different Plant Structures

Virtually any plant structure, be it leaf, seed, fruit, root, or timber, can serve as a repository of silica deposition. Within these structures, deposition can be highly localized in a single kind of tissue, or be distributed throughout the entire plant body. Some plants can have silicon evenly distributed between roots and shoots, while in many others aerial organs accumulate it to a far greater degree than do subterranean organs. As a short preface to this section, it should first be stated that there still remain entire families of plants for which particular structures have not been evaluated for phytolith production. Consequently, it will be many years before anything approaching a more complete synthesis will be possible. This need not dissuade us from recounting the various patterns found in the many species investigated. These patterns are generally very consistent within families and species, and they may eventually be considered significant taxonomic and phylogenetic characters.

Herbacious Plants

Among herbacious plants, the grass family has been the most intensively studied. All portions of the above-ground plant bodies may have discrete deposits of solid silica. In cereals, the inflorescence bracts of wheat, oats, barley, and rye, comprising the glumes, lemmas, and paleas, often have a higher silica content than do the leaf blades, leaf sheaths, and culms. This pattern was not affected by varying the supply of monosilicic acid in the soil solution from 7 to 67 ppm (Jones and Handreck, 1967). Among leaf blades, sheaths, and culms, there is often a decreasing gradient of silica content, with blades having the highest values and culms showing negligible amounts, ranging from only one-tenth to one-fifth of leaf blade content (Lanning *et al.*, 1980; Geis, 1978).

The husks of maize, which are modified leaf sheaths, generally have low silicon concentrations, with amounts varying between 0.1 and 0.9% (Table 2.3). These are values ranging from only one-tenth to one-half of the content of leaf blades. Cross- and dumbbell-shaped phytoliths form a significant part of the solid silica content of husks, but it is highly doubtful that they would achieve good representation in soil profiles. Maize tassels have silica levels varying from one-third to over three-quarters of leaf content, and hence, production is more variable than in either leaves or blades. It would be difficult to predict how visible tassel phytoliths would be in soils, although in Chapter 7 evidence from archaeological sites from Panama is presented indicating that maize phytoliths are primarily from leaf decay with possibly a small amount of tassel and no husk contribution.

The roots and rhizomes of many grasses contain silica in amounts ranging from 0.7 to 15%, but values from species to species are hard to compare because of different ages of the roots assessed (Geis, 1978; Lanning and Eleuterius, 1985; McNaughton *et al.,* 1985). Discrete identifiable phytoliths appear to be in the minority, although Geis (1978) isolated saddle-shaped silica bodies and epidermal long cells from the rhizomes and distinctive pitted plate phytoliths from the roots of three species of grasses. Most of the silica from roots is in the form of small, nodular aggregates from endodermal cell silicification whose morphological significance is difficult to assess (Sangster, 1978).

Bennett and Sangster (1982) reported no phytoliths from the adventitious roots of maize, while Lanning *et al.* (1980) measured significant amounts in roots of some maize varieties from Kansas. Morphology was not described, thus it is difficult to assess if particles were nondescript fragments or morphologically significant. In regard to grass root silicification, Sangster (1978) believes that silica distributional patterns between species may depend on the phylogenetic affinities of the taxon and further that root silicification may be under active or genetic control. He notes that silicon deposition is restricted to the endodermal cell walls in all species of the tribe Andropogoneae studied, a constant pattern irrespective of differences in the basic root anatomical pattern.

Lanning and Eleuterius (1981, 1983) have studied silica distribution in a number of nongrass herbacious silica-accumulating species and found patterns similar to those of grass deposition, with leaves often contributing more than twice as much silica as do stems. They showed that the inflorescences of *Cyperus* (Cyperaceae) and *Juncus* (Juncaeae) have a high silica content, often exceeding amounts in leaf blades (Lanning and Eleuterius, 1985). Culms and stems of species whose leaves did not have a high silicon content correspondingly showed no production either. Bozarth (1985) has isolated distinctive-looking phytoliths from the achenes of sunflower *(Helianthus anuus)*, opening a new area of research for archaeological investigations of this important North American domesticate.

I have examined seeds from several tropical herbacious families, including the Marantaceae, Cannaceae, and Cyclanthaceae. Patterns of production are found in Table 2.3. The Marantaceae and Cannaceae are notable for having high quantities of seed phytoliths, and in the former morphologies are often quite distinctive, as discussed in Chapter 3. Seed phytoliths were often produced by species whose leaves contributed significant amounts, but were never present in species showing no or low phytolith production (i.e., the Cyclanthaceae). This appears to be a fundamental pattern of phytolith production that may be used in designing sampling strategies for modern collections: silicon-accumulating or nonaccumulating species can be defined on the basis of leaf content; that is, if leaves show considerable production other plant structures of that species may show production as well, in varying proportions to leaves; but if leaves do not have significant phytolith quantities, other plant parts most probably will not produce phytoliths at all. [The exception is wood of tropical plants, which shows significant phytolith production in many species considered nonsiliceous (Ter Welle, 1976).]

Nonaerial organs of herbacious, nongraminaceous species have been examined very little. Tomlinson (1969) reports the presence of phytoliths in the roots of *Maranta arundinaceae,* a South American root crop, and indicates that phytoliths are absent from Zingiberaceae roots. Lanning and Eleuterius (1983, 1985) note that several herbacious roots they studied *(Cyperus, Juncus)* had a substantial silica content, often equaling or halving that of leaves, but it is difficult to know whether siliceous particles take the form of morphologically recognizable structures or only nondescript fragments.

Woody Species

Distribution of silica deposition sites has been studied in a limited number of arboreal, woody species. Geis (1983) compared foliage, branches, bark, and bolewood for sugar maple *(Acer saccharum)* and pine *(Pinus resinosa)*. In sugar maple, almost 60% of the silica was concentrated in the foliage but red pine foliage contained only 16% of all the total silica content of the plant. Diagnostic phytoliths represented only a minor component of sugar maple silica and a very minor fraction of red pine silica, and was restricted almost entirely to the foliage. Significant quantities of silica have been isolated from the wood of numerous species of dicotyledons (Amos, 1952; Scurfield *et al.,* 1974; Ter Welle, 1976).

I have examined fruits and seeds from about 100 species of tropical palms and dicotyledons. The pattern is the same as found in herbacious plants; species that accumulate silica in their leaves may also, but not always, do so in reproductive structures, whereas in leaf nonproducers phytoliths are almost never found in fruits or seeds. The single nonaccumulating species found thus far to have seed phytoliths is *Bursera simaruba* (Burseraceae), a tropical forest tree. Among woody plant families so far studied, the Burseraceae, Chrysobalanaceae, Palmae, and Urticaceae are notable for the production of fruit and seed phytoliths. Production is often restricted to the fruit exocarp or mesocarp, where it perhaps functions to protect the propagules of plants from predators and is less common in seeds. As will be seen in Chapter 3, morphology is often highly significant in identifying families and genera of plants. Examples of production patterns for woody species are found in Table 2.4.

Localization of Silicon Deposition in Plant Structures

Just as any plant organ may be silicified, the location of solid silica within particular kinds of tissue or cells may be highly varied. In the above-ground parts of grasses, phytoliths are derived mainly from epidermal cell silicification, and the complex epidermal structure of grasses results in a myriad of phytolith shapes. The same pattern of predominant epidermal silicification takes place in Cyperaceae (sedge) leaves, while in the Palmae and Marantaceae phytoliths are mainly subepidermal in origin. Tomlinson (1969) has provided a useful review of the loci of silicification in monocotyledons. Other patterns compiled from studies can be found in Table 2.3.

Among dicotyledons, epidermal tissue, including hair cells (trichomes) and hair bases is often the main locus of silicification (Table 2.4) Whole families may be characterized by the production of certain kinds of epidermal and hair phytoliths. Phytoliths derived from other kinds of tissue are generally less frequent and more sporadically produced.

On the cellular level, silica can have three loci of deposition; it can form as incrustations of cell walls, in the interior of cells, called lumina, and in intercellular spaces. At the beginning of this century, Hans Solereder (1908, p. 1110) wrote, "there are far fewer records of the occurrence of siliceous material in the lumina of cells than of silicification of the cell wall," a statement that is every bit as true today. We will discuss later in the chapter how an ancient (and still fundamental) function of silica in plants may have been as a compression-resistant element, where it helps to prevent collapse of cell walls when the water contained in them is under tension during transpiration. Thus, it is not surprising that cell wall silicification should be so common an occurrence. Discrete silica occurring as deposits in epidermal cell walls is especially noteworthy of many dicotyledons and resulting phytoliths appear to be much more susceptible to dissolution in soils. Solid lumina deposits of silica are less common in dicots, an unfortunate situation, because these are the types of phytoliths that appear to be very resistant to chemical and mechanical weathering and dissolution over long periods of time.

In monocotyledons, on the other hand, the silicification process more often involves an infilling of the cell lumina, resulting in solid "plugs" of silica of varying shapes. It is not too surprising then that phytoliths belonging to such taxa as the Palmae, Bromeliaceae, Gramineae, Cyperaceae, Marantaceae, and Heliconeaceae have been recovered more frequently from sediments than certain dicots, such as the Moraceae, whose levels of silica may be high, but whose phytolith populations are largely comprised of cell wall incrustations.

Mechanisms of Phytolith Deposition in Soils

Silicified cells are liberated from their organic matrix and deposited in soils by means of a number of pathways. When recovered from soils or isolated from living plant material they are known as phytoliths. The most common circumstance of deposition is through death and decay of the plant, which releases siliceous materials into the A horizons of soil profiles or into archaeological middens. Phytoliths can also be released into the environment via animal droppings or fire.

Because phytoliths are formed predominantly in the vegetative organs of plants and released into soils, not into the air, we can expect that a large proportion of the phytolith record represents a highly localized, *in situ* deposition. However, long-distance transport of phytoliths, while often negligable as compared with pollen grains, should not be disregarded as an unlikely event. It has been shown that phytoliths can move long distances in wind-blown dust (Folger *et al.*, 1967; Twiss *et al.*, 1969; Parmenter and Folger, 1974). They have been recovered from deep-sea cores off the Northwest African coast at distances up to 2000 km from their source areas on the continent (Melia, 1980). The strong Harmattan winds that blow during the dry season and the sparse, arid vegetation that leaves ground cover bare and exposed to dust storms are thought to be responsible for strong movement from the African continent.

The combination of wind and fire can also contribute heavily to phytolith movement in the air (Labouriau, 1983). Deposition away from source areas may also occur in animal droppings and through a decrease in vegetative cover and resultant soil erosion that expose phytoliths to water runoff, especially in areas of high precipitation. Human transport of plants may also cause phytoliths to be deposited in areas removed from original plant growth.

Factors influencing phytolith movement, such as the presence of a well-developed vegetal mat that covers A horizons of soils and binds phytoliths and the degree of subsequent soil erosion and aerial transport, are dependent on various cultural and environmental conditions like rainfall, wind velocity, and the degree of human modification of plant cover. Therefore, as discussed in Chapter 8, paleoecological data pertaining to the amount and nature of phytolith transport as evidenced from stream or other discharge into lakes and occurrence in deep-sea cores can provide valuable paleoenvironmental and climatic information.

Chemical and Physical Characteristics of Phytoliths

A number of studies on the chemical and physical properties of phytoliths have demonstrated several compositionally and structurally related measurable properties. Phytoliths are often referred to as "plant crystals", when, in fact, siliceous secretions are composed mainly of amorphous (noncrystalline) silicon dioxide (SiO_2) with varying amounts of water, usually ranging from 4 to 9%. Lanning (1960, 1961) and Sterling (1967) have reported small amounts of crystalline silica in plants. The real nature and derivation of these crystalline structures was for some time a matter of dispute. Jones and Milne (1963) argued that crystalline particles were artifacts produced by dry ashing (oven ignition at high temperatures) of modern plant material, which supposedly produced physical and chemical changes in amorphous silica. However, Sterling (1967) bypassed potential methodological difficulties of identification by "wet ashing" plants, a process whereby organics are chemically destroyed, and still found measureable levels of crystalline silica in several modern species examined. Several studies suggest that as phytoliths age, their composition is changed into more crystalline forms, which are more durable in soil environments (Wilding *et al.,* 1977).

Significant amounts of occluded, chemisorbed, or solid solution impurities, such as Al, Fe, Ti, Mn, P, Cu, N, and C can be present in phytoliths. Nitrogen and carbon may be a result of occlusion of cytoplasmic material during phytolith formation in living cells, or of silica impregnation of cellulose and lignin found in cellular structures (Wilding *et al.,* 1977).

Biogenic silica of plant origin is optically isotropic, ranges in refractive index from 1.41 to 1.47, has a specific gravity from 1.5 to 2.3, and ranges in color under transmitted light from colorless or light brown to opaque (Jones and Beavers, 1963). Noncolored phytoliths are often transparent, and it is possible to look through them and ascertain three-dimensional characteristics without actually turning them in the mounting medium. Darker forms are related to higher quantities of organic carbon pigmentation occluded within or on the surface of the phytolith and have a lower specific gravity than lighter colored forms (Jones and Beavers, 1963). These are also commonly produced when plants are burned, and thus may serve as an index of the presence and intensity of prehistoric firing of the vegetation.

Wilding *et al.* (1967) have, furthermore, demonstrated that over 50% of the occluded carbon on phytoliths is not readily accessible to oxidation and can be made available for radiocarbon dating. In the past, direct dating of carbon occluded opal phytoliths with the conventional beta decay counters required rather laborious methods. Wilding (1967), for example, had to use over 50 kg of soils to isolate enough carbon for dating purposes, soils which he processed in a commercial cement mixer (L. P. Wilding, personal communication, 1983). The recent advent of the linear accelerator, which requires only extremely small samples of carbon, will make possible the practical application of direct dating of phytoliths.

The utility of phytoliths as a paleoecological tool depends on its stability in soil environments. The latter term encompasses a variety of depositional settings, from highly weathered soils in open and forested temperate or tropical settings, to the relatively protected situations underneath lakes and the overhangs of rockshelters. The degree of phytolith preservation will no doubt vary according to the chemical and physical nature of the environment and the particular taxon that has left silicified remains.

With regard to differential solubility of phytoliths from different taxa, characteristics such as shape and surface area appear to play important roles. Sheetlike phytoliths derived from epidermal cell wall incrustations of some deciduous (temperate, broadleaf) and tropical trees appear to be much less stable than the solid plugs of silica deposited by other plants because they are found infrequently in soils underneath these kinds of forests (Wilding and Drees, 1973, 1974; Special Topic 2: Chapter 6). Wilding and Drees (1974) and Bartoli and Wilding (1980) have provided experimental data on the differential solubility between plant taxa. In the first study, between two-thirds and three-quarters of the total opal isolated from temperate-zone three leaves was dissolved in a 2–5 min digestion in boiling 0.5 N NaOH, whereas only 5% of grass opal would have been dissolved with a similar digestion. Comparisons of deciduous forest, grass, and conifer phytoliths by Bartoli and Wilding (1980) revealed a number of other interesting relationships. In general, deciduous forest phytoliths from beech *(Fagus)* were younger, more hydrated and had lower aluminum contents, factors which all contributed to substantially increased solubility over the other two types in cold- and hot-water dissolution. Surprisingly, coniferous phytoliths from pine and fir were even more resistant to dissolution than grass phytoliths, presumably because they were older, less hydrated, and, most importantly, contained greater amounts of aluminum. The presence of aluminum apparently reduces the surface area of phytoliths, a factor important in rendering them less soluble.

This somewhat negative picture of deciduous tree phytolith durability will probably be ameliorated under natural conditions because the physical and chemical characteristics of soils themselves have important roles to play in phytolith dissolution. A common attribute of soils, especially those in highly weathered environments, that may enhance phytolith durability is the presence of free iron and aluminum oxides. They tend to become chemisorbed to siliceous surfaces, retarding phytolith dissolution (Wilding *et al.,* 1977).

Extreme levels of pH tend to have the opposite effect. They are as important for phytolith as they are for nonbiogenic silica solubility, which is independent of pH from 3 to below 9, but increases rapidly above pH 9 (Wilding *et al.,* 1977). Pease (1967) has reported a corresponding pH solubility curve for phytoliths. In archaeological situations, highly alkaline soils can be expected to occur in shell middens, where deposits are continually saturated with carbonates and pH levels are artificially raised. Tropical shell middens from Panama, whose pH levels measured from 8.9 to 9.1, have been shown to be devoid of phytoliths, except for those with carbon occlusions, which are said to be more resistant to dissolution (Piperno, 1983, 1985b). These two sites were the only contexts out of numerous sediments examined from Panama, archaeological and geological, that did not yield significant amounts of phytoliths (Piperno, 1983, 1985b,c,). The fact that high soil pH levels dissolve phytoliths does not seem to be universally the case. Lewis (1978) isolated large amounts of phytoliths from a bison kill site in Nebraska where soil pH values ranged from 8.2 to 9, although one wonders about preservation if pH values had exceeded 9. Conditions found in tropical shell middens may be particularly inimical because they are compounded by high year-round temperatures and annual rainfall.

In terms of general characteristics of depositional environments, we might envision that protected sediments underneath lakes and rockshelters would be kinder to phytoliths than soils in open, terrestrial environments, which are subjected to constant weathering. Similarly, phytoliths beneath a well-developed layer of litter may survive better than ones lying exposed at the surface of soils. In sum, it is clear that differential preservation of phytoliths in sediments will occur. The overall preservation factor will differ from region to region depending on the nature of the climate, vegetation, soil, and depositional environment. It is probably premature to assume that phytoliths formed from cell wall silicification will always be far less frequent than phytoliths derived from cell lumina. Numerous soils from many areas of the world will require analysis before we can draw firm statements about these relationships.

The Functional Significance of Phytoliths

The occurrence of silicon dioxide in many phyla of the present and past world has led to study and speculation of its adaptive significance, phylogenetic development, and role in the normal growth and development of living species. Among plants that deposit silica diatoms are probably the best studied, and there is no doubt that silica is absolutely essential for a variety of metabolic processes. Studies have shown that it is just as important for cellular metabolism as for cell wall formation (Simpson and Volcani, 1981).

Investigations concerning the role of silicon in higher plant growth have yielded conflicting results (see Iler, 1979; Kaufmann *et al.*, 1985, for reviews). Kaufman *et al.* (1985, p. 487) state the prevailing opinion, noting that "since there is no evidence that silica plays a direct role in the metabolism of grasses, or in other plants which accumulate it in considerable quantities, one cannot conclude that it is an essential element." Yet, many plants accrue a number of benefits to growth and reproduction that seem to be directly related to the presence of silica. In rice it enhances resistance to fungus disease and increases the grain yield. Rice leaf blades are more erect when silica slags are applied, allowing more light to reach the lower leaves and resulting in increased photosynthetic activity. Grass shoots deficient in silica are stunted or weak and several species have shown increased resistance to predation by chewing incests and mammalian herbivores, presumably as a result of their high phytolith content (e.g., Okuda and Takahashi, 1964; Iler, 1979; McNaughton and Tarrants, 1983; McNaughton *et al.,* 1985). Cucumbers, morning glory, and French beans are other plants that may derive increased resistance to pathogenic fungi from the presence of silica (Iler, 1979).

Mechanical support and increased resistance to herbivory and pathogenic fungi have thus all been cited as functions of silica in plants, although it is often difficult to ascertain if the benefits are direct (directly affecting the plant itself) or indirect (merely modifying the environment). An example cited by Iler (1979) is appropriate here. He notes that one study showed how the addition of colloidal silica to rice grown in nutrient solution appeared to make the plant more tolerant of potassium. However, it is equally possible that the silica acted as an ion adsorbent and therefore kept the potassium ions out of solution and away from the plants.

To further add confusion to an already poorly understood situation, a few studies appear to have shown that silica is an indispensable element for the normal growth and development of *Equisetum* (Hoffman and Hillson, 1979; Chen and Lewin, 1969). Shoots of *Equisetum,* a silica accumulator, collapsed when grown in a silica-free medium (Chen and Lewin, 1969). Silica may be a vital element for beet *(Beta vulgaris)* growth as well (Iler, 1979). We are left with a conflicting picture of the roles and necessity of silica in plant growth and development. There is little doubt that in many silica accumulators, such as rice and other grasses, its presence results in a number of benefits to the plant. In addition, it may be an essential element for the normal growth and development of other species, such as the scouring rushes.

Recent studies have addressed the evolutionary history of silicon deposition in higher plants and its role in the long-term adaptation of some animals that feed on silica accumulators. Thomassen's (1983, 1984) and Smiley and Huggins' (1981) observations of phytoliths in Miocene-age grass, sedge, and beech fossils demonstrate that a silicon-accumulating system developed at an early stage of plant evolution. The negligable silica content of some modern-day families can be attributed to a loss of silicification or to its absence throughout their evolutionary history. Dahlgren and Clifford (1982) note that a silica deposition system may have evolved independently in several monocot orders, citing the abundance of phytoliths in the Cyperaceae (Cyperales) but not in the closely allied Juncaceae (Poales).

Raven (1983) feels that an ancient and still fundamental function of silica in plants was as a compression-resisting element, where it helps to prevent collapse of cell walls when the water contained in them is under tension during transpiration. This is a role analagous to that of lignin and may explain why silica deposition in dicotyledons is so often restricted to cell walls. Energetic efficiency studies have shown that it is much "cheaper" for plants to incorporate silica than lignin, and in view of this it is curious why more extant plants do not accumulate silica in significant amounts. The concept of trade-offs is important to those interested in studying adaptive strategies, and as Raven (1983) notes, if a plant does incorporate silica in leaves than it must "pay" for the added density that SiO_2 imparts by adding more silica to cells further down the shoot, possibly a more expensive strategy than simply using lignin, a lighter material.

McNaughton and Torrants (1983), McNaughton *et al.* (1985), and Herrera (1985) have emphasized the coevolutionary relationship between the silica content of African grasses and some anatomic and behavioral characteristics of herbivores native to the East African plains. In grasses silica is deposited in places likely to facilitate mouthpart abrasion of herbivores, and therefore may have exerted considerable selective pressure on the evolution of herbidont cranial anatomy and dentition.

Phytoliths may also be a potent force in human evolution. Recent studies have unraveled a possible link between ingestion of foods high in silica and the etiology of esophageal cancer (Parry and Hodson, 1982; Hodson *et al.*, 1982; Newman and Mackay, 1983). In regions such as northeast Iran and and northern China, the incidence of esophageal cancer is unusually high, and here the diet of local communities is based on grass seeds containing high amounts of silica (for example, *Setaria italica* and *Phalaris* sp). In Iran, silicified hairs from the inflorescence bracts of *Phalaris* occur in the diet as a contaminant of wheat flour. Apparently plant silica parallels some of the mechanical effects of asbestos fibers on cell growth. Silica fragments were found in the esophageal mucosa surrounding tissues of cancer sufferers in nothern China. Siliceous hair cells were shown to be powerful stimulants of the growth of cultured cells, and they promoted skin tumors in mice (Parry *et al.*, 1984).

In cases such as these, cause and effect is always difficult to demonstrate, but there is clearly a correlation between cancer and ingestion of highly siliceous foods. As Parry *et al.* (1984) have noted, the current popularity of coarse, fibrous foods containing crushed cereal grains may, therefore, have some serious implications for human health.

3

Phytolith Morphology

Introduction

The intricacy and diversity of silicified forms found in lower plants such as diatoms have fascinated microscopists for more or less the entire history of microscopy. Ernst Haeckel (1899–1904) left some remarkable drawings of their exquisite structures from the nineteenth century, and many later-day investigators, including the great limnologist G. Evelyn Hutchinson, have wondered and written about their beautiful and manifold patterns. Were they the product of differentiation brought about by the long-term action of natural selection on genetic diversity or merely a result of undecipherable random processes accumulated over time? It seems that the former process may have been the most important because silica is an essential element of diatom metabolism, and diatom classifications are based on the morphology of their silica shells. Still, the processes that underlie their development continue to attract scientists and to elude them.

Such intellectual curiosity and inquiry has seldom been extended to siliceous structures found in the cells of higher plants. Indeed, it is almost superfluous to point out that scientists have not spent a great deal of time wondering about the diversity of phytolith morphology. On the contrary, a considerable amount of literature can be found attesting to the narrow limits of phytolith production (supposedly being confined largely to certain monocotyledons) and the homogeneity of shapes in plants that are silicon accumulators (with taxonomic significance supposedly limited mainly to species in the grass family). A review of the phytolith literature produced during the past 30 years conveys the impression that phytolith-based paleoecological research would be minimally informative, because the identification of specific taxa from an assemblage of fossil phytoliths would be difficult and achievable only on a very limited level. This prevailing and, indeed, pervasive feeling is the reason that when palynological applications in paleoecology blossomed and even when pollen specialists sought ways to improve the accuracy of pollen data phytoliths were given only passing attention by Quaternary paleoecologists.

If one had been alive in Germany during the first third of this century, a period described in Chapter 1 as the botanical period of research, quite a different view of phytolith production and morphology would have been imparted, for there existed myriad published accounts of phytolith formation and shape in dicotyledons and lower vascular plants such as ferns. The significance of phytolith morphology in these plants was well appreciated. As early as 1908, the German plant anatomist Hans Solereder (1908), writing about a class of phytoliths called cystoliths, which are found mainly in the Urticaceae (nettles), Moraceae (fig family), and Boraginaceae, said "they occur only in a few orders [author's note—this may mean family], and are characteristic of whole tribes or genera. Data as to the extent to which cystolith-like structures can be employed in distinguishing species are contained chiefly in Mez's and Priemer's papers, which deal with cystoliths of the Cordiaceae [author's note—probably meaning Boraginaceae], and Ulmaceae, respectively" (pp. 1113–1114).

The subsequent period of phytolith research, termed the period of ecological research, which began circa 1950, saw an emphasis devoted to the survey and description of phytoliths from monocotyledons, especially grasses. Much less attention was devoted to dicotyledons and almost none to lower vascular plants although studies by Geis (1973), Brydon *et al.* (1963), Wilding and Drees (1971, 1973), and others demonstrated notable levels of production and morphological diversity in angiosperm and gymnosperm tree leaves. As a consequence of the discrepancy in the amount of attention devoted to different plant taxa, there resulted a number of serious misconceptions about, and lacunae in, the phytolith morphological data base. Descriptions and comparisons of phytoliths from entire families were never made available to modern researchers. Even among grasses, by far and away the most intensively studied plants, important morphological and metric attributes such as three-dimensional shape, size, and relative frequencies of phytoliths in individual species remained poorly studied. Hence, the widely held notion that phytoliths could not be used to identify genera and species of grasses was derived.

The obvious and pressing need to develop reference phytolith collections for archaeological study regions, with basic information on the kinds of plants that contribute phytoliths to archaeological and geological records and the taxonomic significance of silicified forms, led the author and other archaeobotanists and paleoecologists to recently compile broad modern phytolith comparative collections and to more closely study some attributes of already known phytolith shapes. These results, combined with those of previous investigations, form the body of this chapter. Phytolith morphological studies are now sufficiently advanced to propose classifications of siliceous secretions across a wide range of families. The diversity of phytolith shapes is as equally impressive as the diversity of diatom shapes; so remarkable that someday a future Ernst Haeckel will have cause to wonder why there are so many different kinds of phytoliths.

How Phytolith Shapes Are Formed

We have discussed in Chapter 2 how virtually all tissues of the plant body and each of their cellular components may act as loci for the deposition of solid silica. The particular shape(s) of phytoliths that individual plants may come to produce are determined by two factors: the kind of cell that accumulates silica and its precise location in the plant body. For example, cells forming the surface epidermis, hair cells (also known as trichomes), and hair bases may all be silicified, resulting in mineralized replicas, virtual "casts" conveying many of the structural features of the once-living cell. A second category of phytolith is formed by an incomplete silicification of the cell lumen or interior, whereby shapes are produced that do not conform to the original configuration of the cell. Examples of this type include the familiar dumbbell, saddle, and cross-shaped phytoliths found in grasses, which are collectively known as "short cell phytoliths." In addition, palms and other monocots like the Cyperaceae and Marantaceae have phytoliths formed in unevenly thickened cells called stegmata that do not assume the original shape of the cell.

Phytolith Description and Classification

The terminology and classification systems applied to phytoliths have not as yet been standardized. There are two basic arrangements by which phytoliths could be classified. These are called taxonomic and nontaxonomic schemes. A nontaxonomic classification emphasizes shapes of objects under study with little emphasis on equating shapes back to the organisms that produced them, or tying them into the larger taxonomy of the organism (e.g., to which subfamily, family, or order it belongs). A taxonomic classification, on the other hand, stresses the correspondence between phytolith shape, the species that produced it, and the evolutionary relationship of that taxon with other plants. Some investigators (e.g., Deflandre, 1963; Rapp, 1986) called for a nontaxonomic classification because they prematurely concluded that individual phytoliths are of little utility in the identification of plants from which they are derived, and a single species may contain several different kinds of phytoliths. However, it will be shown that individual phytoliths often characterize particular taxa at various levels (genus, tribe, subfamily, and family), and there is a close correlation between the types of phytoliths and taxonomic affinities of plants containing them. These are factors that obviously call for considerable attention to taxonomic considerations in classification schemes.

With regard to the nomenclature of phytoliths, investigators have preferred either morphological terms, based on shape (Rovner, 1971; Wilding and Drees, 1973, 1974) or generic terms, based on specific cellular origins (Geis, 1973; Geis and Klein, 1978; Piperno, 1985a). Since many phytoliths can be assigned to specific loci of origin on the basis of their morphology I prefer a system that describes and names major categories of phytoliths on the basis of their origin in living tissue. Morphological categories can then be added to this system as needed, in cases where cellular origin cannot be specified after removal from living plant tissue or where subcategorization and further description is required. For example, individual phytoliths can further be characterized by size, shape (both two- and three-dimensional), ornamentation of walls, thickness of cell walls, orientation of cells, and relation to surrounding silicified cells. With this system phytolith nomenclature can be more easily standardized, and jargon is avoided.

In the compilation of descriptions and comparisons of phytoliths between taxa, it is extremely important to consider carefully which attributes to choose for closer morphological examination and subcategorization. An adequate classification should incorporate categories that are discrete, consistently identifiable from sample to sample, and demonstrate little variation within species. We could have a field day describing the minute particulars of the diverse assemblages isolated from grasses, but many phytolith attributes, such as the particulars of end lobes of dumbbells, are without significance in grass identification, and so setting up forms for forms sake adds little progress to the development of phytolith classifications (see Twiss, 1986a, for some very pertinent comments along these lines). In the construction of regional phytolith descriptions and keys, I believe it will prove beneficial to go with fewer rather than many basic forms and then very prudently choose those attributes with which to express maximum phytolith detail.

The Major Classes of Phytoliths

The following types of plant tissues and cells are commonly silicified and produce discrete phytolith types: epidermis, including hair cells, hair bases, and stomata; hypodermis; mesophyll; schlerenchyma; and vascular. They produce the major classes of phytoliths whose basic shapes and features are now described. The term epidermis denotes the outermost layer of cells on the primary plant body. The main mass of cells includes the epidermal cells proper, which vary in shape; but in most plants epidermally derived phytoliths can be divided into two broad classes. These are anticlinal, which have sinnuate or wavy margins, and polyhedral (Fig. 3.1a, b). The latter may have from four to eight sides and a square to rectangular overall

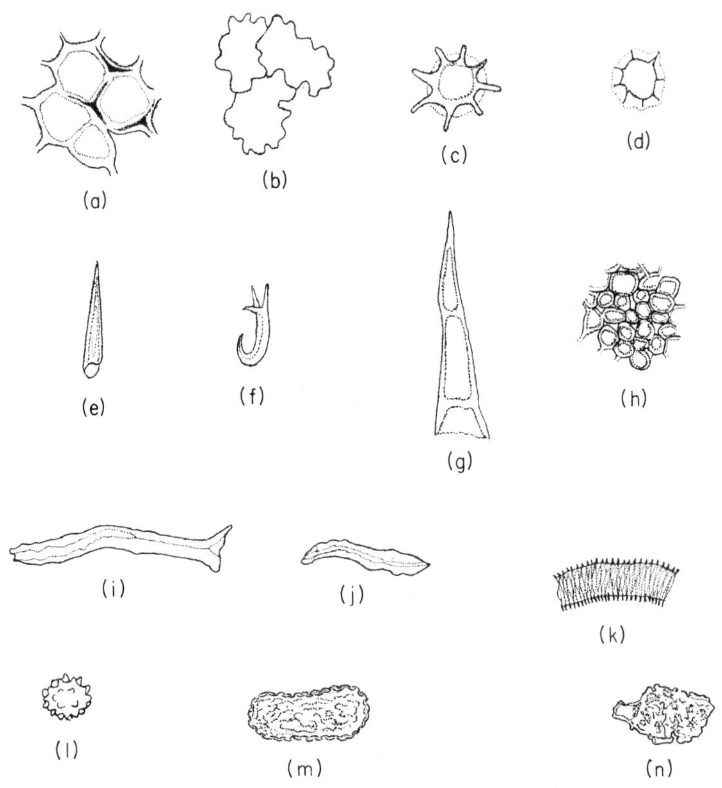

Figure 3.1 Major classes of dicotyledon phytoliths: (a) polyhedral epidermis; (b) anticlinal epidermis; (c) hair base; (d) hair base; (e) nonsegmented hair; (f) nonsegmented hair; (g) segmented hair; (h) mesophyll; (i) sclereid; (j) sclereid; (k) tracheid; (l) cystolith; (m)cystolith; and (n) cystolith.

appearance. Anticlinal epidermal phytoliths are found in many dicotyledons and a few spore-bearing plants, including the Polypodiaceae (ferns), Cyatheaceae, and Schizeaceae. Silicified polyhedral epidermis is found widely throughout the dicotyledonae division of angiosperms and in the monocotyledon families Commelinaceae and Gramineae.

A variety of other kinds of epidermal phytoliths are found in monocotyledons. The specialized nature of the grass epidermis gives rise to a complex array of phytoliths. Best known among these are the dumbbell, cross-shaped, saddle-shaped, and circular to acicular forms [some of the latter have been reclassified as trapezoids by Brown (1984) and Mulholland (1986)] derived from idioblast cells, which are primarily located over the veins of the leaf epidermis (Fig. 3.2a–f). These are collectively called short cell phytoliths. They have restricted distributions within the Gramineae. Dumbbells (now happily called bilobates by Brown, 1984) and cross

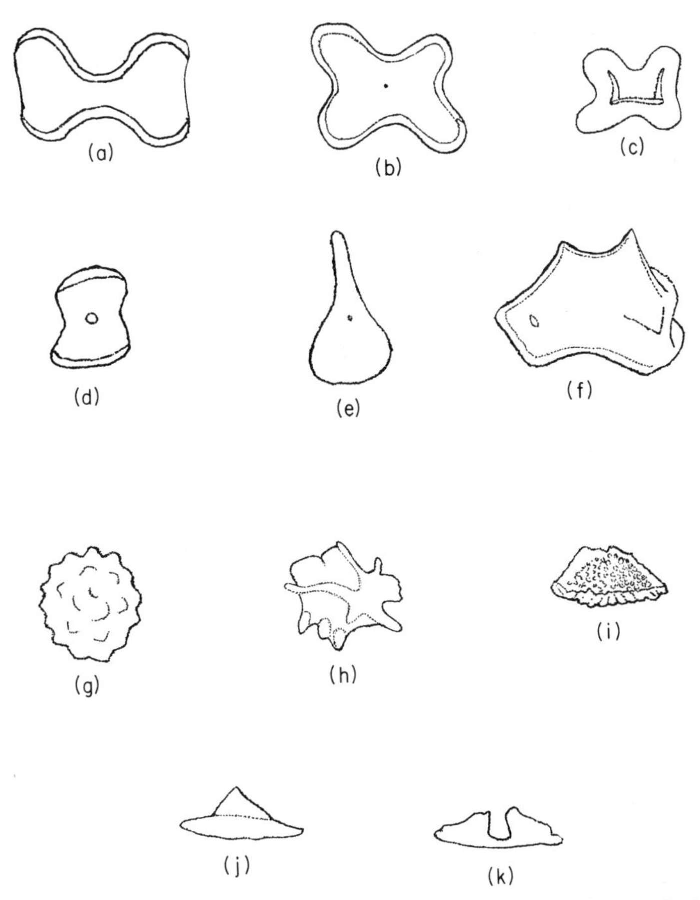

Figure 3.2 Major classes of monocotyledon phytoliths: (a)–(f) grass short cells; (g) spherical phytolith from the Palmae; (h) phytolith with irregularly angled to folded surface from the Marantaceae; (i) conical to hat-shaped phytolith from the Palmae; (j) conical to hat-shaped phytolith from the Cyperaceae; (k) silica body with trough from *Heliconia*.

shapes occur primarily in the Panicoideae or tall grass subfamily and in some bamboos. Saddle shapes occur abundantly in the Eragrostoideae, which includes the short, Chloridoid grasses, and in many bamboos. In the Eragrostoideae bilobates are uncommon, and cross shapes are not produced. Circular to acicular forms, which form the primary silica body in many Festucoid grasses, are produced in small numbers in other subfamilies, while other kinds of phytoliths are restricted to the Pooideae (Twiss' Festucoid subfamily), which lacks bilobates and cross shapes (Brown, 1984, 1986; Piperno, 1987a). Some lesser known short-cell phytolith types include those produced by tropical bamboos like *Chusquea* that may be genus specific (Piperno, 1987a) (Fig. 3.2f). They certainly have very restricted distributions in the Gramineae, being found thus far only in *Chusquea* out of 135 species, comprising 75 genera, tested from the New World tropics.

Hair cells are appendages on the leaf epidermis that serve various protective and secretory functions. Silicified hair cells are often triangular, with flat, spherical, or elliptical bases (Fig. 3.1e–g). The base of the phytolith is the widest part which fits into the hair base (described later), while the apex is the free end of the hair cell. Infrequently, hair-cell phytoliths have nontapered (nontriangular) shafts and rounded apices. Hair-cell phytoliths can be further subdivided into two broad categories that conform to their origin in living tissue—segmented (multicellular), which have a number of distinct divisions across the phytolith, and nonsegmented (unicellular), which have no such divisions (Fig. 3.1e–g). Further differentiation of hair-cell phytoliths is based on such characteristics as surface features, shape, and size; these will be described later. They are found primarily in dicotyledons, where they have considerable taxonomic significance, and in a few monocotyledons such as grasses and sedges.

Hair bases are cells in the epidermis from which the hair cells originate. Derived phytoliths are spherical, often with a circular mark in the center that is actually a siliceous protuberance into subepidermal tissue (Fig. 3.1c and d). They can have a variety of designs on the surface or on the perimeter of an optical section of the cell. Hair-base phytoliths are very common in dicotyledons, where they can be used in discriminating some taxa. They have not been observed in monocotyledons or lower vascular plants.

Some epidermal cells, especially in the Moraceae, Urticaceae, Boraginaceae, and Acanthaceae, may have specialized cells called lithocysts which contain cystoliths. Cystoliths are outgrowths of the cell wall impregnated with silica and/or calcium carbonate. They sometimes extend into the ground tissue of the leaf. All forms described here consist of silicon dioxide, since possible calcereous secretions were removed by treatment with dilute HCl. Cystoliths are usually large, distinctive phytoliths exhibiting various kinds of surface features (Fig. 3.1l–n).

Mesophyll is the ground tissue of the leaf enclosed within the epidermis. It is usually specialized as a photosynthetic tissue. Two types of mesophyll tissue may be silicified: palisade mesophyll, which as the name implies consists of cells elongated at right angles to the epidermis and arranged like a row of stakes, and spongy mesophyll, which has a less regular appearance (Fig. 3.1h). All dicotyledon mesophyll tissue is much the same in appearance and derived phytoliths have little taxonomic value.

In many monocotyledons, like palms, sedges, the Marantaceae, Musaceae, and the Orchidaceae, specialized cells with unevenly thickened walls called stegmata, found often in the mesophyll and less frequently in the epidermis, give rise to a number of different phytolith shapes; conical to hat-shaped, spherical to aspherical, bodies with troughs, and various irregular shapes (Fig. 3.2g–k). Often these shapes are taxonomically significant, and in addition, each exhibits a variety of surface features that are specific to families or subfamilies of plants. They will be described in detail later.

Schlerenchyma cells are strengthening elements of mature plant structures. They have thick, often lignified cell walls. Two forms of cells are distinguished by plant anatomists, sclereids and fibers. Sclereids are most commonly silicified, resulting in a number of shapes that are easily identified (Fig. 3.1i and j). Their morphology is often one and the same between quite unrelated taxa, but as we shall see later they may be valuable in the separation of herbacious and arboreal taxa.

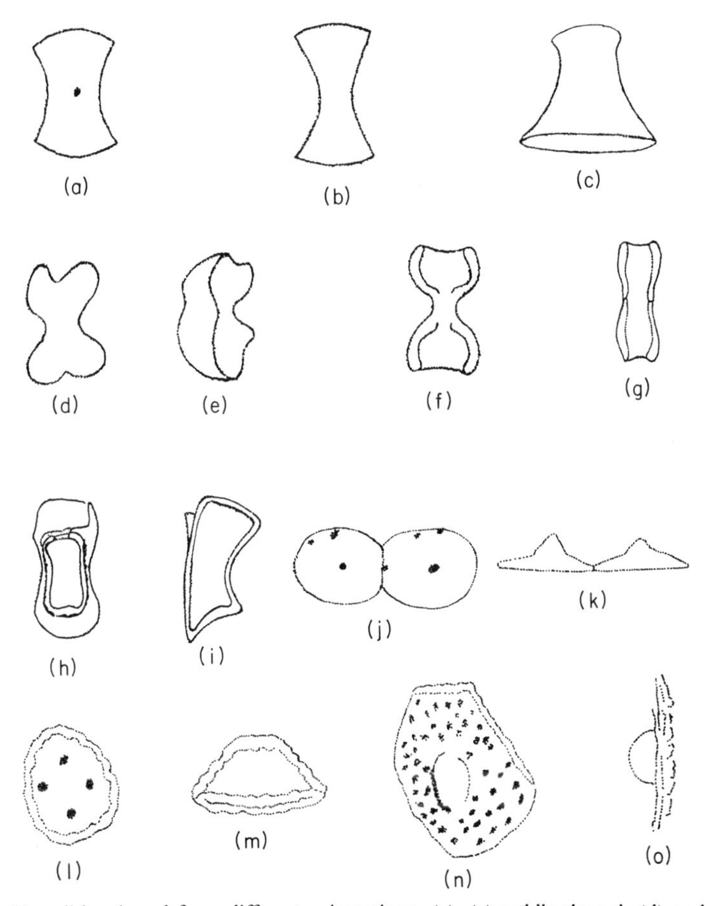

Figure 3.3 Phytoliths viewed from different orientations: (a)–(c) saddle shaped; (d) and (e) cross shaped; (f) and (g) dumbbell shaped; (h) and (i) bamboo short cell phytolith; (j) and (k) Cyperaceae conical shapes; (l) and (m) Palmae conical to hat shaped; (n) and (o) dicotyledon seed phytolith.

Vascular tissue, comprising the xylem and phloem, is concerned with conduction of water, storage of food, and support. The elements of vascular tissue silicified are tracheids, part of the xylem complex, which has a primary water conducting function. They can be elongated or irregular in outline and are often solidly silicified (Fig. 3.1k). For the most part they have no value in plant identification.

The phytoliths depicted in Figs. 3.1 and 3.2 have been outlined in only one orientation. With some types it is important to be able to turn the body in order to enable viewing from different geometric orientations because the same phytolith may have two different shapes when viewed from different angles, making identification somewhat tricky. Also, the determination of phytolith three-dimensional structure is an important taxonomic character in identifying certain monocotyledons. For example, circular and acicular silica bodies from Festucoid grasses (Twiss *et al.* 1969) are conical to hat shaped or trapezoidal when turned on their side. Studies of bilobate and cross-shaped three-dimensional structures have shown that considerable variation exists between Panicoid grass genera, offering a basis for the discrimination of certain taxa like maize (*Zea mays* L.). (Piperno, 1984; Piperno and Starczak, 1985; Special Topic 1 in this chapter). Several wild grass genera have been found to have possible genus-specific, three-dimensional short-cell phytolith variants (Piperno, 1984). These will be described in detail later. Figure 3.3 shows several different kinds of phytoliths and the different shapes they assume when viewed from various geometric orientations.

Phytolith Surface Ornamentation

The surfaces of phytoliths show varied and intricate details, many of which are important in identification. Before introducing them, we must briefly engage again in terminology and definitions. Scanning electron microscopy will be used for documentation in some cases, but it should be emphasized that many of the features discussed are visible and differentiable with the conventional light microscope at powers of 400× and 1000×, making practical the day-to-day identification of phytoliths from plants and soils.

The kinds of surface ornamentation that I have observed and defined are (1) spinulose, (2) nodular, (3) rugulose, (4) smooth, (5) irregularly angled or folded, (6) verrucose, (7) tuberculate, (8) stippled, (9) armed, and (10) nonarmed (Piperno, 1985a). The last two apply to segmented and nonsegmented hairs found in dicotyledons. Of the first five, rugulose and smooth surfaces are found both in monocots and dicots, usually on spherical shapes, but sometimes on different classes of phytoliths. Spinulose, nodular, and irregularly angled surfaces occur solely in monocotyledons. Verrucose and tuberculate surfaces occur only on cystoliths, which are produced by dicotyledons, and stippled surfaces are found solely on dicotyledon fruit and seed phytoliths.

A spinulose surface decoration is a pattern in which small projections (spinules) are fairly large, regular, and evenly distributed on the body of the phytolith. Spinulose surfaces have been found in two monocot families, the Palmae (Plate 1) and the Bromeliaceae (Plate 2). Many species in the Palmae have spinules that are larger, more frequent, and better defined than Bromeliad spinules, and these characteristics serve as factors distinguishing the two families. Palm genera may have predominantly pointed or rounded spinules. The greater ease in distinguishing Palmae spinules may be related partly to phytolith size as bromeliad spherical phytoliths are often much smaller than palms, ranging only from less than 2 to 8 μm in diameter, whereas palm phytoliths vary from about 6 to 25 μm in diameter. Some Palmae genera, such as *Cocus* (coconut) contribute phytoliths whose sizes and surface features are more like the Bromeliaceae.

A nodular surface decoration, found only in members of the Marantaceae, a small tropical family of herbacious perennials, is defined by the presence of many (or sometimes fewer) small prominences that are unevenly distributed over the face of the phytolith (Plates 3 and 4). The sizes, shapes, and irregular distribution of the protuberances make them clearly differentiable from the spinulated forms. Nodular surfaces are to be found both on spherical and conical Marantaceae silica bodies.

A rugulose surface pattern refers to a rugged or rough surface where the presence of spinules, nodules, or bulges is not clearly evident (Plate 5). Phytoliths of this kind occur in the seeds and leaves of a very limited number of arboreal dicotyledons (*Licania* and *Hirtella*) and also in the vegetative structures of several herbacious monocotyledons (the Cannaceae, Heliconiaceae, and Marantaceae). There are significant size differences in phytoliths bearing these surface patterns. Those from monocots very often range from 9 to 25 μm in diameter, while dicot shapes are much smaller, varying from 3 to 9 μm in size.

As the name implies, a smooth surface decoration refers to phytoliths that bear no pattern on the siliceous body (Plate 6). As a formal category, the description applies to spherical phytoliths found commonly in the seeds of one arboreal dicot, *Hirtella,* and on rare occasions in the leaves of a few species in the Cannaceae and Marantaceae. Size differences again appear to be a means of discrimination between the monocot and dicot forms, with smooth phytoliths from the two groups having the same ranges as described for rugulose forms.

As can be imagined, the distinction between rugulose and smooth may sometimes be difficult to make, particularly on phytoliths that are less than 6 μm in diameter. In these situations, we at least try to be consistent with the identifications. More scanning electron microscopy is clearly needed for a closer investigation of the smaller phytoliths.

Phytoliths having irregularly angled to folded surfaces occur in three monocotyledon families, the Marantaceae, Cannaceae, and Zingiberaceae. This distribution makes sense taxonomically because these families share the same order of plants, the Zingiberales (Plate 7).

Surface patterns called verrucose and tuberculate denote the wart-like projections found on cystoliths of the Urticaceae and Acanthaceae, respectively (Plates 65 and 70). A stippled surface refers to a particular pattern found thus far only on the fruit and seed phytoliths of the Burseraceae (Plate 83). In certain dicotyledon families, such as the Boraginaceae, Moraceae, and Urticaceae, segmented and non-segmented hairs may have short surficial spines distributed either across the whole phytolith, or confined to a certain portion of the hair shaft. These are called armed hairs and partially armed hairs, respectively (Plate 8).

Phytolith Variation in a Single Species

The silicification of different structures of a plant body and of numerous kinds of tissues within a single structure frequently leads to the occurrence of several different phytolith types in a single species. Four or five disparate shapes are not uncommonly found in dicotyledon leaves, a situation that develops when all of the various tissues (e.g., tracheids, sclereids, epidermis) are impregnated with silica to one degree or another. This is a factor that complicates the establishment of classification systems and subsequent identifications of discrete phytoliths in soils. It is, however, far from being an insurmountable situation. When more than one phytolith is found in a plant they often have varying levels of taxonomic significance, and with enough comparative work it becomes possible to specify which are shared by many families; which occur in a few families; or which are family, genus, or species specific. Phytolith keys and identifications are then made accordingly and modified as the collection is expanded. Dicotyledon leaf epidermal phytoliths, for example, are very often the same no matter from what family they derive, and the same holds true for tracheid and mesophyll phytoliths. Cystoliths, hair cells, and hair bases, on the other hand, demonstrate a far greater degree of diversity and taxonomic significance. With the establishment of good descriptive systematics and keys and the proper amount of familiarization with them, the investigator can often quickly decide whether a particular phytolith encountered from a plant or soil assemblage is worth some added special attention or if it should be quickly relegated to the "redundant phytolith types" list.

Some or all of the totality of phytolith shapes found in a single species, called phytolith assemblages, may also be employed to yield significant taxonomic information. When it is known or surmised that a phytolith assemblage from a given plant is only partially held in common by others, either related or unrelated, then the construction and statistical comparison of morphological type frequencies may be useful in plant identification. Morphological type frequencies entail the compilation of all morphological variants and their sizes in a species and a tabulation of their incidence or percentage frequency.

The grass family is a good example of where this kind of analysis is useful. Among monocotyledons, grasses present the extreme situation of having more than one kind of phytolith. They have a very varied epidermal structure which is almost entirely silicified, thus producing a menagerie of phytolith types. Under the traditional grass classification system, identification is often possible only to the subfamily level because compendia of shapes, sizes, and frequencies are not tabulated in species or statistically compared to others. Examples will be presented later of how such analyses are carried out and result in genus- and species-level identifications.

We have so far layed out some of the basic facts of phytolith shape, size, surface decoration, and distribution. We will now discuss in detail phytolith morphology as it is presently known in the major groups of the plant kingdom. Major plant domesticates of the Old and New World will be presented first, followed by wild monocotyledons, dicotyledons, gymnosperms, and ferns.

Phytoliths in Old and New World Domesticates

The careful, detailed, and systematic analysis of phytoliths from major domesticates of the Old and New World has only just begun. Only a few, such as maize *(Zea mays)*, manioc *(Manihot esculenta)*, squash *(Cucurbita)*, barley *(Hordeum distichum)*, wheat (*Triticum monococcum* and *T. dicoccum*), and achira *(Canna edulis)* have been studied in any detail (Bozarth, 1986; Pearsall, 1978, 1979; Piperno, 1983, 1984, 1985a; Piperno and Starczak, 1985; Miller, 1980; Miller-Rosen 1985). Many, especially those comprising the native cultigen inventory of Africa and North America, have gone completely unexamined. This section reviews the present state of knowledge concerning the contribution of phytoliths to studies of agricultural origins and dispersals and offers some suggestions and areas for future research that may prove profitable.

Tables 2.3 and 2.4 show patterns of production and silicification in 21 plant domesticates that have been examined. Maize will be discussed separately in detail in the special topic to follow. Domesticate phytolith production is unfortunately rare to nonexistent in many major cultivated plants so far studied. Phytoliths have not been observed in the leaves of mamey *(Mammea americana)*, papaya *(Carica papaya)* and chile peppers *(Capsicum annuum)*, the leaves and tubers of otoy *(Xanthosoma sagittifolium)*, taro (*Colocasia* sp.) yams (*Dioscorea* spp.) and sweet potato *(Ipomoea batatas)*, tubers and bark of manioc, the pods and fiber of cotton *(Gossypium barbadense)* or the fruit rinds of avocado *(Persea americana)*. As discussed in Chapter 2, it is certainly possible that repeated extractions of additional material from these plants would reveal the occasional epidermal, mesophyll, or tracheid phytolith, but it is highly unlikely that such phytoliths would be diagnostic, and it is improbable that they would occur in substantial amounts.

Plant Domesticates Producing Nondistinctive Phytoliths

Nondiagnostic phytoliths occur in small to moderate amounts in several plant domesticates. Manioc *(Manihot esculenta)*, a tropical root crop of major importance, produces extremely small numbers of phytoliths derived from hair cells (Plate 9). They also occur in wild Euphorbiaceae such as *Acalypha diversifolia*. The value of 0.42% silica obtained from manioc leaves (Table 2.4) was predominantly in the form of nondescript fragments. Therefore, there is little chance that manioc hair cell phytoliths would be isolated from archaeological soils. No phytoliths were observed in extractions of manioc tubers, wood, and bark.

The leaves and fruit exocarp of pineapple *(Ananas comosus)* contain abundant numbers of small (3–6-μm) spherical spinulose phytoliths of a kind found in many wild Bromeliads and some of the Palmae. There is little chance that they will provide independent evidence for the presence of pineapple; however, if the plant is suspected on other paleoecological grounds phytoliths of a kind just described should be isolated from soils in large numbers.

Segmented hair cell phytoliths with flat, spherical, or ellipsoidal bases (two or three segments) and tapered shafts and apices were isolated from bottle gourd leaves *(Lagenaria siceraria)* (Plate 10). These types are also found in other Cucurbitaceae genera, such as *Cayaponia, Sicyos,* and *Citrullis,* and somewhat similar forms occur in the Compositae. The rinds of bottle gourd produce a kind of slightly scalloped spherical phytolith to be described in detail later when *Cucurbita* is discussed. Bozarth (1986), who first described these phytoliths, doubts that they will be useful in identifying archaeological gourd.

The pods of *Phaseolus vulgaris* and *P. lunatus* produce hook-shaped hair cells (Bozarth, 1986), that I have described as thin and curved nonsegmented hairs (Plate 11) (Piperno, 1985a). They occur in a large number of related and unrelated plants (see phytolith key, p. 253 of the appendix) and cannot be used to identify archaeological *Phaseolus*. The heads of domesticated sunflower (*Helianthus annuus* var. *macrocarpus*) produce segmented hair cell and hair base phytoliths (Bozarth, 1986) that have been found in various tropical wild Compositae (Piperno, 1985a), hence taxonomic significance of sunflower phytoliths is presently unclear.

Phytoliths isolated from the leaves of cacao *(Theobroma cacao)* (tracheids and sclereids), cotton *(Gossypium barbadense)* (mesophyll and tracheids), avocado *(Persea americana)* (tracheids and stomata), and tomatoes *(Lycopersicum)* (tracheids) are very redundant in form with those from many other plants and have no taxonomic value (Plates 12 and 13).

Despite the lack of production or repetitive shapes found in many crop plants, there is still an enormous amount of productive and important work to be done and a large potential flow of new information concerning phytoliths and their relation to agricultural origins and dispersals.

The most distinctive kinds of phytoliths have been found in the following plants and plant structures: the leaves and husks of maize, the leaves and rinds of squashes (*Cucurbita* spp.), the leaves of banana (*Musa* spp.), the glumes of wheat (*Triticum monococcum* and *T. dicoccum*) and barley *(Hordeum distichum)*, the leaves and glumes of rice *(Oryza sativa)*, the leaves of achira *(Canna edulis)*, and the leaves, seeds, and seed bracts of arrowroot *(Maranta arundinaceae)*.

New World Domesticates Producing Distinctive Phytoliths

What appear to be genus-specific phytoliths occur in the rinds of squashes *(Cucurbita)* (Plate 14). They were first described by Stephen Bozarth (1986) from wild *(Cucurbita foetidissima)* and cultivated *(C. pepo, C. maxima)* species. They are spherical, range in size from about 48 to 87 μm, and have a distinctive scalloped surface decoration formed by contiguous concavities. The author carried out an intensive investigation of the possible occurrence of similar forms outside of *Cucurbita* by isolating phytoliths from the leaves, tendrils, and stems of 36 wild non-*Cucurbita* species of the Cucurbitaceae. They did not produce similar phytoliths; therefore, we can state with a high degree of confidence that the spherical, scalloped phytoliths from *Cucurbita* are distinctive at the level of genus. It is possible that a means of discrimination between wild and domesticated forms might be found on the basis of such features as the size and depth of the individual depressions or scallops and the mean phytolith size.

There appears to be a considerable amount of capriciousness in the production of scalloped phytoliths by *Cucurbita*. S. Bozarth (personal communication, 1985) initially tried a wild buffalo gourd *(Cucurbita foetidissima)* from the American Southwest and found nothing. He then tested a different plant of the same species grown under similar conditions and isolated the scalloped form. I extracted several Central American varieties of *Cucurbita pepo,* which yielded no scalloped phytoliths whatsoever, and then found such phytoliths in 3000 year old archaeological sediments from Panama. Questions that need to be addressed are the differential occurrence of these phytoliths in different cultivars and species of wild and domesticated *Cucurbita* and their sometimes spotty distribution in the same population.

The leaves of *Cucurbita* also contribute small numbers of apparent distinctive kinds of phytoliths which are derived from hair cells and consist of three types. One kind is short and wide with two segments, a wide (>40 μm) basal segment and a small, round apical or last segment (Plate 15). Somewhat similar phytoliths occur on rare occasions in *Cayaponia citrullifolia,* a wild member of the Cucurbitaceae, but *Cucurbita* hairs can be differentiated because they are much wider. Another type is also short and wide but with three segments. The basal and intermediate segments are slightly nontapered and the apical segment is small and round (Plate 16). Their sizes range from 99 to 231 μm in length and from 59 to 112 μm in width.

The last type has two segments, a shaft that is either slightly tapered or not tapered at all and a rounded apex or apical segment. Their sizes range from 99 to 135 μm in length and 59 to 69 μm in width. Hair cells of this type were also observed on rare occasions in *Cayaponia citrullifolia,* but they were not as wide as in *Cucurbita,* not exceeding 40 μm at their widest point.

In addition to these, segmented hair cell phytoliths were observed that have flat or rounded bases and tapered shafts and spices, as well as other kinds of hair cell types. These are all held in common with wild members of the Cucurbitaceae (Plate 17). None, including the distinctive kinds, are solidly silicified, having only their cell walls encrusted with silica; so it is questionable how durable they are once deposited into soils.

In the Cucurbitaceae, a characteristic surface pattern occurs on hair cells, one that when viewed with the light microscope appears to consist of short, fine striations. Scanning electron microscopy shows that they are actually small spines (Plates 17 and 18). This pattern thus far appears unique to the squash family. In *Cucurbita* all of the hair cells described previously may possess this pattern, providing another clue that they are from the Cucurbitaceae.

Many of the *Cucurbita* hairs of all types are very wide in comparison to others of their kind from related wild species, and it may be that width alone may allow discrimination of some. More comparative metric and morphological studies of hair cell phytoliths from this family are needed before *Cucurbita* can be confidently identified in archaeological soils on the basis of their characteristics.

In addition to the hair cell phytoliths, hair base and anticlinal epidermal phytoliths occur in cultivated squashes, but the hair bases are of a kind found commonly in wild members of the family (Plate 19), and anticlinal epidermis occurs throughout the dicotyledonae.

Regrettably, only a few of the New World tropical plants that were domesticated primarily for their roots or tubers have phytoliths with taxonomic significance. They are achira *(Canna edulis)* and arrowroot *(Maranta arundinaceae)*. Both are South American root crops. In the leaves of achira, Pearsall (1979) has documented the presence of large, spherical spined phytoliths that occur in chains or strings (Plate 20). I believe the spiney surface pattern that Pearsall refers to is a designation for what I am classifying as rugulose. Such phytoliths also occur in wild Cannaceae, the Marantaceae, and Heliconiaceae but have not been observed in long strings (Piperno, 1985a). These phytoliths may therefore be characteristic of achira. Achira roots have not been tested for phytolith production, but they require examination.

The leaves of arrowroot contribute abundant numbers of spherical, nodular phytoliths, a type limited in occurrence to the arrowroot family, Marantaceae (Plate 3). They range from 9 to 18 μm in diameter. If this distribution holds, an ecological argument for the presence of arrowroot may be used in archaeological regions removed from the areas of natural distribution of the Marantaceae, such as the dry coasts of Peru and southwest Ecuador. The seeds and seed bracts of arrowroot also contribute large numbers of phytoliths that are distinctive to at least the family level. Their morphologies are completely distinct from those found in Marantaceae leaves and have shapes ranging from roughly spherical to rectanguloid. They often have undulating or decorated surfaces and can show the presence of a well-defined stalk that presumably attaches to the walls of the cells in which they are formed. These are the forms that I initially called irregular, with and without stalk when they were still unknown phytoliths isolated from Panamanian archaeological soils (Piperno, 1985b). *Maranta* seed phytoliths may also be of great potential in the identification of archaeological arrowroot, for similar forms from only one other Marantaceae genus *(Stromanthe)* have been isolated, and discrimination between *Maranta* and *Stromanthe* seems possible.

Tomlinson (1969) reports that the tubers of arrowroot have phytoliths; hence more possibility exists for differentiation, especially since roots were of greater economic importance than seeds to prehistoric and historic-period populations.

Arrowroot leaves have three other kinds of phytoliths that can be found in both related and nonrelated plants. Conical to hat-shaped phytoliths are somewhat similar to a kind that occur in some of the Palmae (Plate 21): spherical to aspherical rugulose bodies occur in the Cannaceae and Heliconiaceae; and bodies with irregularly angled or folded surfaces are found in wild Marantaceae, the Cannaceae, and Zingiberaceae.

Achira is a root crop whose importance in prehistoric economic systems of South America might well have been underestimated. The present day consensus on arrowroot is that it was not grown for its tubers until the Historic period. Achira and arrowroot phytoliths may provide indications of where and how extensively they were grown in pre-Colombian and Historic-period South America.

Old World Domesticates Producing Distinctive Phytoliths

The leaves of bananas and plantains (*Musa* spp.) contribute phytoliths with troughs that appear to be at least genus specific (Plate 22). Their size range is about 15–21 μm in length and 4–6 μm in width. They have a variety of shapes from roughly rectangular to having both ends pointed. The troughs are shallow, often positioned laterally on the phytolith, and can be on an elevated part of the phytolith. These characteristics all serve to separate *Musa* phytoliths from those of *Heliconia* in the related family Heliconiaceae (cf. Plates 22 and 43). The implications of the *Musa* phytoliths for tracing the development of the important food crops bananas and plantains are obvious.

The glumes of wheat (*Triticum dicoccum* and *Triticum monococcum*) and barley *(Hordeum distichum)* possess a considerable phytolith content, often two to three times as high as in leaves. Miller-Rosen (1985) has conducted the first exacting comparative analysis and developed a preliminary classification scheme to identify wheat and barley phytoliths in archaeological soils. Figure 3.4 is a schematic representation of the different features of the glume epidermal long cell walls in each plant.

The distal and medial sections of single glumes from each species are different, and generally the distal glume contained the most diagnostic characteristics. In barley, the thin cell edges are serrated and end in sharp or knobbed points. The thickened cell walls are sometimes square-like and usually occur in waves of even amplitude. The distal glume also contains adjacent short cells and microhairs of uniform size. The distal glumes of emmer and einkorn wheat contain epidermal long cells with very low, blunted waves, which are often squared in the thin cells. The thin as well as the thick long cell walls in the medial glume of both types of wheat have rounded, erratic waves of irregular amplitude.

The glume epidermis of wheat and barley have a tendency toward very strong silicification, resulting in multicelled sections of epidermal tissue that preserve many of these distinctive characteristics of living cells. One reason for this pattern may be the hot and arid habitats of these plants, which promote intensive transpiration and then silicification due to water loss and the creation of supersaturated levels of soluble silica. Miller-Rosen cautions that wild grasses must be studied to ensure

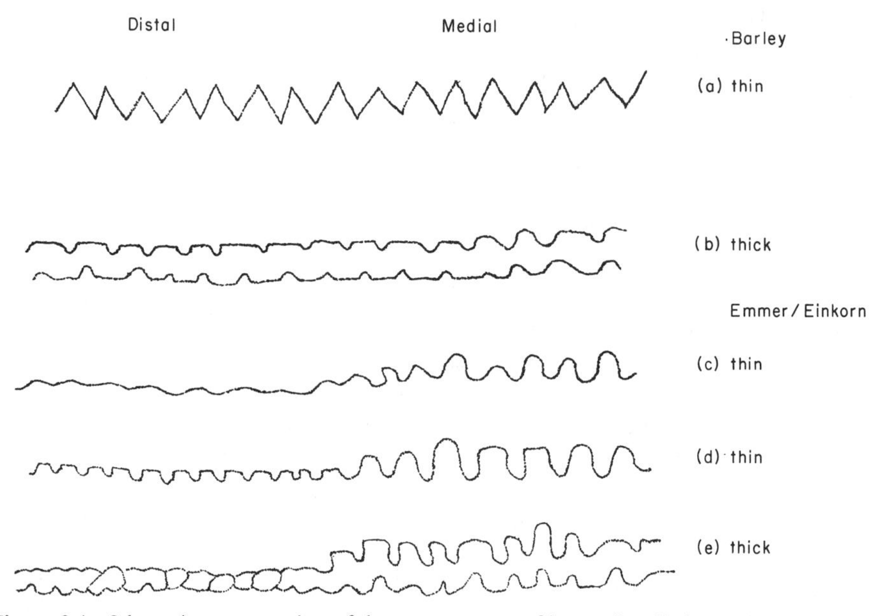

Figure 3.4 Schematic representation of the wave patterns of long cell walls from wheat and barley glumes. [From Miller-Rosen (1985).]

that no overlap exists in the discriminating features. Also, large collections of a single species must be built to measure the effect of interpopulation variability on phytolith shape. Clearly, a considerable amount of potential exists for the archaeological documentation of wheat and barley with phytolith data.

A great deal of money, energy, and time has been expended in southeast Asia tracing the origins and early dispersals of rice *(Oryza sativa)*. It appears that such investigations may ultimately benefit from phytolith analysis. Phytoliths from rice have been studied since the early days of the botanical period of research (Grob, 1896; Edman and Söderberg, 1929; Formanek, 1899). An assessment of their taxonomic significance is provided by Watanabe (1968). He focused on phytoliths from two structures of the plant, the glume subtending the grain and the leaf. Watanabe believes that the shape and arrangement of long cell phytoliths in the epidermis of the glumes are highly diagnostic as are short cell phytoliths in the leaf epidermis.

The glumes are described as having hollow swellings with acute tips and deeply sinuated or sharply dentated walls. Watanabe considers the swellings in and of themselves to be a rice character. He also notes that leaf short cell phytoliths were of shapes consisting of "pairs of ovals joined side by side at the end of their short axces," a description quite probably of cross-shaped phytoliths. He correctly pointed out that many grasses do not have cross-shaped types but gives no explanation of why he considered *Oryza* types distinctive. Metcalfe (1960), however, has also commented on differences of rice cross-shaped phytoliths, calling them the Oryzoid type. He lists only 11 other species of grasses in his extensive survey as having this kind of cross shape. Another distinctive feature of short cell phytoliths in rice is their alignment with long axes perpendicular to the long axis of the leaf. Judging from Metcalfe's survey of grasses, this pattern appears to occur only in the rice tribe Oryzeae, which in the Old World includes the genera *Leersia* and *Maltebrunia*. Hence, if multicelled sections of leaf epidermis occur in archaeological situations, it may be possible to make rapid identifications of *Oryza*. Once again, there is a great deal of exacting comparative work to be carried out with wild members of the rice genus and other wild grasses in order to rigorously assess the taxonomic significance of rice phytoliths.[1] Measurements must be made of short cell and other phytoliths and their three-dimensional characteristics described. Clearly an enormous amount of potential exists for the archaeological documentation of early rice domestication with phytoliths, which will be far more resistant to destruction than macroremains in the often inimical wet and humid southeast Asian environments.

[1]I have recently come upon a reference (Fujiwara *et al.*, 1985) that deals with phytolith work on rice in tropical Australia, where the authors isolated wild *Oryza* silica bodies from a 1000 year old archaeological site. I have not seen the paper and can not say which discriminatory attributes they are using, but the authors are confident in their identifications, as were the early German botanists who first worked with archaeological rice phytoliths and Watanabe.

Prospects for Future Research

The results reviewed here have only touched on the magnitude of work yet to be carried out in relation to phytoliths and domesticated plants. The list of plants for which data have been presented can be at least tripled by the numbers of important food crops whose phytoliths have not been investigated or have been only superfically explored. Such crops include the North American domesticates sunflower *(Helianthus annuus)*, sumpweed *(Iva annua)*, and maygrass *(Phalaris caroliana)*; the African domesticates sorghum *(Sorghum bicolor)*, finger millet *(Eleusine caracana)*, watermelon *(Citrullus lanatus)*, other melons *(Cucumis)*, and ensete *(Ensete ventricosa)*; the Near Eastern domesticates oats (*Avena* sp.) and rye *(Secale cereale)*; and the Asian domesticates foxtail millet *(Setaria italica)* and sugar cane *(Sacharrum officinarum)*. The probability that these crops have phytoliths is high because they belong to families that are characteristically strong silica accumulators. We can expect a flood of new information when the appropriate comparative analyses are carried out. At the same time, we can anticipate that such domesticates as potatoes (*Solanum* spp.), peas *(Pisum sativum)*, beans (*Phaseolus* and *Canavalia* spp.), African yams (*Colocasia* spp.), and peanuts *(Arachis hypogaea)* will not yield a phytolith record of value, either because of poor production or the absence of shapes having taxonomic significance.

Special Topic 1: Some Special Attributes of Maize Phytoliths: A Basis for Their Identification in Archaeological Soils

Introduction

One of the first modern applications of phytolith research to archaeological subsistence reconstructions was done by Deborah Pearsall (1978). She proposed a method, based on phytolith size, to differentiate maize from wild grasses. It was an extremely important study because maize was one of the preeminent pre-Colombian food staples. The implications of Pearsall's work caused it to be controversial at the outset among botanists and archaeologists, for perhaps no other subject has evoked their interest, excitement, and antipathy as has the origin and early dispersals of maize.

Academic battles have been waged for years over the ancestry, development, and early diffusion of corn (for example, Beadle, 1972, 1980; Mangelsdorf, 1974, 1986; Galinat, 1971; Iltis, 1983; Doebley, 1983; Doebley *et al.,* 1984; Iltis and Doebley, 1984). The archaeological macrobotanical and pollen record is full of temporal and spatial lacunae for a number of reasons, including poor preservation, poor dispersal of pollen from prehistoric fields and gardens into sites, and the relatively small quantities of maize that must have been utilized when it was a primitive and incidentally used crop plant.

The reemergence of teosinte *(Zea mexicana)* as the probable wild ancestor of maize (Beadle, 1972; Galinat, 1971) made it likely that Mesoamerica was the cradle of maize domestication because there is no evidence that teosinte is naturally distributed south of Honduras. Furthermore, if teosinte were corn's wild ancestor the earliest cobs uncovered in the Tehuacan deposits (MacNeish, 1967) were completely domesticated, having already undergone considerable morphological and perhaps genetic modification by 7000 years ago.

Isoenzymatic work with annual and perennial teosinte and maize has provided intriguing hints into maize's origins. The teosinte populations called Balsas by Wilkes (1967), and now classified as *Zea mays* ssp. *parviglumis* by Doebley and Iltis (1980) and Iltis and Doebley (1980), are genetically closest to maize, even though morphologically they are most distant (Doebley *et al.,* 1984). This finding has led Iltis and Doebley (1984) to conclude that Balsas teosinte, native to the warm and seasonally dry mid-altitude elevations of the Balsas watershed of western Mexico, is a likely candidate for the ancestor of maize. The Tehuacan Valley and indeed all of the central highland area, once thought to be the hearth, may have been marginal to maize's earliest evolutionary history.

Botanists (e.g., Pickersgill and Heiser, 1977), citing the differences between contemporary varieties of primitive corn in Mexico and South America, have argued that primitive maize was dispersed at an early date (about 7000 B.P.) to South America. Archaeologists have supported (Lathrap, 1975; Pearsall, 1977–1978; Zevallos *et al.,* 1977; Piperno, 1983; Piperno *et al.*, 1985) and rejected (Roosevelt, 1980) this idea. Some argue that highly productive races of maize were first developed in South America and then dispersed to Mesoamerica (Zevallos *et al.*, 1977).

Pearsall's early work with maize phytoliths (1978, 1979) entered the heart of this controversial matter, for her data provided the first, hard botanical evidence that the early ceramic Valdivia cultures (3000 B.C.–1500 B.C.) of southwestern Ecuador were indeed growing maize. This, of course, meant that lower Central American peoples were cultivating primitive corn and were intimately involved in the early diffusion of the plant southward.

Dunn (1983) attempted a critique of Pearsall's work, but it was not grounded on any real refuting evidence, and it inexplicably presented little relevant data at all. Dunn cited the occurrence of some large-sized cross shapes in *Tripsacum* (which Pearsall had noted) and *Sorghum* (an Old World grass), while offering no information on their frequency of occurrence; the point, which as Pearsall and I (1984) have consistently emphasized, is the factor separating maize. Furthermore, it is not clear which axis of the cross shape was measured. Since Pearsall's and my studies are based on short-axis dimensions, Dunn's results may not be comparable. Dunn then went on to find problems inherent in her discovery that *Tripsacum* produces fewer cross shapes than maize (if true—and it is not always—then archaeologically it would be advantageous, since *Tripsacum* would be less likely to be well represented in fossil deposits) and somehow concluded that since four different taxa, *Tripsacum, Sorghum,* annual teosinte, and maize, appear to have different frequencies of cross shapes (no numbers were provided) such differences may be environmentally, not genetically, modulated.

Roosevelt's (1984) commentary on the problems of archaeological maize phytolith evidence was based largely on Dunn's paper. Roosevelt concludes that the identification of maize by phytolith size is a problem because (citing Dunn) "cross-shaped phytoliths are found in several different species, and size is rarely diagnostic at the species level, since it varies greatly within a plant according to the location of a phytolith-bearing cell on the plant body." Dunn carried out no such studies on the size of cross shapes from various plant structures. While it is by no means unusual to find hasty conclusions and questions about new applications of a technique, premature impressions of phytoliths by commentators having little or no practical experience lead to misconceptions at a time when the promise, variables, and problems of phytolith analysis must be defined and controlled by systematic and careful research.

The importance of maize, the demands of my own work on agricultural origins in Panama, and the questions surrounding the use of phytoliths to identify maize archaeologically, made necessary very detailed comparisons of maize and wild grass phytoliths, using a large number of maize races and wild species, analysis of different structures of the same plant, measurements of interpopulation variability, and criteria other than phytolith size not previously investigated. Results of my first intensive investigations are presented here. Phytoliths may become a key element in unravelling the mysteries of early maize, for it appears that maize phytoliths can be separated on the basis of sound quantitative and qualitative criteria from wild tropical grasses, and much potential even exists for discrimination between phytoliths from maize and teosinte.

The identification of phytoliths from maize relies on a number of attributes of cross-shaped phytoliths. They are (1) short axis measurement, (2) three-dimensional structure, and (3) the percentage of cross shapes measured against the sum of dumbbells and cross shapes in the same species. Cross-shaped phytoliths are focused on in maize discrimination for a number of reasons. They are found only in the Panicoid subfamily of grasses and a few bamboos. They tend to be much more common in maize than in wild non-*Zea* Panicoid grasses, where dumbbells often predominate. Most importantly, they are formed in short cells of grasses (Prat, 1948). As shown later, short cell phytoliths demonstrate little within-species variations but considerable between-species variation and appear to be under active genetic control, attributes which impart reliability to paleoecological studies based on their characteristics.

Cross-shaped phytoliths are easy to recognize. They consist of three or four lobes attached to a central body, the lobes being demarcated by three or four indentations (Plate 23). They tend strongly to be almost as long or only up to 9 μm longer than wide. A few grasses have small percentages of elongated phytoliths with indented ends and a constricted body, which could be called transitional between the classic dumbbell and cross-shaped forms, though in appearance they much more closely resemble dumbbells (Plate 33). Pearsall (1978) chose to disregard the transitional types by defining cross shapes as being not more than 9 μm longer than wide, a useful distinction that I have retained in my work.

An important aspect of maize identification lies in short cell phytolith three-dimensional morphology, a characteristic virtually ignored by previous investigations. It is highly variable between species and important in discriminating grass taxa. The original maize identification scheme of Pearsall (1978, 1979) relied on a size typology that separated cross shapes into four categories; small (6.87–11.40 μm), medium (11.45–15.98 μm), large (16.03–20.56 μm), and extra large (20.61–25.19 μm). Large-sized cross shapes were far more common in maize than in wild Ecuadorian grasses, and Pearsall used this factor to identify maize in archaeological deposits from that region. My initial analysis of tropical wild species (1984) indicated that a few had cross shapes as large as maize, but they and many others had cross-shaped, three-dimensional morphologies very different from those found in corn. I have defined eight different kinds of three-dimensional structures from analyses of 23 races of maize, 8 varieties of teosinte, and 39 wild Panicoid grasses and bamboos native to the grass cover of Panama, but largely pan-tropical as well. The three-dimensional structures defined by the morphology of the face opposite the primary cross-shaped face are

1. plain, that is, mirror-image cross-shaped (Plate 23);
2. tentlike arch (Plates 24 and 25);
3. large nodules on each corner (Plate 26);
4. covered by a thin, elongate plate;
5. two elevated pieces of silica along the long axes that form the outline of a near dumbbell;
6. irregularly trapezoidal to rectangular (Plates 24, 27, and 28);[2]
7. dumbbell-shaped; and
8. near-cross with a conical projection at each corner.

Each of these has been described in detail by Piperno (1984). Their distributions in grasses, expressed as percentages of the sum of all three-dimensional variants found in each species, are presented in Tables 3.1 and 3.2, along with measurements

[2]As described in Piperno (1984), Variant 6 phytoliths encompass a variety of shapes that the author preferred to place in a single category, since differences between them can be quite subtle and difficult to define, and they may all occur in a single species. Also, Variant 5 cross shapes can be difficult to distinguish from Variant 6 forms and may eventually be subsumed within this group.

Table 3.1
Characteristics of Cross-Shaped Phytoliths in Tropical Wild Grasses

	% Variant						Size variant			% Cross shapes	Number of cross shapes	Number of dumbbells
	1	2	3	6	7	8	1	2	6			
Andropogon brevifolius	95			5			12.1		11.6	1.3	20	192
A. bicornis	87			13			12.2		10.8	1.3	75	1347
A. leucostachyus	100						12.3			0.4	4	965
A. scoparius	88	6		6			11.7	6	9	6.3	18	270
A. virgatus										0	0	200
Andropogon sp.	90	6		4			10.6		9	17.2	25	120
Axonopus compressus	90	4		6			11.5	8.1	10.9	34	160	310
Cryptochloa sp.			100							96	50	2
Bambusoid Grass #2	100						12.1			4.7	50	700
Cenchrus echinatus	25	25		50			13.7	10.6	12.7	37	402	684
C. pilosus	5	38		56	1		11.9	10.6	12.1	45	45	55
Digitaria sanguinalis	100						10.5			0.8	4	496
Echinochloa crus-gulli										0	0	50
Eragrostis ciliaris										0	0	30
Hymenache amplexicaulis	87	5		8			10.9	7.5	8.5	15.9	194	1021
Imperata cordata	100						10.3			3.3	10	290
Isachne arundinaceae										0	0	100
Olyra latifolia						100				56	40	31

Oplismenus burmanii			100				4	5	120
Oplismenus hirtellus	28		72	13.8		13.4	6.9	87	1163
Panicum fasciculatum							0	0	100
Paspalum paniculatum	7		93	10.8		9.8	9.1	15	150
Paspalum plicatulum	49	17	34	12.8	10.5	11.4	31	379	844
Paspalum virgatum	87	4	9	11.3	7.4	11.0	4.2	69	1577
Paspaium sp.	10	70	20	9	7.5	10.4	9	20	200
Pharus latifolius							0	0	100
Pharus parvifolius							0	0	100
Setaria scheelei							0	0	200
Setaria sp.	22	28	50	12.3	9	11.2	25	50	150
Streptochaeta sodiroana							0	0	100
Streptochaeta spicata							0	0	100
Trachypogon moutiflora							0	0	100
Tripsacum andersonii	0	10	90		9	12.5	10	10	90
Tripsacum dactyloides	5	10	85	12	9	11.2	10	20	180
Tripsacum lanceolatum	0	18	82		11.1	12.5	54	40	34

Table 3.2
Characteristics of Cross-Shaped Phytoliths in Different Races of Maize[a]

	% Variant				Size variant				Number of
Zea mays L.	1	2	6	7	1	2	6	% Cross shapes	cross shapes
Maiz Dulce	70	17	13		14.8	12	13.5	67	30
Pepitilla	63	23	14		15.9	12.1	12.3	63	40
Pardo	77		20	3	13.7	13.5	12.5	56	50
Amapito	77	7	16		14.2	9.4	11.3	32	30
Pisankala	33	20	47		12.8	10.0	10.5	14	20
Jala	98		2		12.9		15.8	53	40
Chapalote	66	16	16	2	12.6	9.5	13.5	83	60
Palomero Toluqueno	70	15	15		13.9	9.9	10.0	35	30
Tepecintle	60	32	8		13.8	11.2	12.5	62	50
Cororco	93		7		14.2		13.5	46	30
Nal-tel	50	34	16		13.8	12.0	13.5	56	50
Cuzco	87	7	6		13.6	11.3	11.3	33	30
Jalisco	54	13	23	10	14.9	9.8	13.6	29	30
Reventador	80	7	13		12.4	12.0	11.7	54	40
Cacahuacintile	40	27	27	6	13.0	10.4	11.0	43	40
Puya	61	23	16		12.7	11.2	11.4	39	120
Pira	60	10	25	5	13.6	9.0	11.9	17	30
Chechi Sara	63	17	20		13.8	13.5	13.2	25	40
Zapalote Chico	52	29	19		13.8	12.0	13.0	31	33
Cateto	85	5	10		13.0	11.8	11.7	31	125
St.Croix	73	14	12		13.9	11.9	11.7	29	200
Pororo	79	7	14		12.8	10.6	10.3	28	115
Canario de Ocho	80	12	8		14.2	11.8	11.6	27	150

[a]Mean and standard deviation of size variants: Variant 1, 13.7±0.8; Variant 2, 11.2±1.3, and Variant 6, 12.2±1.3.

of the two other variables employed in maize identification; mean sizes of cross shapes, based on short-axis measurements, and the percentage of cross shapes in each species. In documenting phytolith size, actual sizes in micrometers are recorded instead of the four-tiered classification system of small, medium, large, and extra large. Phytolith measurements in micrometers provides a sounder method of quantification and one more amenable to statistical verification.

Three-dimensional structures of cross shapes are easily recognized with the light microscope at powers of 400×. The mirror-image cross shapes of Variant 1s, for example, show no other structure on the opposite side of the phytolith (Plate 23), while the trapezoidal/rectangular structures opposite the cross-shaped side of Variant 6s are usually well defined (Plates 27 and 28). Phytoliths are transparent, hence it is often possible to discern a structure on the reverse side without actually turning the sample (Plates 25 and 29). The tentlike arch of Variant 2 forms, for example, is easily recognized in surface view as a line running the length of the center of the cross shape (Plate 25), and the rectanguloid–trapezoidal structures of Variant 6s are often readily visible as well (Plates 27 and 28).

It is apparent from Tables 3.1 and 3.2 that significant differences in the sizes and percentages of three-dimensional variants exist between grass species. However, before discussing the differences between maize and wild grass phytoliths and their implications in interpreting the fossil record, the significance of interpopulation phytolith variability must be assessed.

Phytolith Interpopulation Variability

Studies of interpopulation phytolith variability seldom have been made, but in comparisons of domesticate and wild phytoliths they were urgently needed to gain a measure of its effect and determine if phytolith characteristics held constant under varying environmental regimes. For eight wild grasses in Table 3.1 where cross shapes or dumbbells were most common, three to five replicate samples (two leaves from each plant) were analyzed; each was drawn from a different region of Panama. Often, one additional replicate was taken from a different tropical country. For five races of maize, four replicates each of leaves, husks, and tassels drawn from the same population were studied. Tables 3.3 and 3.4 present the results of this analysis.

The replicated wild samples displayed a notable constancy of phytolith size and morphology, as ranges for cross-shaped sizes (especially Variant 1) and three-dimensional frequencies were usually quite narrow. Grasses that in the original study of three-dimensional structure (Piperno, 1984) displayed high percentages of non-Variant 1 cross shapes, such as *Cenchrus echinatus* and *Oplismenus hirtellus*, displayed this same characteristic in each of the replicates. The same constancy held true for such species as *Axonopus compressus* and *Hymenache amplexicaulis*, which displayed high percentages of Variant 1 cross shapes in every specimen. Grasses such as *Paspalum plicatulum* and *Oplismenus hirtellus* that produced kinds of genus-specific phytoliths in the original study (irregularly shaped three-lobed short cell phytoliths and dumbbells with thick rectanguloid structures on their opposite face, respectively) produced them in significant quantities in every sample (Table 3.5).

There are many wild Panicoid grasses that produce extremely few cross-shaped phytoliths and dominant numbers of dumbbells, a feature demonstrated by the many species in Table 3.1 with very low frequencies (less than 5%, of cross shapes). That this is an unvarying characteristic of many grasses is made clear by the very high frequencies of dumbbells counted in every specimen of two different species of *Andropogon*.

Phytolith size also varied little among the replicates. Size ranges within species are usually quite small, again even in plants sampled as far apart as Panama and Mexico. *Cenchrus echinatus* and *Oplismenus hirtellus* are producers of cross-shaped

Table 3.3
Replicated Analysis of Cross-Shaped Phytoliths from Wild Grasses[a]

Species	$\overline{X}$% Variant			Range: Variant			$\overline{X}$ Size variant			Range: Variant		$\overline{X}$% Cross shapes	Range	Number of populations
	1	2	6	1	2	6	1	2	6	1	6			
Oplismenus hirtellus	28	—	72	15–38	—	62–85	13.8	—	13.4	13.7–14.1	11.2–15.1	6.9	4.5–8.5	3
O. hirtellus (Mexico)	0	—	100				—	—	14.9			12.5		1
Paspalum plicatulum	49	17	34	35–60	—	25–52	12.8	10.5	11.4	11.8–13.9	10.8–12.1	31	23–39	4
P. plicatulum (Mexico)	53	7	40				12.5	—	10.5			20		1
Cenchrus echinatus	25	25	50	7–34	—	42–57	13.7	10.6	12.7	13.3–14.0	12.3–13.3	37	29–51	4
C. echinatus (Belize)	13	30	57				15.1	—	13.6			20		1
Axonopus compressus	85	2	13	70–96	—	4–30	11.5	8.1	10.9	11.2–12.0	9.4–13.2	34	21–41	3
A. compressus (Mexico)	82	10	6				10.4	—	12.0			30		1
Hymenache amplexicaulis	87	5	8	79–98	—	1–14	10.9	7.5	8.5	10.5–11.3	7.5–9.3	15.9	9–28	3
H. amplexicaulis (Mexico)	94	6	—				11.7					33		1
Paspalum virgatum	87	7	6	80–100	—	0–12	11.3	7.8	11.0	9.6–12.4	9–11.5	4.2	2.6–6.2	4
Andropogon leucostacyus	100	—	—	—	—	—	12.3	—	—	12.0–12.5	—	0.4	0–2	3
A. leucostrachyus (Mexico)	—	—	—	—	—	—	—	—	—	—	—	0		1
Andropogon bicornis	87	—	13	76–95	—	0–20	12.2	—	10.8	10.8–12.8	9–12	1.3	0–2.2	4
A. bicornis (Guatemala)	—	—	—	—	—	—	—	—	—	—	—	0		1

[a]All sizes given in micrometers. Ranges for Variant 2 measurements are not listed.

Table 3.4

Replicated Analysis of Cross-Shaped Phytoliths from Maize[a]

Race	$\bar{X}$% Variant			Range: Variant			$\bar{X}$ Size variant			Range: Variant		$\bar{X}$% Cross shapes	Range	Number of different plants
	1	2	6	1	2	6	1	2	6	1	6			
Puya														
Leaf	61	23	16	54–66	—	5–24	12.7	11.2	11.4	12.5–12.9	9–13.6	39	26–44	3
Husk	25	—	75	—	—	—	15	—	13	—	—	7	—	3
Tassel	85	3	12	77–100	—	0–20	12.1	11.5	13.3	11.4–12.5	13–13.5	70	68–72	3
Canario de Ocho														
Leaf	80	12	8	72–90	—	4–16	14.2	11.8	11.6	13.0–15.5	10.4–12.0	27	25–40	4
Husk	39	15	46	17–75	—	12–60	16.4	12.9	14.9	15.1–18.4	13.8–15.8	39	30–51	4
Tassel	55	16	29	28–82	—	10–46	13.5	10.9	12.1	13.3–13.7	11.6–12.4	57	46–69	4
Pororo														
Leaf	79	7	14	65–88	—	4–35	12.8	10.6	10.3	12.2–13.4	9–12.8	28	20–32	4
Husk				Cross-shaped phytoliths occurred rarely in each of four specimens										
Tassel				Cross-shaped phytoliths occurred rarely in each of four specimens										
St. Croix														
Leaf	73	14	12	64–90	—	4–18	13.9	11.9	11.7	13.4–14.4	9.8–12.8	29	17–40	4
Husk	26	26	48	8–35	—	32–60	16.8	15.4	16.5	15.3–18.8	14.9–18.1	34	21–49	4
Tassel	64	2	34	62–65	—	34–35	12.7	—	—	12.0–13.4	—	69	63–74	4
Cateto														
Leaf	85	5	10	72–100	—	0–20	13.0	11.8	11.7	11.8–14.6	10.2–13.0	31	21–37	4
Husk	48	5	45	36–58	—	42–52	14.4	12.0	13.1	12.5–15.8	12.4–14.6	40	34–50	4
Tassel	53	9	38	44–68	—	28–44	11.4	8.3	11.0	11.2–11.7	10.0–11.6	60	40–78	4

[a]All sizes given in micrometers. Ranges for Variant 2 measurements are not listed.

Table 3.5
Percentage of Distinctive Non-Cross-Shaped Phytoliths in Two Wild Grasses[a]

O. hirtellus dumbbells	# 9268	30%
	# 8331	33%
	# 1417	49%
P. plicatulum irregulars	# 5216	4.2%
	# 5100	6.5%
	# 5072	15%
	# SC-245	5%

[a]Of all short cell phytoliths.

phytoliths whose size overlap maize, while *Hymenache amplexicaulis, Axonopus compressus* and *Paspalum virgatum* have smaller cross shapes. These results are also in agreement with previous phytolith size studies (Piperno, 1984), which were based on the analysis of a single specimen. It is important to emphasize that individual samples were drawn from habitats having very different soils and average annual rainfall, factors that would favor variability in phytolith attributes should they be environmentally modulated. Particularly striking is the similarity of specimens from Panama and other tropical countries. It appears that phytoliths formed in short cells of the grass epidermis are under a considerable degree of active genetic control, a factor that imparts validity to their use in paleoecological reconstruction.

Table 3.4 shows the results of the replicate analysis of maize leaves, husks, and tassels. Results of quantitative silica analyses in these structures has been discussed in Chapter 2. Husks often have values ranging only from less than one-tenth to one-third of the content of leaves, while tassels may contribute larger amounts, ranging from one-third to four-fifths of the leaf content. It is doubtful that the small quantities produced by husks of many races and tassels of certain races are sufficient to achieve good representation in fossil assemblages.

In replicated maize leaves, three-dimensional phytolith structure shows little variation, as phytoliths are predominantly Variant 1 in every specimen. Mean size of Variant 1 cross shapes does not fall below 12.7 μm. It is clear from these data and analysis of single specimens of many more races (Table 3.2) that the production of numerous Variant 1 cross shapes with mean sizes between 12.7 and 15 μm is a fundamental characteristic of Central and South American maize leaves. As we will discuss shortly, these attributes in combination discriminate maize from wild grasses. Phytoliths from husks and tassels have not been described elsewhere; hence, it is appropriate to do so here in some detail.

Husk phytoliths, in contrast to silicified cells from leaves, take the form of only three primary shapes: dumbbells, cross shapes, and trichomes. Husk cross shapes are predominantly Variant 6 (Plates 28 and 29) and in Canario de Ocho displayed a considerable degree of variability from sample to sample. Some of this may relate to the fact that three-dimensional morphologies in this race deviated from the classic Variant 6 or Variant 1 types and presented difficulties of classification. Variant 6 cross shapes from husks are larger than those of leaves and are extremely thick (Plate 29), with this dimension averaging about 10.8 μm.

Variant 1 cross shapes are also large and thick (average = 7.3 μm) when compared to leaf phytoliths. Husks also contribute a high proportion of irregular, atypical short cell phytoliths broadly classifiable as dumbbells or cross shapes but which have extra lobes or other irregularities. In many characteristics then, especially shape, size, and thickness, husk phytoliths differ markedly from those of maize and wild grass leaves, and may be identifiable in soil profiles if recovered in any number.

Phytoliths from tassels are on the whole more similar to leaf silica bodies, though their sizes tend to be smaller, in some races probably small enough to leave a record indistinguishable from wild grasses. Three-dimensional morphologies are predominantly Variant 1, though not as markedly so as in leaves. It is therefore questionable if tassel phytoliths recovered from soils would classify as maize in the identification scheme about to be proposed. An interesting aspect of tassel phytolith production is the very high frequency of cross-shaped phytoliths found in every specimen analyzed.

In sum, analysis of replicate maize leaves, husks, and tassels from five different races demonstrates several clear and consistent patterns in the production of phytoliths from each structure, another finding pointing to the genetic control of short cell phytolith attributes. With intra- and interpopulation variability measured and shown to play a negligable role in determining short cell phytolith characteristics, it is possible to proceed forward and systematically compare phytoliths from wild grasses and maize.

Comparisons of Maize and Wild Grass Phytoliths

Three variables are used in this analysis: (1) percentage of Variant 1 cross-shaped phytoliths measured against the sum of all kinds of three-dimensional variants, (2) short axis size for each cross-shaped variant, and (3) the percentage of cross-shaped phytoliths measured against the sum of dumbbells and cross shapes. Only leaf phytoliths are considered.[3] A survey of Table 3.2 shows that only two maize races out of 23 contribute less than 50% of Variant 1, while a considerable

[3]A preliminary analysis of phytoliths from the floral bracts (glumes, lemmas, and paleas) of five of the wild grasses (*Oplismenus hirtellus, Cenchrus echinatus, Andropogon leuchostachyus, Paspalum plicatulum* and *Axonopus compressus*) indicates that cross-shaped phytoliths are produced in very low frequencies by these structures and in every species are much smaller than leaf cross shapes. Therefore, the inclusion of bract phytoliths in this analysis would confirm and strengthen the results and conclusions. Also, investigation of 12 additional Panicoid and Bambusoid grasses, comprising 9 different genera, shows that they do not contribute any attributes that are potentially confusable with maize (Piperno, 1987a).

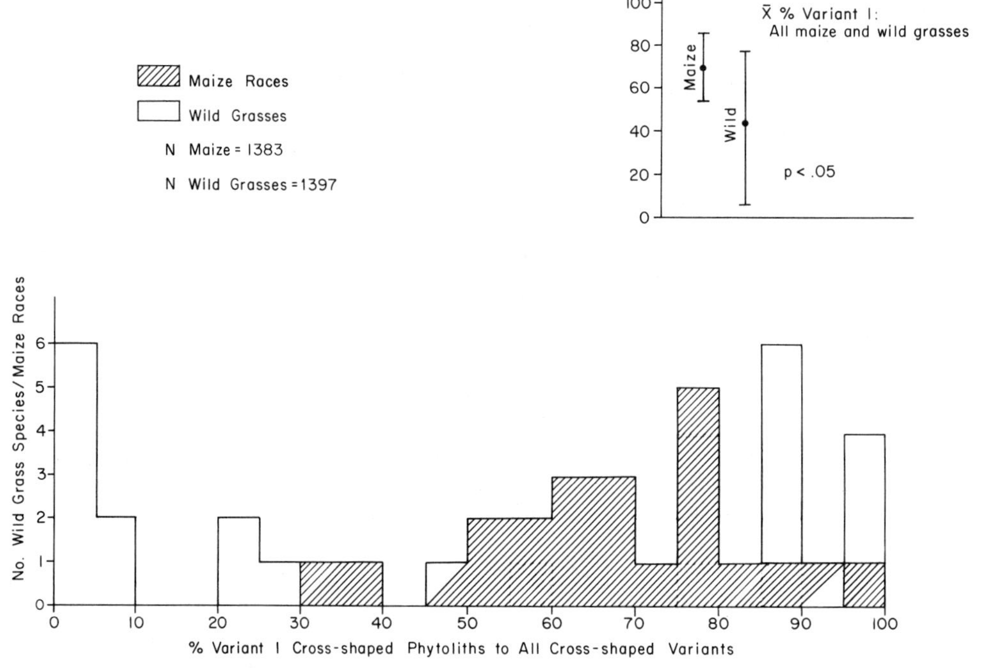

Figure 3.5 The percentage of Variant 1 cross-shaped phytoliths in tropical grasses and maize.

number of wild grasses contribute far fewer. These are the Variant 2/6 producers, such as *Oplismenus, Cenchrus,* and *Tripsacum*. A few wild grasses, namely *Axonopus, Hymenache,* and *Andropogan,* like maize, are dominant producers of Variant 1 types.

The histogram in Fig. 3.5 summarizes these differences. Maize is shown in cross-hatch and wild grasses in an open bar. The box in the upper right-hand corner shows the mean and standard deviation for the two groups (maize and wild) when all representatives of each category are combined, along with results of the *t* test. In the histogram, maize values cluster starting around 60%, while many wild values are much lower.

Short-axis measurements of maize Variant 1 (as well as Variant 2 and 6 cross shapes) are very often significantly larger than those of wild grasses, as shown by the histogram in Fig. 3.6. In maize, mean size of Variant 1s is 13.7 μm as compared to 11.9 μm for wild grasses, a difference significant at the 0.001 level. Maize races cluster starting at a mean size of between 12.5 and 13 μm, while the great majority of wild grasses have smaller sizes. On the basis of this analysis, grasses can be divided into two groups which differ in their mean size of Variant 1 cross-shaped phytoliths. Many wild grasses contribute cross shapes of sizes less than 12.5 μm. Many maize races have cross shapes of sizes between 13 and 15 μm. The two wild

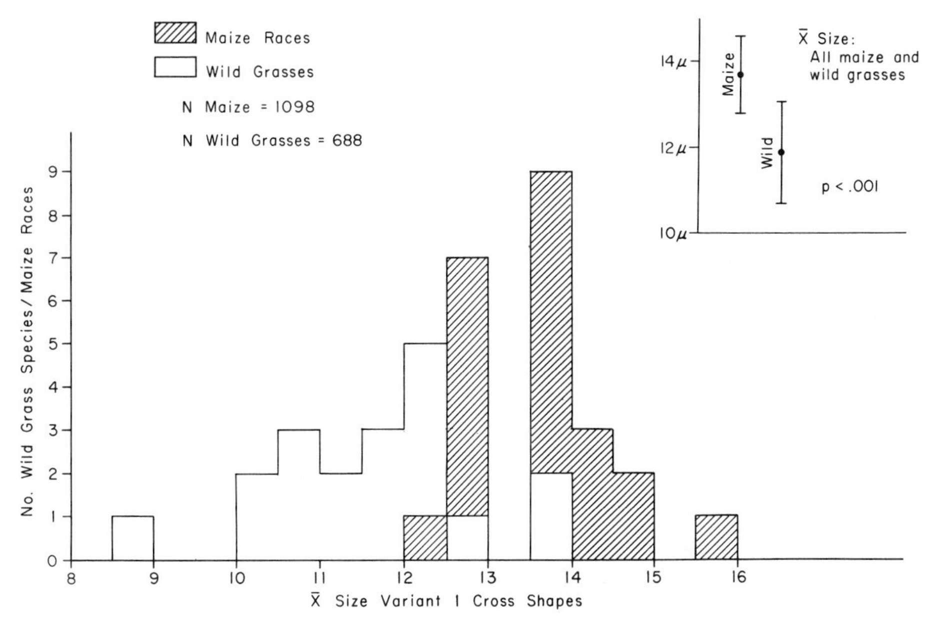

Figure 3.6 The mean size of Variant 1 cross-shaped phytoliths in tropical grasses and maize.

grasses contributing cross shapes measuring between 13.5 and 14 μm are *Oplismenus hirtellus* and *Cenchrus echinatus,* and they, besides having dominant percentages of Variant 2 and 6s, also have other kinds of durable, distinct phytoliths that identify their presence in soil profiles. This is an important qualitative consideration in grass identification that is discussed on page 90. In addition, not a single Variant 1 cross shape exceeding 20.6 μm (Pearsall's extra large category) has been observed in wild grasses, whereas they occur in maize. Thus, there appears also to be an absolute size difference between the two groups.

Another differentiating attribute of maize is a character I previously called the dumbbell to cross-shaped phytolith ratio, computed by dividing the number of cross shapes counted in a species into the number of dumbbells counted. Maize very often contributes proportionally many more cross shapes, as shown by Fig. 3.7. In this analysis, ratios were converted into percentages because the latter are easier to evaluate statistically. These results and Brown's (1984) study of Central North American grasses show that cross-shaped phytoliths are a very uncommon component of wild non-*Zea* Panicoid grasses, which produce predominantly dumbbells.

The result of this investigation is a series of characters displayed by maize and wild grasses whose means are considerably different, but whose distributions overlap, sometimes considerably. On the basis of one character, we could not with any

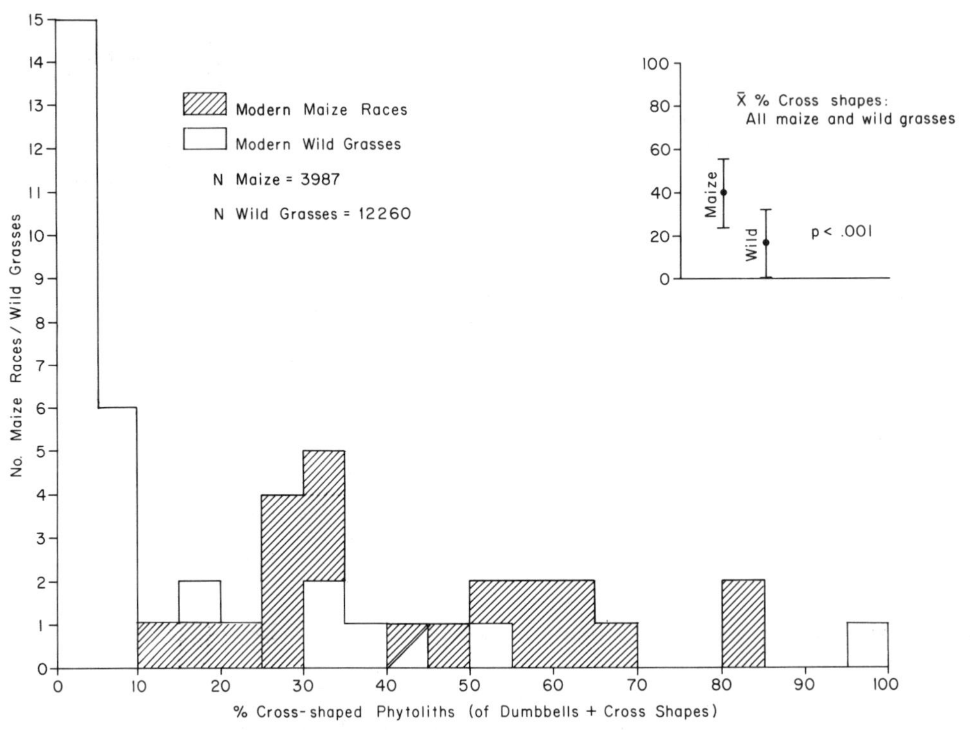

Figure 3.7 The percentage of cross-shaped phytoliths in tropical grasses and maize.

degree of confidence accurately identify an unknown specimen from an archaeological sample as belonging to a maize or wild population. It is possible to visually inspect the data and rely on intuition to judge that only maize combines the attributes of large cross-shaped size, high percentages of Variant 1 cross shapes, and high percentages of cross shapes of all types measured against dumbbells. An intuitive argument was made to identify archaeological maize phytoliths in Panamanian sediments (Piperno, 1984). A more formal, statistical approach was obviously desirable and this was accomplished by means of Fishers Linear Discriminant Function analysis, a multivariant technique (Piperno and Starczak, 1985).

Discriminant function values were calculated for each wild grass and maize race using three and five variables; the results are plotted in the histograms of Figs. 3.8 and 3.9. The histograms show that maize and wild grasses can be separated into two populations based on the measurements made, as two discrete clusters are present in each graph. As will be shown in Chapter 7, cross-shaped phytoliths from archaeological sites can be identified as wild or maize with the same kind of analysis. Although the five-variable analysis, which includes the variable percentage of cross shapes, achieves a somewhat better separation of modern phytoliths, we

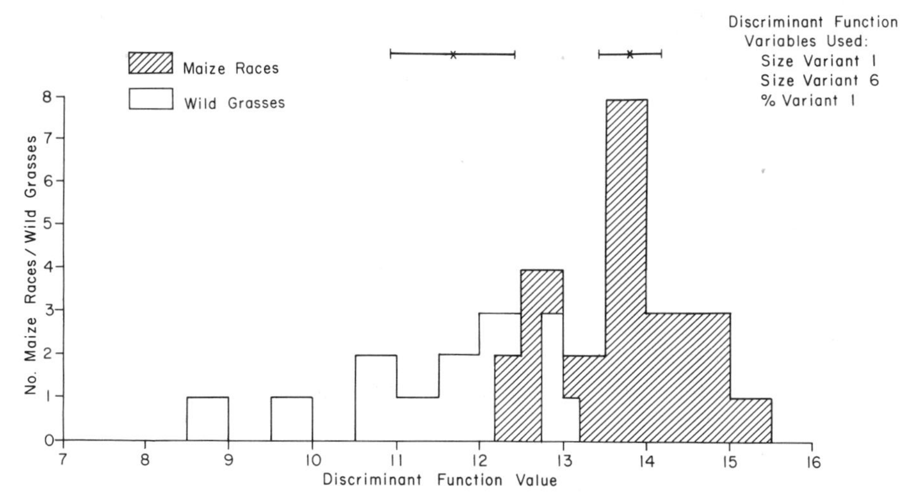

Figure 3.8 A three-variable discriminant function analysis of tropical grasses and maize.

will find it is often not appropriate for use with archaeological data because of inflation of archaeological deposits by dumbbells from wild grasses.

Only wild grasses contributing both Variant 1 and 6 cross shapes were included in the multivariate analysis. Inclusion of wild species not run because of lack of production of Variant 6s such as *Andropogon leucostachyus, Imperata cordata,* and Bambusoid grass #2 would not have affected the population segregation because

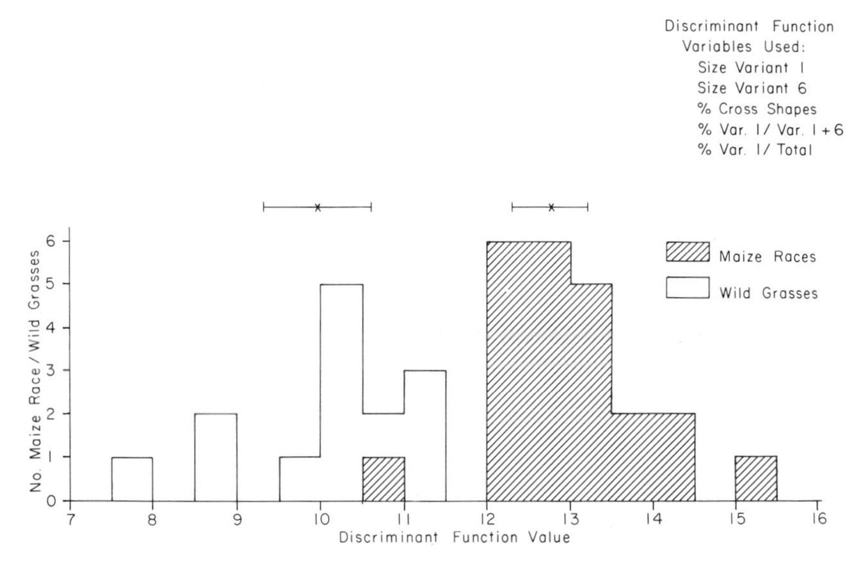

Figure 3.9 A five-variable discriminant function analysis of tropical grasses and maize.

they have small Variant 1 crosses (Table 3.1). In sum, it appears that a multivariate approach to short cell phytolith attributes effectively separates many wild taxa from maize. Such an analysis should be extended in the future to include variables such as cross-shape thickness and dumbbell three-dimensional structure. It should be stressed that wild grasses forming the native cover of other archaeological study regions (e.g., eastern and southwestern North America) must be examined before this identification procedure can be widely applied.

In addition to the approach just described, a presence/absence analysis of archaeological phytolith assemblages can aid in determining the identification of cross-shaped phytoliths. Several cross-shaped producing wild grasses can be ruled out from possible representation in fossil deposits because they have high numbers of apparent genus-specific phytoliths or phytoliths confined to a narrow range of genera, not including *Zea*. They include *Paspalum plicatulum*, *Oplismenus*, and *Cenchrus*. Their shapes are described later in the section on wild grass phytoliths. It is important to emphasize again that replicate studies showed they were consistently produced in large numbers, independent of the environmental conditions of plant growth.

Some Characteristics of Teosinte Phytoliths

So far, teosinte has been left out of the discussion, not because phytoliths show no promise, for preliminary investigations suggest that teosinte phytoliths have a great deal to offer in way of elucidating the early history of maize cultivation. Table 3.6 presents the sizes and percentage frequencies of cross-shaped phytoliths in leaves of eight varieties of teosinte, based on replicate studies of from two to seven populations of each variety.

Races such as Balsas and *Zea luxurians* and *Z. diploperennis* contribute high frequencies of Variant 2 and/or Variant 6 cross-shapes, and therefore have three-dimensional patterns very unlike maize. Mean sizes of cross shapes vary, with Balsas and Chalco producing very small phytoliths, *Zea luxurians* Variant 1s even larger than maize, and Huehuetenango and Central Plateau having phytolith sizes that overlap many races of maize. There as well can be a great deal of interpopulation variability between specimens of the same race. One sample of Race Nobogame, for example, produced a frequency of 88% Variant 1 cross shapes, while in another population a percentage of 35% was determined. However, all populations of *Zea luxurians, Z. diploperennis,* Huehuetenango, and Balsas displayed a constancy of three-dimensional attributes. One wonders if differing degrees of hybridization with maize and maize introgression into teosinte had a role to play in the interpopulation variability noted in some populations.

If discriminant function values are calculated from the means of each variable, Fig. 3.10 results. It is seen that only *Zea luxurians* has a discriminant function

Table 3.6

Characteristics of Cross-Shaped Phytoliths in Teosinte

	N	% Variant			Range Variant			$\overline{X}$% Size variant			Range			$\overline{X}$%		Number of
	Populations	1	2	6	1	2	6	1	2	6	1	2	6	Cross shapes	Range	cross shapes
Balsas	7	24	47	26	15–48	28–73	12–43	11.9	8.8	10.5	10.8–12.7	7.6–10.7	8–10.9	33	13–64	195
Central Plateau	5	49	14	36	17–89	8–30	3–60	13	10.2	11.1	11.3–14	7.9–12	9–12.4	48	38–53	195
Chalco	4	47	27	24	27–75	8–36	5–46	12	9	11.6	11.8–12.5	8.3–9.7	10.2–12.8	30	29–33	135
Nobogame	2	62	22	16	35–88	4–40	8–25	12.5	9.9	10.7	11.4–13.5	9–10.7	9–12.4	21	15–26	65
Huehuetenango	3	74	9	17	68–82	8–12	12–20	13	8.8	9.6	12.1–13.9	7–10.5	9.3–10.2	38	32–51	100
Zea luxurians	5	12	26	62	0–22	8–43	54–80	14.9	11.1	12.0	12.9–17	9.9–12.8	11.4–12.7	61	43–83	195
Zea perennis	3	29	16	55	5–76	2–35	22–85	11.1	9.1	11.3	9–13.5	9–9.2	9.6–12.6	26	20–29	90
Zea diploperennis	3	30	22	48	25–35	0–40	35–70	12.7	9.4	11.2	11.7–13.4	9.2–9.6	9.8–12.5	60	47–77	80

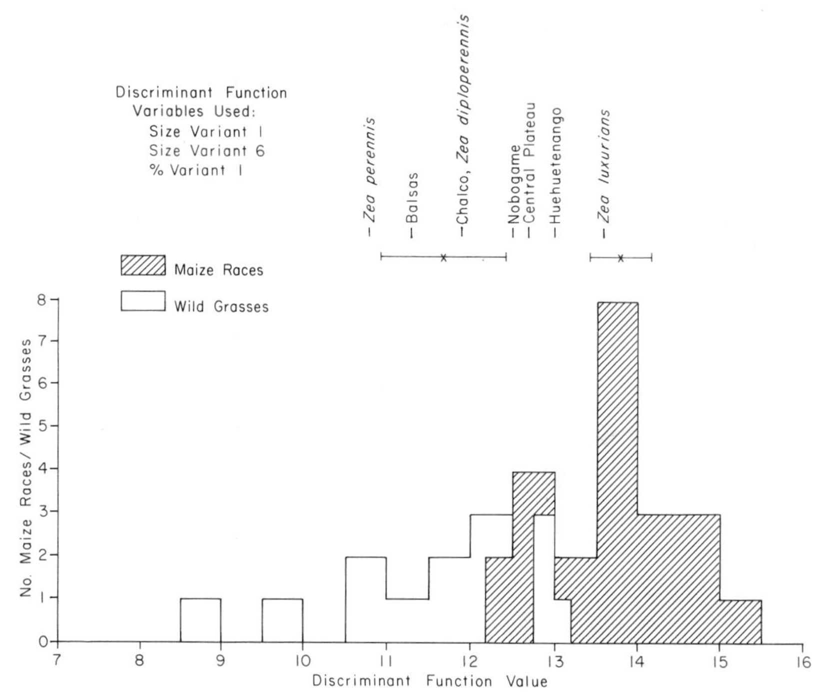

Figure 3.10 A three-variable discriminant function analysis of teosinte.

(DF) value falling between the maize confidence intervals, due to its extremely large Variant 1 cross shapes, which occur in slim numbers. If recovered from soils, the high frequency of Variant 6 phytoliths in this plant would preclude an identification of maize. Huehuetenango, Nobogame, and Central Plateau teosintes fall into the murky area of maize/wild overlap, while Chalco, *Zea diploperennis, Z. perennis,* and Balsas fall clearly into the wild grass cluster.

The really intriguing and exciting aspect of these data lies in the phytolith attributes of Balsas teosinte, a large, truly wild population native to the seasonally dry and warm western slopes of southern Mexico, which has been little affected by hybridization with maize (Wilkes, 1967; Doebley, 1983). Phytolith spectra from all seven Balsas populations exhibited (1) low frequencies of Variant 1 cross shapes, whose sizes are significantly smaller than Variant 1s from maize ($p < 0.05$),[4] (2) high frequencies of Variant 6 cross shapes, and (3) high frequencies of Variant

[4]Russ and Rovner (1987) presented preliminary results from a stereology and computer-assisted image analysis of maize and teosinte, which suggested that Balsas cross shapes are not significantly smaller than maize. However, measurements were taken on the long axis of phytoliths; hence, results are not comparable to those presented here.

2 cross shapes, whose sizes are also significantly smaller than maize Variant 2s ($p < 0.05$) (significance of size differences tested with the t test). In fact, the mean value of 47% Variant 2 cross shapes in Balsas teosinte is the highest yet found in any grass.

Isoenzymatic data are indicating that Balsas populations are genetically closest to maize (Doebley, 1983), and Iltis and Doebley (1984) believe that they were maize's wild ancestor. If true, then during the course of domestication, maize phytoliths have undergone major structural changes (from Variant 2 to Variant 1 cross shapes) and size modification (toward significantly larger cross shapes), changes that might well be documented in an archaeological phytolith record from western Mexico.

That changes under domestication might be manifest in the phytolith record is not a surprising revelation. It has been continually emphasized how many phytolith characters appear to be under genetic control. The dramatic differences apparent in some morphological attributes of maize and teosinte are an expression of genetic changes resulting from the selective pressures of domestication. Phytoliths are formed in plant cells. There is little reason to suppose that genetic changes are any less expressed at the cellular level than at the macromorphological level. Such changes might therefore contribute to significant modification of phytolith morphology and size. We will stress the large investment in fundamental research needed to further assess phytolith characteristics in teosinte and maize, and we will continue to emphasize that after necessary comparative research is carried out, phytolith applications to problems of the earliest origins and dispersals of maize may prove to be extremely rewarding.

Some Factors Governing Size and Morphological Variation in Grass Silica Bodies

What are the reasons behind the dramatic differences seen in phytolith size and morphology in maize, teosinte, and non-*Zea* grasses? One explanation for the differences in three-dimensional morphology seems to involve the location of phytoliths in the leaf. Some of the extracted grasses, including maize, *Setaria,* and *Andropogon,* still had phytoliths embedded in organic tissue, and it was possible to see the orientation of phytoliths in the leaf. Almost invariably, Variant 1 cross shapes were in cells lying over the veins and Variants 2 and 6 were in cells located in between the veins. It is presently unclear why such a relationship should exist. The proportions of cross-shaped, three-dimensional types in these grasses were the result of the number of cross-shaped phytoliths located over versus between the veins.

The location of phytoliths in the leaf also affects their size. Phytoliths in cells over the veins are very often larger than phytoliths in cells between the veins. This may not be the whole story, however. Part of the explanation for phytolith size may also lie in a value called the leaf venation index. This is the ratio of the number of veins in the leaf to the width of the leaf. Grasses with higher indexes (more veins per surface area) than maize like Balsas teosinte have smaller phytoliths than maize. The possible relationship between phytolith size and leaf venation index may be a functional one; if a grass has relatively fewer veins across its leaf, they may have to be wider to transport sufficient quantities of water and minerals to the plant. Remember that cross-shaped phytoliths are sized by their width, not their length.

But this is still not the whole story. Phytolith size and shape proportions will also vary depending on the number of rows of silica-bearing short cells located over the leaf veins of each species. In Panicoid grasses, the number usually varies from 3 to 5 rows (Metcalfe, 1960). Cross-shaped phytolith size and three-dimensional structure, then, are complex and little-understood results of the interaction of at least three factors: the location of phytoliths in the leaf, the leaf venation index, and the number of rows of silica bearing cells over the veins in the leaf. Further investigations of these characters are needed to better understand the variation found in silica bodies of different grass taxa.

Phytoliths in Wild Monocotyledons

Phytolith morphology is immensely complex. This fact has no doubt already been impressed upon the reader from descriptions and discussions accompanying the few plants discussed thus far. It is not my intention to describe and classify all of the phytolith shapes that have been isolated from modern plants. Atlases and catalogs intended for this purpose should soon be forthcoming and there would not be enough space. Still, because phytolith morphology is so little understood and descriptions and illustrations currently so difficult to find in published literature, the following sections offer a fairly wide survey and large number of photographs. Some characteristic shapes and sizes of phytoliths will be presented as they occur in major groups of plants and species with different kinds of growth habits.

The Gramineae

We will start with that most familiar of families, the Gramineae. Descriptions and illustrations of grass phytoliths abound in the literature. Some basic types and their distributions have been discussed previously and illustrated in Fig. 3.2. Some remarks on lesser known aspects of morphology and distribution of forms compiled by recent investigations are presented here.

The well-known classification system of Twiss *et al.* (1969) that correlates the four major short cell shapes of grass leaves (bilobate, cross, saddle, and circular to acicular) with three subfamilies (Panicoid, Chloridoid, and Festucoid) is still a very workable system. Modifications have been made to accomodate changes in the higher level taxonomy of grasses and there can be an annoying lack of correspondence between phytolith shape taxonomy and grass taxonomy that must be considered, depending on the region where archaeological studies are carried out. Five subfamilies of grasses are now generally recognized. They are the Pooideae, which includes the Festucoid grasses; the Eragrostoideae, subsuming the Chloridoid grasses; the Panicoideae, the Panicoids; the Arundinoideae, which includes such genera as *Phragmites, Danthonia,* and *Aristida;* and the Bambusoideae, bamboos. Some basic kinds of short cell phytoliths that typify a particular subfamily, such as Panicoideae dumbbells, are more broadly dispersed, a factor to be evaluated in making taxonomic assignments of fossil grass phytoliths.

Brown's (1984) extensive and excellent survey of phytoliths in grasses common to central North America (see also examples of his phytolith keys in the appendix) provides examples of overlap of shape distributions and offers changes in phytolith terminology. For example, some genera in the subfamily Arundinoideae, including *Phragmites,* produce saddle shapes, and all *Danthonia* and *Aristida* species (also Arundinoideae) produce bilobates, formerly called dumbbells (Brown 1984, 1986). Certain circular to acicular Festucoid-type (Pooideae) phytoliths of Twiss *et al.* (which Brown calls nonsinuous short trapezoids) are produced in low numbers in other subfamilies, a feature which Twiss *et al.* (1969) noted. Other Festucoid phytoliths, such as the oblong sinuous class of Twiss *et al.* (Brown's "woven" or "wavy" trapezoids) are restricted to the Pooideae (Brown, 1984; Piperno, 1987a). The necessary deviations from the original scheme of Twiss *et al.* must be worked out region by region through intensive analyses of grasses comprising the native cover.

Still more detailed analysis in a regional context may determine that finer attributes of phytoliths can distinguish tribes and genera or are found in a narrow range of genera. This provides a basis for discriminating and ruling out the presence of certain taxa in fossil phytolith assemblages, both of which are important in interpretation. For example, Brown (1986) has noted that Arundinoideae bilobates tend to be much larger with slender and longer shafts and differ markedly from Panicoideae bilobates. These features may help eliminate potential confuser taxa in determining whether phytoliths in fossil sediments are primarily from C3 (members of the Arundinoideae) or C4 (Panicoid) grasses (see discussion in Chapters 7 and 8). Brown found several shapes to be specific at the tribal level and isolated a particular kind of trichome only from the genus *Stipa* out of 112 taxa examined.

New World tropical bamboos examined by the author demonstrate a number of taxonomically significant shapes. For example, irregular mesophyll (?) cell phytoliths (Plate 34) occur only in certain Bambusoid genera such as *Olyra, Bambusa, Cryptochloa,* and *Lithachne*. Those in the latter two genera differ markedly from the *Olyra* and *Bambusa* types. Cross-shaped phytoliths having Variant 3, three-dimensional structures (Plate 26) have been isolated only from the genera *Cryptochloa* and *Lithachne,* while the similar Variant 8 types occur in *Olyra*. Variant 8 types have also been isolated in small numbers from a few races of maize. *Olyra* and *Lithachne* cross shapes also appear to have different general shape characteristics than Panicoideae cross shapes, their lobes being larger, with a wider arc.

Three genera of bamboos, *Chusquea, Pharus,* and *Streptochaeta,* contribute short cell phytoliths with three-dimensional structures thus far not found in any other grass examined; hence, the possibility of genus-specific separation is real (Plates 35, 36, 37) (see the phytolith key in the appendix for details). Bamboo phytoliths will bear much importance in reconstructions of tropical vegetation. Since they are adapted mainly to the shady, cool understory of forests, they permit identification of forest versus nonforest grasses, something that cannot be accomplished with a pollen record. They will contribute to the identification of forested vegetation in tropical paleoecological profiles, and judging from preliminary results of studies in modern forests in Panama and Costa Rica to be presented in Chapter 6, may help investigators to discriminate between dry and wet tropical forests in fossil phytolith sequences.

We have already seen in Special Topic 1 how size and three-dimensional attributes of short cell phytolith morphology can be used to identify maize in tropical sediments. Clearly, three-dimensional morphology of phytoliths is a subject that deserves a great deal of attention in future investigations. In addition to maize and bamboos, a number of other tropical grasses have been found to have possible genus-specific variants or forms found in a narrow range of genera. They include *Paspalum,* with irregular three- and four-lobed phytoliths that are Variant 6 in three-dimensional structure (Plates 30 and 31); *Oplismenus,* with dumbbells having a thick rectanguloid structure on their opposite face (Plate 32); and *Cenchrus* and *Pennisetum,* which produce long Variant 5/6 dumbbells (Plate 33).

In addition, there appears to be a clear segregation of Panicoid grasses with regard to the type of three-dimensional forms taken by short cell phytoliths (both cross shapes and bilobates). Such genera as *Tripsacum, Cenchrus, Oplismenus,* and others are predominantly Variant 6 producers, while *Andropogon, Axonopus,* and others are predominantly Variant 1 producers. It is important to emphasize again that these distinctive short cell phytoliths were consistently produced in high numbers in my replicate studies of the various taxa, using plants sampled from regions having very different environmental conditions of growth. When all of the phytolith categories just discussed, three-dimensional structure, size, and thickness, have been examined closely for the various kinds of short cell phytoliths, I feel that tribal and genus-level identifications of grass phytoliths recovered from soils will be far more common than at present.

Other kinds of grass phytoliths, such as fundamental or long cells, bulliform cells, and interstomatal cells, are less useful in identifying taxa below the level of family. Their production is also more haphazard and random than silicification of short cells, and is possibly correlated more with the environmental conditions of growth, such as available levels of silica and soil moisture, rather than with a genetically programmed response. Hence, caution is required in their use as taxonomic markers in paleoecological contexts.

Square to rectangular chunks of silica, trichomes, and elongate phytoliths are frequently recurring grass phytoliths; however, all but the elongate, spiny class have been isolated from the Cyperaceae (Piperno, 1983, 1985a). Elongates with smooth outlines and square to rectangular chunks are produced by a number of nongrass monocots and a few dicots, while bulliform-like phytoliths occur in some dicots such as *Piper*. The classic fan-shaped and many of the keystone-shaped bulliform phytoliths can, however, be easily discriminated from the latter forms.

Phytoliths in Nonleaf Parts of Grasses

Phytoliths thus far discussed can be found in the leaves, sheaths, culms, floral bracts, and rhizomes of grasses, depending on the particular species and its subfamily affilation. Exceptions include the characteristic elongate forms of leaves that apparently are not found in floral bracts. There is also a much greater divergence of epidermal patterns in the leaf blade than in the leaf sheath, leading to a corresponding greater divergence in blade phytoliths.

Phytoliths found in floral bracts—the glumes, lemmas, and paleas—deserve particular mention because they are produced in great quantity, appear to be highly diagnostic of these structures, and may ultimately provide significant data on archaeological seed processing. Their potential for identifying some Old World cereals has been discussed previously. Thorough reviews on the ocurrence and morphology of floral bract phytoliths can be found in Hayward and Parry (1980) and Parry and Smithson (1966). Two shapes appear to be confined solely to bracts, dendriform and asteriform opal (Fig. 3.11). Dendriform opals are described as cylindrical rods of varying length that have protrusions or spines radiating from a central core. They are derived from long cells. Asteriform opals are roughly spherical spiky phytoliths. Also, scutiform opals are saucer-shaped bodies that have a unique, slanted apex and occur much more frequently in bracts than in leaves.

A number of different patterns of long cell phytoliths are described from different grasses, an extension of the variability documented by Miller-Rosen (1985) from wheat and barley. It seems also from Miller-Rosen's work that configurations of different kinds of silicified cells may identify grass genera.

I have examined glumes, lemmas, and paleas from several species of Panicoid grasses. Scutiform, dendriform, and asteriform phytoliths are commonly produced

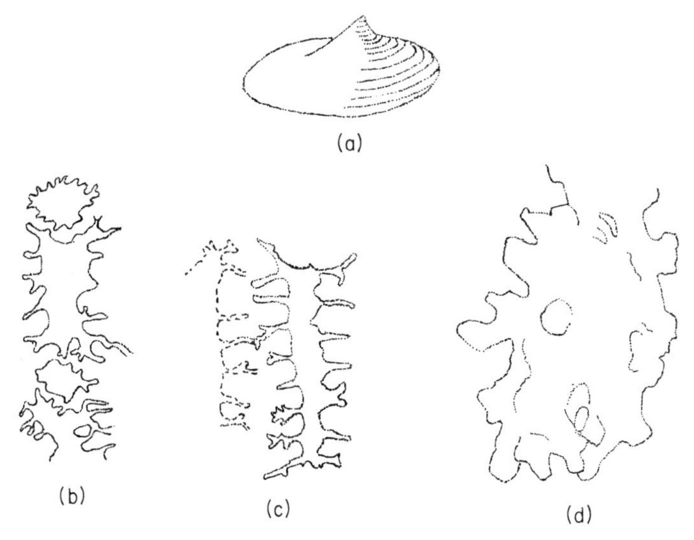

Figure 3.11 Phytoliths from floral bracts: (a) scutiform opal; (b) and (c) dendriform opal; and (d) asteriform opal.

and did not occur in the leaves of any of the species examined. Needless to say, floral bract phytoliths are yet another area that deserves a considerable amount of future attention.

In comparison to aerial structures of grasses, roots and rhizomes have received only incidental attention, but discrete phytoliths appear to be much less common than in the aerial portions. In some Panicoid species, rhizomes produce saddle-shaped phytoliths and long cells, while in roots phytolith shapes are those characteristic of vascular and intercellular tissue. Roots also produced a possible distinctive pitted, square to rectangular plate phytolith containing a central, spherical concentration of silica (Geis, 1978). As reviewed by Sangster (1978) some genera, mainly in the tribe Andropogoneae, contribute spherical silica aggregates from roots that appear similar to some deposits from dicotyledon woods. These, then, probably have little value in grass identification.

Nongrass Monocotyledon Morphology

The survey that follows of nongrass monocotyledons takes a different approach. Major headings will be organized according to major classes of phytoliths instead of plant taxa. This will permit the most distinctive kinds of phytoliths to be presented efficiently, and at the same time identify areas of broad overlap between families and genera. There are many phytoliths that have limited distributions in plants, and these will be discussed in detail as well. Phytoliths to be described next have been isolated from the leaves of plants, with exceptions as noted.

Conical to Hat-Shaped Phytoliths

A major class of monocotyledon phytolith is the conical to hat-shaped type which is found in four families, the Palmae, Cyperaceae, Marantaceae, and Orchidaceae. Many are family specific, the differentiating criteria being based on surface decoration, shape, or both. Some of their characteristics can be quite subtle, requiring the use of high magnifications (1000 ×) for discrimination. Their sizes are small, ranging from 3 to 25 μm in diameter.

In the Cyperaceae, conical-shaped phytoliths have pointed apices and smooth or nearly smooth surfaces (Fig. 3.3k). These two characteristics, taken alone or in combination, serve to separate them from conical shapes in other families. Also, individual cones may have more than one peak or, more commonly, one main peak may be encircled by others that are less tall and clearest in lateral view. A surface view of this type shows a larger central body encircled by several particles that Metcalfe has named satellites (Metcalfe and Gregory, 1964; Metcalfe, 1971) (Fig. 3.3j).

Andrejko and Cohen (1984) have implied that these structural details are identifiable only when viewed with the scanning electron microscope, SEM (a situation that is not true, as Metcalfe originally defined them and I have characterized them by using a standard light microscope, and especially oil immersion). Andrejko and Cohen (1984) have also characterized the apices of some Cyperaceae conical shapes as rounded, mainly to show that some sedge genera can be separated from others. However, for interfamilial comparison I would continue to describe all sedge conical shapes as having pointed apices.

All species in the Cyperaceae contribute these distinctive phytoliths, often in high numbers. Andrejko and Cohen (1984) feel that genus-level identification may be possible based on details (size, number, location) of the satellites surrounding the main conical body, a factor that would be of considerable ecological importance. This is the aspect that may require scanning electron microscopy.

There are other, less diagnostic phytoliths found in sedges. They include trichomes, elongates, and square to rectangular shapes, all of which overlap the grass family. Descriptions can be found in Metcalfe (1971) and Piperno (1983).

In the Palmae, hat to conical-shaped bodies [they are usually described as hat shapes; see Tomlinson (1961)] are one of two kinds that occur, the other being spherical shapes. It is highly unusual to find other types of phytoliths in this family. With the exception of two genera, *Latania* and *Socratea,* a species will contribute either hat shapes or sphericals, but not both. Copious amounts of these types are to be found in leaves, petioles, seeds, fruits, and wood. The Palmae is, in fact, unique in producing the same kind of phytolith in all of its plant structures.

All of the Palmae hat-shaped phytoliths have, to one degree or another, a surface decoration described by Tomlinson (1961) as spinulose, and I have kept this designation in my own terminology (Plate 38). Its meaning has been described. Size and morphological overlap exist between some of the Palmae hat shapes and those from the Marantaceae, where at least one genus in the Marantaceae, *Maranta,* produces conical shapes with surface decorations very similar to Palmae genera *Cocus* and *Acrocomia,* among others (Compare Plates 21 and 38). The problem here is that many phytoliths are very small and difficult to describe without the scanning electron microscope, whose use will be fundamental in this area.

However, hat shapes from other Palmae genera such as *Geonoma, Chaemadorea, Socratea,* and *Bactris* appear to be at least family specific because of the shape of the body or the number of spinules lying on the platform of the phytolith (Piperno, 1983, 1985a). *Geonoma* phytoliths, for example, are bulkier than other palm conical shapes and have a more rugulose surface decoration (Plate 39). In *Chaemadorea,* phytoliths have from 7 to 17 distinct spinules arranged in no particular pattern on the top of the structure, a feature not noted in other palms examined thus far (Plate 40).

Many Marantaceae conical shapes contributed by such genera as *Calathea* and *Ichnosiphon* are family specific due to their nodular surface patterns (Plate 41). Also, many have no flatish structure or platform on the top of the phytolith as do many Palmae phytoliths. The Orchidaceae contribute conical shapes that may be family specific, as they have a rounded cone and a smooth surface.

Spherical to Aspherical Phytoliths

Spherical to aspherical phytoliths are the second major class produced by monocotyledons. They are found in large numbers in the Palmae, Marantaceae, Bromeliaceae, Cannaceae, and Heliconiaceae and can be divided into three classes based on surface decoration: spherical spinulose, rugulose, and smooth. Those having a spinulose surface occur in the Palmae and Bromeliaceae; however, the presence of larger, more numerous, and better defined spinules in many of the Palmae permits this family to be identified (Plates 1 and 2). Some Palmae genera, including *Cocus,* have phytoliths with fewer and more indistinct spinules like those found in the Bromeliaceae. Size ranges for spherical spinulose phytoliths are from less than 3 to about 25 μm in diameter.

Spherical phytoliths with a rugulose surface decoration are found in the Cannaceae, Marantaceae, and Heliconiaceae. Their size ranges are from 9 to 24 μm in diameter (Plate 5). Spherical smooth phytoliths are rare in monocotyledons, being found in small numbers in the Cannaceae and Heliconiaceae. Size distributions are from 12 to 24 μm. Those in *Canna* appear to have poorly developed folds. Spherical nodular phytoliths occur only in the Marantaceae. They are readily distinguished from spinulose forms by their smaller protuberances, which are often unevenly distributed over the surface and can be fewer in number (Plates 3 and 4).

Lanning (1972) has also reported spherical phytoliths from the leaves of *Juncus* (Juncaceae). No information was given on size and surface patterns, but in published photographs they appear similar to those isolated from the Bromeliacee. Stant (1973) reported and illustrated with SEM small, roughly spherical phytoliths from *Gibasis* (Commelinaceae), although they were still embedded in tissue and their shape was difficult to discern.

Irregularly Angled to Folded Surfaces

A third major class of monocotyledon siliceous body is that having an irregularly angled to folded surface (Plates 7 and 42). It occurs in the Zingiberaceae (where it is the only phytolith thus far described) Marantaceae, and Cannaceae. This distribution makes sense taxonomically because these three families share the same order of plants, the Zingiberales. There may be some basis for differentiation of some of the Zingiberaceae phytoliths due to the presence of a more definable body (Plate 42).

Phytoliths with Troughs

The last major class of monocotyledon phytolith includes those with troughs. They are found solely in the Musaceae and Heliconiaceae and are specific to family and genus. *Heliconia* phytoliths (Plate 43) have troughs that are deep and centrally located, points that differentiate them from similar phytoliths in the related genus *Musa* (Plate 22).

Before concluding this section, a few other kinds of phytoliths should be mentioned because they are taxonomically important. Distinctive tongue-shaped phytoliths occur only in the Commelinaceae, a family of annual or perennial herbs that generally prefer damp conditions (Plate 44). Elongate phytoliths with undulating margins have been isolated from only *Heliconia* and *Calathea*.

Phytoliths in Monocotyledon Fruits and Seeds

This is an area that needs a great deal of investigation on a fundamental level. Phytoliths from monocotyledon fruits and seeds are providing a new and extremely promising area of research, for they often offer an array of shapes not found in leaves that can be morphologically unique to families, and possibly genera, of plants. These phytoliths are to be found in a fewer number of species than leaf silica. So far, they have not been found in plants whose leaves do not contribute significant amounts of phytoliths, but do not occur in all plants where leaf production is high. I have found substantial seed or fruit production, high enough to leave an archaeological record, in three monocot families, the Palmae, Marantaceae, and Cannaceae. In palms, the characteristic spherical or conical-shaped spinulose phytoliths are present, and they tend to be slightly larger in size than leaf phytoliths. *Socratea* is the only genus so far examined that produces different fruit and leaf phytolith shapes (conical in the leaf and spherical in the fruit). In palms, production is much more characteristic of fruit exocarp and mesocarp than kernel.

Marantaceae seed phytoliths have a variety of shapes, substructures, and surface decorations that are often restricted to individual genera (Plate 45). Cannaceae seed phytoliths are very small (1–3 μm) spherical forms that require examination with the scanning electron microscope.

Other Plant Parts

The inflorescence bracts of the Marantaceae can produce phytoliths very similar to those found in seeds, but more frequently they are conical with nodular surfaces and, hence, resemble leaf phytoliths. There is little doubt that other families will form phytoliths in these structures as well. As mentioned previously the petioles and stems of palms contribute numerous conical-shaped or spherical phytoliths of a kind found in leaves. Root phytoliths do not appear to occur in many species (for reviews see Tomlinson, 1969; Sangster, 1978), although many taxa have not been tested for phytolith production.

Summary: Monocotyledon Phytolith Morphology

In way of a summary on monocotyledon phytolith morphology it is pertinent to first review again the numerous families that are not accumulators of silica. They include the Araceae, Cyclanthaceae, Dioscoreaceae, Xyridaceae, Amaryllidaceae, Liliaceae, and Alismataceae. Phytoliths reported for yucca (Bryant, 1974) are in fact deposits of calcium oxalate. It seems that phytolith formation in monocotyledons is not as common as previously thought.

Monocotyledons having deposits of solid silica yielded a significant number of morphological forms with taxonomic significance. Phytoliths minimally distinctive at the family level have been isolated from the Palmae, Cyperaceae, Marantaceae, Commelinaceae, probably Orchidaceae, and Heliconiaceae. Differentiation below the family level is certainly possible in at least the Palmae, Marantaceae, and Heliconiaceae, with considerable potential also for the Cyperaceae. Some kinds of phytoliths occurred in a limited number of families, for example, the irregularly angled to folded surfaces from the Marantaceae, Cannaceae, and Zingiberaceae. Ecological considerations may permit the exclusion of many alternatives in the fossil phytolith record.

Monocotyledon phytoliths will have considerable importance in paleoecological reconstruction. Sedges are frequent components of secondary vegetation throughout the world and *Heliconia* is an excellent indicator of disturbed plant associations in the humid, lowland tropics. Their presence, along with grasses, in fossil phytolith assemblages will provide evidence for anthropogenically disturbed vegetation and slash-and-burn agriculture. The Palmae have great economic significance in tropical areas and they, along with the Marantaceae, are important components of tropical forests. Clearly, their presence in phytolith assemblages will provide considerable information on the nature of past tropical vegetation.

Notice that the production of a single major type of phytolith is common in monocotyledon leaves, this situation occurring in the Palmae, Orchidaceae, Bromeliaceae, and Zingiberaceae (Table 2.3). The number of shapes contributed by a plant corresponds to the number of kinds of tissue silicified and the degree of cellular differentiation within that tissue. Palmae phytoliths are all of mesophyll origin with only rare epidermal silicification occurring in some plants. It is not surprising then to find a uniformity of phytolith morphology here. The Gramineae, which among monocots represents the other extreme, has an extremely varied epidermal structure that, when silicified, produces a range of phytolith shapes.

It has been demonstrated that many monocotyledons produce phytoliths having taxonomic significance. Hopefully, the fixation on studying grasses to the exclusion of other plants in the belief that phytolith analysis should be most productive in the grass category is gone.

Phytoliths in Wild Dicotyledons

As with monocots, this section is organized according to major classes of phytoliths, with an emphasis devoted to the most and least distinctive types. There is, especially in dicotyledons, a very large grey area of morphology in between the two extremes that will be made more clear only after extensive shape descriptions and measurements become available (see Piperno, 1985a). In this section there will also be some discussion and segregation of shapes according to whether they occur in woody or herbacious plants. This will start us on a track toward assessing the paleoecological significance of dicotyledon phytoliths.

As with monocots, much of the following information comes from results of my own work with tropical dicotyledons, and discussions of dicot shapes can be found in the works of Wilding and Drees (1971, 1973, 1974), Rovner (1971), Geis (1973), Wilding, *et al.* (1977), Robinson (1979, 1982), and Bozarth (1985, 1986). Discussion is mostly oriented toward leaf phytoliths with a section at the end devoted to lesser-known siliceous forms from seeds, fruits, and other plant structures.

Hair Cell Phytoliths

Phytoliths derived from hair cells are probably the most common type found in dicotyledons. As with most types, their morphology is significant at varying taxonomic levels. They can be produced both in leaves and seeds, but are much more common in the former and demonstrate a most impressive degree of variation within and between taxa, a feature that makes description and comparison difficult. Whole families may be characterized by the presence of predominantly segmented or nonsegmented silicified hairs (Table 2.4). The great majority are not solidly silicified, having only cell walls impregnated with silica. I have observed a host of different kinds of hair-cell phytoliths in dicotyledons. What follows is a summary of the most distinctive and redundant types.

The most widely shared hair-cell phytolith among nonrelated plants is the nonsegmented hair, often a simple V-shaped form with no distinguishing substructure or surface decoration (Plate 46). These are to be found in many plants, both herbacious and woody (see phytolith keys in the Appendix).

Some kinds of nonsegmented hairs are more limited in occurence and useful in discriminating taxa. *Morus rubra* produces abundant numbers of hook-shaped hair cells with a spherical protuberance of solid silica on the upper surface (Geis, 1973). The Dilleniaceae and *Trattinickia* (Burseraceae) produce solidly silicified short, deltoid-shaped hairs that often have prongs, and in the Dilleniaceae, lateral projections at the base (Plate 47). *Trattinickia* hairs are often thinner than those in the Dilleniaceae and have more prongs projecting from the base of the hair. Short, very thin hair cell phytoliths are found in *Aristolochia* (Aristolochiaceae).

In *Davilla aspera* (Dilleniaceae) and a few other plants longer hairs are found having fiberlike lines the length of the cell that probably represent an incomplete silicification of the inside of the hair. Some also have prongs and lateral projections at the base. Solidly silicified nonsegmented hairs with elongated or spherical flat and indented bases occur only in few genera in the Moraceae, especially in *Trophis racemosa* (Plates 48 and 49).

Several kinds of nonsegmented hairs occur in a limited number of families. Thin, curved hairs (Plate 11) are found commonly in the Moraceae, Urticaceae, and Aristolochiaceae, one genus each in the Boraginaceae *(Bourreria)* and Menispermaceae *(Odontocarya)*, and a few genera in the Leguminoseae. Hair cells with rounded apices occur in a limited number of genera in the Moraceae, Piperaceae, Labiatae, and Urticaceae (Plate 50). Some differentiation of these forms based on phytolith size and shape seems possible. Long, threadlike hairs occur in the Compositae (two genera), Malvaceae (two genera), Urticaceae, and Ulmaceae (Plates 51 and 52). They are commonly found in the last two families.

Nonsegmented hair-cell phytoliths with surficial spines, called armed hairs, occur much less commonly than unarmed hairs. They have been observed thus far in only four families; the Boraginaceae, Urticaceae, Moraceae, and Ulmaceae (Plate 53). Some differentiation seems possible based on such characteristics as whether spines extend the length of the cell (partially armed versus entire shaft armed), shape, and size. Armed hairs from the Boraginaceae may be family specific as they are often shorter and wider than those from other taxa.

Segmented hair-cell phytoliths often show a greater diversity of shape and surface decoration. A number of families of considerable importance to economic and environmental reconstruction contribute distinctive kinds. In the Compositae, one of the largest families of flowering plants with numerous herbacious members, a number of phytoliths presently specific to family, genus, and species are encountered. For example, three kinds of segmented, partially armed hairs were isolated from *Eclipta alba, Melanthera hastata,* and *Wulffia baccata* (Plates 8 and 54). Hairs with squarish segments and thick cell walls are common in composites, especially in *Bidens* (Plate 55). This is a type shared only by some of the Cucurbitaceae, and ecological considerations will permit the exclusion of this family from many temperate zone phytolith assemblages, where the Compositae are important constituents of disturbed vegetational formations.

Many other Compositae hair-cell phytoliths have flat or spherical bases, tapered shafts and apices, and thick cell walls. Differentiation of some of these to at least the family level seems possible (Plate 56). Compositae pollen has long been used in pollen diagrams to document cultural vegetational disturbance in association with agricultural activity. Their many distinctive phytoliths may serve the same role in future paleoecological studies.

The Cucurbitaceae, a family of mainly climbing plants, also contributes a number of distinctive phytoliths derived from segmented hairs. They include forms with circular segments (Plate 57), very long (> 90 μm) and thin segments, bell-shaped surficial designs, and dark, spherical inclusions. Each of these is restricted to one or a few genera in the family.

Cucurbitaceae hair-cell phytoliths also have distinctive surface features. They consist of faint spines described previously for *Cucurbita* (Plates 17 and 18) and long, fine striations that traverse the length of the cell. These patterns occurred in a limited number of genera, and some differentiation with respect to genus should be possible based on the particular kind of hair having the surface decorations.

Highly distintive hair-cell phytoliths occur in a number of other taxa (Plate 58). A large number of segmented hair phytoliths have not been subcategorized with respect to segment shape and size, and much work is needed here with these and other phytoliths not so clearly distinctive of or shared with other families.

Hair Base Phytoliths

Hair bases are another common phytolith-producing plant cell in dicotyledons. They have been observed only in leaves and like most phytoliths can either be very useful in discrimination of taxa or not very useful at all. The most widely shared forms are spherical, sometimes with a circular mark in the center, and with no distinguishing surface or peripheral decoration.

Some taxa contribute highly distinctive forms. In the Dilleniaceae, multicelled hair bases are found in *Curatella americana* and *Tetracera* (Plate 59). Those in *Curatella* often have 5 to 10 cells and are surrounded by polyhedral epidermis. In *Tetracera* the presence of more than 4 cells is rare, and hair bases are surrounded by anticlinal (sinnuate) epidermis. This combination of attachment of hair base to epidermis permits generic, if not species-specific, identification of these plants.

A number of distinctive hair base phytoliths are found in the Boraginaceae, for example, in *Heliotropium, Cordia lutea,* and *Tournefortia volubilis* (Plates 60 and 61). A number of other types have limited distributions in dicotyledons. Several species of *Ficus* produce hair bases that enclose an irregular pattern of striations radiating from the center to the periphery of the cell (Plate 62). Only forms from *Chlorophora tinctoria* and *Ficus* are stellate shaped and enclose striations that emanate from the inside (but not the center) to the periphery of the cell (Plate 63). These phytoliths from the Moraceae will be important in the identification of tropical forests in paleoecological profiles.

The *Chlorophora* hair bases are surrounded by silicified epidermis with spherical inclusions (Plate 63), a feature found only in a few families including the Compositae, Cucurbitaceae, Boraginaceae, Dilleniaceae, Moraceae, Sterculiaceae, and Verbenaceae. Some genera and species can be differentiated by variation in the kinds of epidermis, inclusions, and hair bases that are jointly silicified.

Hair bases with remnants of spines from armed hairs occur infrequently in five families, the Compositae, Cucurbitaceae, Moraceae, Ulmaceae, and Boraginaceae (Plate 19). Hair bases consisting of two half-spheres joined together occur in the Compositae and Cucurbitaceae, while those with four half-spheres or complete spheres joined together were observed only in the Cucurbitaceae (Plate 19). Hair bases with short lines projecting from the perimeter of the cell (remnants of adjoining epidermis) occur very commonly in the Moraceae and Ulmaceae (Plate 64).

Cystoliths

Cystoliths are kinds of specialized phytoliths found in a limited number of families; when they occur in plants they are quite often characteristic of families and genera. I have isolated and described cystoliths from four families; these are the Urticaceae, Moraceae, Boraginaceae, and Acanthaceae. Solereder (1908) noted that they could also be found in the Hernandiaceae, Gesneraceae, and Olacinaceae. Cystoliths are outgrowths of the cell walls of specialized cells called lithocysts, which occur in the epidermis and ground tissue of leaves. They consist of silica and/or calcium carbonate. It does not appear to be commonly known that in a great many cystoliths silica is the dominant constituent of the body, resulting in highly durable forms.

In the Urticaceae, a family whose great importance in paleoecological reconstruction is demonstrated by their significance in pollen profiles, cystoliths are one of the dominant classes of phytoliths. They are minimally specific at the family level and are typically (1) spherical with a verrucose (warty) surface pattern, (2) curved with a verrucose surface pattern, (3) elongate with densely distributed surface verrucae, or (4) roughly spherical to irregularly shaped with densely distributed verrucae (Plates 65–69). These phytoliths have a disjunct distribution in the family and the possibility for genus-level identification of certain kinds needs to be explored.

In the Moraceae, cystoliths are not as widely distributed and in the few genera where they have been observed (*Ficus, Perebea,* and *Poulsenia*) they do not constitute a significant portion of the phytolith picture. However, they show considerable variation in form, for example, torpedo and elliptical shapes with a rough surface decoration that have not been observed in other families. Torpedo shapes with a stalk attached [a "true" cystolith in the sense used by Solereder (1908) since the point of attachment to the cell wall is evident] have been observed only in *Ficus*. Solereder (1908) also speaks of hair cystoliths, those which originate from the lateral wall at the base of the hair or form a continuation of the body of the hair. Many of the Moraceae and Boraginaceae cystoliths are of this type.

In the Acanthaceae, a large tropical family consisting mainly of herbs and shrubs, cystoliths are usually elongate in shape with two blunt ends, or one tapered and one blunt end (Plates 70 and 71). They sometimes have a surface pattern described by Karlstrom (1978, 1980) as tuberculate, although more often a surface decoration is not well developed. Cystoliths do not form a conspicuous component of silicified cells in this family, and in many species occur only rarely. Their shapes may in some cases approximate some of the Urticaceae, but there is a clear difference in the surface decoration, the Urticaceous forms usually having densely distributed verrucae in contrast to the less clustered tubercles of the Acanthaceae. The latter also have a "seashell" look to them in surface design if the tubercle pattern is not well developed, a pattern which does not occur outside of this family (Plate 71).

If forms not displaying these surface features are found in soils, they should be classified as Urticaceae/Acanthaceae.

Boraginaceae cystoliths can be roughly spherical to elliptical to irregularly shaped with a rough surface pattern. They can occur quite frequently in species of *Tournefortia* and *Cordia* and are probably specific at the level of family.

Epidermal Phytoliths

The main mass of epidermal tissue in plants is made up of the epidermal cells proper, which form a continuous layer extending over the surface of the plant. These cells are quite commonly silicified, but unfortunately they result in shapes that are identical between widely divergent taxa. Epidermal phytoliths have been separated into two broad classes, polyhedral and anticlinal. Examples of each can be found in Plates 72 and 73.

Polyhedral epidermal phytoliths are the cup assemblages, individual cups, and cup fragments of Wilding and Drees (1973) and Wilding *et al.* (1977). They do indeed look like cups when viewed with the scanning electron microscope (Plate 74) because each cell consists of a silicified depression. Wilding and Drees felt they were assignable to forest species because they were found in great quantities in deciduous tree taxa such as sugar maple *(Acer saccharum)*, oaks (*Quercus* spp.), and elms (*Ulmus* sp.). However, they have been found to occur in great quantities in numerous herbacious plants (Piperno, 1985a) and therefore cannot be used as markers of arboreal vegetation. Anticlinal epidermis, as well, occurs in woody and herbacious vegetation and has even been isolated from ferns. Geis (1983) points out that mesophytic deciduous forests produce more anticlinal epidermis than do drier forests, hence a potential exists for suggesting precipitation and soil moisture changes in fossil sequences if forest profiles with shifts in frequencies of different epidermis can be identified.

Occasionally, especially in families such as the Piperaceae and Euphorbiaceae, epidermal cells are heavily and solidly silicified, resulting in polyhedral chunks of silica that can resemble square to rectangular and bulliform grass phytoliths. In soils where grasses have played a negligible role in phytolith deposition, the contribution of these forms might still be substantial (see Special Topic 2, Chapter 6), and it is important to attempt discrimination of grass and nongrass shapes. This can be accomplished, in part, by assuming that they are not grass derived if percentages of grass short cell phytoliths are very low in phytolith assemblages. It has been mentioned already that the classic fan-shaped kind of bulliform cell phytolith has not been observed outside of the Gramineae.

In New World tropical forest species such as *Guatteria, Unonopsis pittieri,* and *Protium panamense,* what looks to be epidermal silicification results in distinctive multifaceted shapes (Plates 75–77). *Guatteria* produces chunky to elongated forms;

Unonopsis produces roughly spherical to aspherical shapes; and *Protium panamense* produces elliptical phytoliths. Elongated phytoliths similar to those from *Guatteria amplifolium* have also been described by Kondo and Peason (1981) from *Michelia compressa*. These phytoliths have not been observed in herbacious plants or lianas, at present are family- or genus specific, and may be excellent indicators of arboreal vegetation in soil profiles.

Uncommonly occurring in herbacious and woody plants are epidermal phytoliths that when viewed from above, appear to have spherical inclusions (Plate 63). When turned on their side it is seen that they bear small protuberances originating from the center of the cell into subepidermal tissue.

Sclerenchyma Phytoliths

There are a number of irregularly shaped phytoliths that the author has isolated in great frequency from predominantly woody species of tropical plants. They have also been observed in Japanese tree leaves (Kondo and Peason, 1981) and in *Magnolia grandiflora* (Postek, 1981). Many of these appear to be derived from vein sheath parenchyma (which in terms of tissue type would technically be schlerenchyma). They consist of elongate and other more characteristic shapes (Plates 78 and 79) and are very refractile and smooth in surface appearance. Some may also have a multifaceted surface. Schlerenchyma-derived phytoliths can be quite common in certain families. Sclereid morphology is often one and the same in quite unrelated species, but here a basis appears to exist for discrimination between woody and herbacious plants because, at least in tropical species, they are very common in some trees and are seldom found in herbs and vines. Out of approximately 500 species of tropical dicots, 46 (12%) have produced sclereids. Of these, 37 (80%) are trees, 4 are lianas, and 5 are vines.

Tracheid Phytoliths

Silicified vascular tissue in the form of tracheids with helical wall thickenings occurs in numerous dicotyledons and has little taxonomic value (Plate 13).

We have so far relied on a nomenclature that reflects a generical basis for the classification of siliceous forms. This is an outcome of the fact that many phytoliths can be assigned to specific loci or origin in plants on the basis of their morphology. There are some cases in which such assignment is difficult, or when bodies in which tissue derivation is known do not conform to the shape of that tissue. These phytoliths will now be described.

Spherical Phytoliths

Recall that spherical to aspherical phytoliths are a major class in monocotyledons. They also occur in dicots but much less frequently and often without the variety of surface decorations displayed by monocots. The leaves of deciduous angiosperms produce significant numbers of smooth, spherical phytoliths that range in size from less than 1 to 50 μm (Wilding and Drees, 1973; Geis, 1973). They tend to assume almost perfect spherical forms and arise as vesicular infillings of epidermal and other cells. Most lack surface detail but may have slight indentations or protrusions which probably mark sites of attachment to host structures.

Spherical smooth phytoliths are produced very rarely in a few tropical herbacious monocots and in considerable amounts by a very few tropical arboreal dicot leaves and seeds. In these assemblages, very regular spheres are an inconspicuous component. It seems, therefore, that in temperate zone assemblages, spheres described by Wilding, Drees, and Geis can be considered a reliable indicator of arboreal vegetation.

Spherical to aspherical rugulose phytoliths have been isolated in great quantity from the leaves and seeds of tropical Chrysobalanaceae (all arboreal) and seem to not occur in dicots outside of this family. In the tropics, size differentiation may permit separation of arboreal from herbacious spherical phytoliths. Those found in monocots (i.e., *Canna, Heliconia,* and the Marantaceae) are almost always from 9 to 25 μm in diameter, whereas those in the Chrysobalanaceae range from 3 to 15 μm, with many occurring in the 3–9-μm size range.

Phytoliths Whose Origin in Plant Tissue Is Unknown

There are a few plants that contribute highly distinctive phytoliths whose origin in plant tissue is unknown. *Licania hypoleuca,* a tropical tree in the Chrysobalanaceae, has phytoliths that assume a shape like a double saucer (Plate 80). *Sloanea terniflora* (Elaeocarpaceae), another tropical tree, contributes rectangular to aspherical forms with cavities of various shapes. The Podostemaceae, a family of aquatic plants that grow adhered to submerged rocks in rapidly flowing water, have phytoliths covered with prominent fingerlike projections. They were first described by Bertoldi de Pomar (1972). The Loranthaceae and Menispermaceae produce large, irregularly pointed bodies (Plates 81 and 82).

Phytoliths from Dicotyledon Reproductive Structures

Phytoliths from the fruits and seeds of dicotyledons are providing a new and extremely promising area of research, for they offer an array of shapes not found in leaves and appear unique to genera and species of some plants. They have been recovered from many families (Table 3.4). Phytoliths are very often derived from epidermis and display a plethora of shapes from spherical to pentagonal and even umbrella-like. They often have a stippled surface pattern and a protuberance emanating from the center of the bottom wall of the cell, which once extended into the subepidermal tissue. The combination of shape, surface decoration, and protuberance imparts an appearance characteristic solely of seed- or fruit-derived opal.

Seed and fruit phytolith production can be quite high. It almost never occurs in plants whose leaves do not contribute high amounts of phytoliths, and in species where leaf production is significant, it occurs in about one-quarter of plants so far examined. Phytoliths are very often found in the fruit exocarp or mesocarp (not the kernel), where perhaps they function to inhibit attack from larvae and other seed predators.

In arboreal tropical species, shapes permitting discrimination of taxa are so far quite common. For example, only in the Burseraceae four- to seven-sided fruit and seed epidermal phytoliths are found, which in surface view have stippled walls and a central, spherical mark (Plate 83). In side view the spherical mark can be seen to be a protuberance from the cell that extended into subepidermal tissue. In *Tetragastris panamensis* phytoliths in side view have sloping, umbrellalike margins and long smooth, stalks (Plate 84), while in *Protium panamense,* the epidermis is thick and nonsloping (Plate 85). In *Protium tanuuifolium* the central protuberance emanantes from a smooth, dome-shaped structure. *Bursera simaruba* produces seed phytoliths with a faintly stippled surface and a small protuberance that is inconspicuous in surface view (Plate 86), while those of *Trattinickia aspera* have jagged edges and small nodules on the surface (Plate 87). This makes five distinct, discrete shapes isolated from five different species in the Burseraceae, a very important family of tropical trees.

Seed epidermal phytoliths with numerous small projections on the cell wall have been isolated only from *Celtis schippi* (Moraceae). The seeds of *Mendoncia retusa* (Acanthaceae) contribute spherical chunks of silica having a surficial pattern of short striations on one side only (Plate 88). The Chrysobalanaceae *(Hirtella)* is unique in contributing spherical smooth and rugulose seed phytoliths (Plate 6). As mentioned previously, these can probably be differentiated from similar forms in monocot leaves by their smaller size and smooth surface.

Other taxa can produce seed and fruit phytoliths whose morphologies are similar to their leaf silicified cells (examples are some of the Compositae, Urticaceae, and Dilleniaceae), and in few taxa (e.g., *Ficus*) hair cells were found that overlapped leaf phytoliths from other genera in the same family. In no taxa examined were seed phytoliths observed that overlapped forms previously described (Piperno, 1985a) as being distinctive to other plant families. As will be discussed fully in Chapters 6, 7, and 8, seed and fruit phytoliths are commonly found both in modern and fossil tropical soils thousands of years old and offer a tremendous potential for retrieving fine-scale paleoecological information on past tropical vegetation.

Among temperate zone dicotyledons, Bozarth (1985) has described promising phytoliths from the achenes of sunflower *(Helianthus annuus)* and nuts of black walnut *(Juglans nigra)*. The latter are described as pitted and multicelled and the former as large and opaque with unperforated surfaces.

Floral bracts and flowers have been studied in far fewer dicotyledons. However, distinctive opaque perforated plates have been recovered from Compositae inflorescences (Plate 89). They were first observed in modern plants by Bozarth (1985) and appear to be the perforated root platelets of Wilding *et al.* (1977), which occurred as unknown phytoliths in many temperate zone soil assemblages.

Phytoliths in Wood and Roots

Phytoliths are found in quantity in numerous species of dicotyledonous woods as intracellular inclusions. Thorough reviews of production and morphology can be found in Amos (1952), Scurfield *et al.* (1974), and Ter Welle (1976). The morphological significance of wood phytoliths is unclear, but many bodies appear rather amorphous. They have been called aggregate grains by Scurfield *et al.*, and they and Ter Welle document shapes that are roughly aspherical with granular surfaces. Size variation in these types can be great, ranging from less than 5 to over 50 μm. Other shapes and surface decorations can occur but none appear to be confusable with the various kinds of morphologies presented above from other plant structures.

Very little is known about root phytoliths in dicotyledons, but we would expect that in many species production is not particularly impressive.

Phytoliths in Gymnosperms

Gymnosperms are probably a less promising area of study than flowering plants. They are generally low accumulators of silica (Table 2.2), having contents very often below 0.5% of dry plant weight. However, phytoliths that do form are diagnostic. Klein and Geis (1978), Rovner (1971), and Norgren (1973) have reviewed the morphology of phytoliths found in the Pinaceae. There are several forms with taxonomic significance, including transfusion tissue tracheids with bordered pit impressions and tapering intrusive ends seen in *Pinus, Picea,* and *Pseudotsuga* (Klein and Geis, 1978), and the well-known asterosclereids produced apparently only by species of *Pseudotsuga,* including Douglas Fir (Brydon *et al.,* 1963). Rovner (1971) isolated square, nettinglike fragments from *Larex* (larch). Significantly, Rovner (1985) has isolated Pinaceae phytoliths from archaeological soils, indicating that aspects of Pinaceae paleoecology are amenable to study with the phytolith record. Silicified endodermal cells abundant in some species of *Picea* (Spruce) are potentially confusable with some bulliform cell phytoliths from grasses (Klein and Geis, 1978).

Phytoliths in Lower Vascular Plants

The leaves of ferns, horsetails, and other spore-bearing plants can exhibit a considerable degree of silicification, and phytoliths display an impressive number of distinctive shapes. Phytoliths have been isolated from about half of all Polypodiaceae (fern) species examined. The most common shape is a very elongated form possessing a flat base and a surface with two undulating ridges parallel to each other (Plate 90). Their lengths range from 70 to greater than 1000 μm. Occasionally, only one side of the phytolith has the undulating ridge, but these may be broken versions of the two-sided forms. These phytoliths appear to be family specific, as they have not been found outside of the Polypodiaceae. Other Polypodiaceae phytoliths include long, thin, loosely silicified forms, anticlinal epidermis, and tracheids.

The Hymenophyllaceae, a family of homosporous epiphytic or terrestrial herbs, contribute large numbers of shallow, roughly bowl-shaped phytoliths that are specific to at least the family level (Plate 91). Their diameters range from 15 to 39 μm. The Hymenophyllaceae are virtually confined to moist forest habitats in the tropics, and hence, their phytoliths may serve as markers of such vegetation in paleobotanical profiles.

The Selaginellaceae is a family of heterosporous annual or perennial herbs also confined to moist forest habitats in the tropics. They produce abundant numbers of elongated epidermal phytoliths with small, conical-shaped projections on the surface (Plate 92). The conical projections are often aligned in two rows extending along the long axis of either edge of the phytolith. Clearly, phytoliths from ferns may provide very important data on paleovegetation. Fern spores cannot often do so because of morphological redundancies.

The Equisetaceae (horsetails) contribute large amounts of silicified cells that appear specific at the level of the family.

Summary: Dicotyledon, Gymnosperm, and Pteridophyte Phytolith Morphology

The taxonomic significance of phytolith morphology in dicotyledons, gymnosperms, and pteridophytes is no longer open to question. The remarkable diversity of shapes isolated from the various taxa defines a new and unique type of microfossil plant community. Phytoliths minimally specific at the family level have been isolated from the Polypodiaceae, Selaginellaceae, Equisetaceae, Hymenophyllaceae, Pinaceae, Acanthaceae, Annonaceae, Boraginaceae, Burseraceae, Chrysobalanaceae, Compositae, Cucurbitaceae, Dilleniaceae, Elaeocarpaceae, Euphorbiaceae, Loranthaceae, Moraceae, Piperaceae, Podostemaceae, and Urticaceae. Differentiation below the level of family is possible in all of the Pinaceae and dicotyledons.

Descriptions of phytoliths and constructions of phytolith keys will be more complicated than in pollen analysis because of production patterns that result in a multiplicity of shapes in a single species and redundancies between several species. Even so, the numbers of distinct phytoliths isolated thus far indicate that identification of many, specific plants given a collection of phytoliths extracted from soils will be possible. We stress again that basic patterns of production and morphology appear little affected by environmental variables. In over 100 species of tropical plants I carried out replicate analyses of leaves, fruits, and seeds from different plants, which showed that shapes of phytoliths are constant in and characteristic of species in which they occur. Exceptions were the silicification of leaf tracheids, and polyhedral and anticlinal epidermis, suggesting that production in these areas may be more influenced by the environmental conditions of growth, such as soil pH and levels of available silica.

Several examples of phytolith keys are presented and discussed in the appendix. They are the ultimate result of a descriptive and classificatory analysis and provide a means of access to the classification system.

4

Field Techniques and Research Design

Introduction

The soil and plant samples that are ultimately analyzed in the laboratory are a result of collection strategies pursued during the course of archaeological field work. We anticipate that such samples represent uncontaminated specimens taken from contexts that permit specific archaeological problems to be addressed with phytolith data. This can be accomplished by preexcavation and on-site planning. It will be emphasized that the particular sampling strategy or strategies chosen depend on the goals of the analysis and on the kind of site that is under investigation, i.e., a rock shelter, shell midden, open-air campsite, nucleated village, or remnant of a prehistoric field.

In addition to an archaeological sampling program, coring programs should be initiated in archaeological study regions whereby geological samples are retrieved from lakes, swamps, or bogs. These provide valuable information on past vegetational distributions and man–land interactions, data that can be extracted from archaeological samples only on a limited and somewhat biased level because of the influence of human choice on plant deposition into sites. The extraction and sampling of geological samples are also discussed in this chapter.

It is perhaps premature to embark on an extensive discussion of sampling methods and strategies in archaelogical phytolith analysis because most relevant work has been carried out only during the last five to eight years. Published results of sampling strategies and resulting patterns of phytolith frequencies from sites are limited to a few areas of the world, and it is a compendium of such results that ordinarily dictate what methods of collection might be the most appropriate in various situations. Yet, because phytoliths are microfossil remains, strategies for removal from sites will not often vary considerably from palynological sampling.

Some of the sampling strategies presented here are accordingly modeled after palynological techniques, as some of the same depositional, taphonomic, and post-depositional considerations should apply to phytoliths as well as pollen. Major differences between the two result from their production and dispersion characteristics. As argued in Chapter 2, much of the archaeological phytolith record should represent a highly localized, *in-situ* decay of plants, whereas more of the pollen record may have blown in from outside of sites or was attached to utilized macrofossil material and, hence, may not represent a product of direct plant usage. Archaeological phytolith sampling should then chiefly vary in the number and location of contexts chosen for analysis.

The Basic Phytolith Sampling Procedures

There are three basic kinds of sampling procedures employed at archaeological sites: (1) column sampling from an exposed and profiled wall, (2) canal sampling from an exposed and profiled wall, and (3) selective horizontal sampling of deposits as the excavation is in progress. The first and third are the easiest, most common, and least labor-intensive. Canal sampling requires somewhat more time and labor expenditure than the other two but yields samples that preserve the stratigraphy of a site after it has been filled in.

Column sampling

Column samples are taken from an exposed and profiled wall after excavations are essentially completed. They involve demarcating a subarea along the wall of a test pit or trench and removing samples at specified intervals from the top to the bottom. The horizontal length of the column need not be longer than 30 cm, as even fine-scale vertical intervals from columns of such a size will yield enough soil for several phytolith (and pollen) extractions with substantial amounts left over for possible future work; 100–200 g of soil are normally all that is needed for an adequate phytolith sample. Ideally, the columns should be from walls that have the clearest stratigraphy and least amount of disturbance, and where some absolute chronology has been established by radiocarbon dating. In practice it is often difficult to find such situations, and of course intrusive deposits may not be visible on profile. In some cases two columns taken side by side can provide good control. One does the best possible under the circumstances by using a liberal amount of common sense and good judgment.

The removal of samples poses no problem. One simply demarks the area and starting from the top removes soil with a trough in prescribed intervals. In sites such as rock shelters where living surfaces comprise a relatively small area, deposits are deep, and long chronology evident, column samples yield good records of plant usage and their changes over time. Phytoliths recovered from columns can describe subsistence for major occupations of the site and define similarities and differences between the occupations. In rock shelter situations it is a good idea to take soil columns both from inside and outside of the dripline. Although deposits from the latter may be more difficult to interpret in terms of stratigraphy and associations, they may incorporate good accumulation of cultural debris which had been swept out of the living area by site occupants.

Sampling intervals within stratigraphic units will depend on site contexts and the nature of midden buildup; as a general rule of thumb intervals should not be more than 5–10 cm. They may be more than 5 cm if it is suspected that deposits have accumulated very rapidly and only a few centimeters if microstratigraphy is evident. The relative merits of contiguous and noncontiguous sampling of columns also depends on site specific contexts. Spaced sampling, say every 10 cm, within stratigraphic units will decrease substantially the number of samples to be analyzed. In some cases taking only one sample from each recognizable stratigraphic level may be tried. The question of how many column samples to take from sites is always open to debate. More than one is obviously desirable, even from small sites, if only to establish that consistent results can be obtained from duplicated samples. We would like to see some critical investigation along these lines, for they have seldom been worked out even for archaeological pollen analysis, which has a much longer history than phytolith studies. In determining how many columns are sufficient, a general rule of thumb is the often stated "the more the better" allowing for constraints of time, labor, and money. The size of sites will always be a factor involved in such a decision, in addition to particular spatial and stratigraphic characteristics that will only be revealed once fieldwork is underway.

Proper techniques of column sampling include scraping the exposed walls prior to soil removal to make sure that local, modern wind-blown phytoliths are eliminated, cleaning the trowel after removal of each section, and placing the soil in secure, well-labeled bags or vials. Surface soils on top of sites and sterile deposits at the bottom of sites should be part of column sampling, as these will provide controls on surface and subsurface mixing.

Horizontal Sampling

The use of column samples will go hand-in-hand with a horizontal sampling strategy, even in small sites. Horizontal samples should provide a better comparison between contemporaneous but spatially discrete areas of a site than do single or multiple column samples. In addition, they are more likely to represent the average phytolith composition of the area being sampled. The extent of horizontal sampling of deposits will depend on the size and nature of the site and the goals of the analysis. Tightly defined features such as hearths, garbage pits, and storage areas should be sampled for the detection of cultigens and other plants that may have been important in the diet. Indeed, phytolith records from such contexts are likely to yield evidence on the function of specialized areas uncovered during excavation.

For example, we can envision a sampling strategy designed to test competing hypotheses about the function of 3-m^2 areas delimited by four postholes commonly found in Late Iron Age sites in Britain (Hodder, 1982, p.120). Were they granaries with raised floors, as some have suggested, or defensive structures, as Hodder believes? The isolation of numerous phytoliths from the caryopses and seed bracts of cereals from soils taken in the feature area would support the granary hypothesis, while the absence of such evidence would tend to support other ideas.

We can immediately think of a number of other questions relating to site function, technology, and intrasite variability that would benefit from horizontal phytolith sampling. House floors and other areas of roof collapse may yield evidence for materials used in construction and thatching or matting and beds. Dirt adhering to lithic artifacts and the surfaces of the artifacts themselves may provide evidence for stone tool functions. The surfaces of stone tools should be considered as candidates for phytolith "washes," whereby the presumed used facet of the tool is rinsed with a steady stream of water in the hope of extracting phytoliths from the pores. The same treatment may be applied to ceramic vessels, metates, and manos. Soils from bone beds in Paleo-Indian kill and butchering sites may yield information about what plants formed the diet of ancient large mammals. Soils from burial contexts and otherwise unidentifiable grave goods and wrappings may yield identifiable phytoliths.

From all of these contexts control samples should also be taken from areas outside primary feature distributions, for example; soils near to or underneath (but not inside of) pits, adjacent to (but not inside of) house floors, and in proximity to soil forming the matrix adhering to stone tools. These will allow the investigator to decide whether patterns isolated are unique to the context sampled, or just a part of the normal cultural residue. Again, the large bulk samples required for macrobotanical analysis need not be taken for phytolith analysis. Rather a series of "pinch" samples that cover the entire desired sampling area and which, when combined into a composite sample, total about 100–200 g should be sufficient in most soils for adequate phytolith recovery. When sampling always remember the characteristics of phytolith taphonomy and preservation—highly localized, if not *in-situ* decay, and remarkable durability.

Large and nucleated villages and other complex sites will require a more intensive sampling program and a broader strategy. Soils should be taken from large numbers of features across a significant extent of the site so that comparisons between contemporaneous but spatially discrete areas of sites can be made. It may be possible to make inferences suggesting that different activities involving plants took place in different areas of houses and in separate residences. Functional distinctions may be drawn between storage and other structures. Possible relationships between occupational specialization, social stratification, and other aspects of prehistoric lifeways with subsistence patterns, house construction materials, and living conditions might then be evaluated. Overlap between modern customs and prehistoric practices may be documented (for example, in burial practices) providing an important body of evidence relating to social organization.

In the reconstruction of prehistoric pathways and the social dimensions of archaeological evidence, phytoliths may prove more useful than the more sporadically preserved pollen and macrobotanical remains. The links between subsistence, technology, the organization of space, and the organization of social relations are strong, and it has been emphasized how a properly conceived sampling strategy may elucidate the aspects of these relationships that leave phytolith residues. Analysis of phytoliths, pollen, and macrobotanical remains in tandem may generate even more significant results.

Canal Sampling

Canal samples are a kind of variant of column samples. In canal sampling a section of a wall approximately 5–10 cm wide is actually cut out into a three-sided aluminum container, wrapped tightly in plastic, and stored in the laboratory for later sampling. Canal samples offer advantages over column samples in that the stratigraphy and soil characteristics of the sampling area remain intact for later visual examination and, if necessary, resampling. Canal samples were taken and used in the phytolith analysis of prehistoric ridged fields from the Calima Valley, Colombia (Piperno, 1985d). In this case, soils for phytolith studies were not removed until two years after canal removal. Stratigraphy was very clear in the canals which had been stored in cool conditions, even clearer than during the original excavations, when wet conditions and a rising water table necessitated a hastier close of the field work than was really desired (Bray *et al.*, 1985). Ideally, both column and canal samples could be removed from sites given sufficient time, money, and man power. If analysis of column soils reveals inconsistencies or possible contamination problems, canals could then be unwrapped, stratigraphy reinvestigated, and some analysis repeated.

Other Considerations of Soil Sampling

In Chapter 2, the subject of poor phytolith preservation in soils of high pH (9 and above) was addressed. We know from Lewis' (1981) studies of bison kill and processing sites from the high plains of Wyoming, Nebraska, and Colorado that alkaline soils with pH values in the range of 8.2 to 9 do not always adversely affect phytolith durability. Tropical shell middens having pH values measured from 8.8 to 9.2 showed depauperate phytolith assemblages with only carbon-occluded forms remaining (Piperno, 1983, 1985b). Here, conditions were compounded by year-round high temperatures and soil leaching, which probably exacerbated the negative effects of alkalinity. In the tropics archaeological deposits of high pH should be considered suspect and, if possible, sampling should be undertaken in areas of sites where less alkaline conditions might occur. In these situations it might be profitable to carry pH paper into the field and determine soil levels. This is an easy procedure, requiring only some water, which is mixed with soil in a 3 : 1 water to soil ratio. The paper is then saturated with the water and approximate pH is determined by comparing the color of the paper to a chart provided in the pH kit.

In thinking about designing strategies for archaeological sites, it is important to stress again that many excavated materials other than soils are appropriate for phytolith analysis. In fact, the first archaeological phytolith studies were carried out on ash heaps and ash isolated from artifactual materials such as burnt pottery (Schellenberg, 1908; Netolitzky, 1914). Techniques for isolation of phytoliths from soils were developed later and not put to use by archaeobotanists until fairly recently. Phytoliths have been found in the teeth of ungulates (Armitage, 1975) and on the surface of stone tools (Anderson, 1980). Fibers and other normally perishable materials, such as those recovered from classic Mayan tombs in the Rio Azul, Guatemala, may be identifiable through phytolith analysis. Clays and muds used for prehistoric structures often were mixed with grasses which may leave distinguishable phytoliths. Coprolites should be suitable for the recovery of silicified remains as they have shown to be for calcium oxalate plant secretions (Bryant, 1974; Bryant and Williams-Dean, 1976).

We have so far discussed sampling design with the goal of carrying out large-scale, intensive phytolith studies. Rovner (1986a) has emphasized that phytoliths also have a significant role to play in cultural resource management, where very quick and limited sampling of sites to assess phytolith content may influence decisions as to which sites should be mitigated, preserved, or written off. Here, analysis of a few samples taken from obvious areas of midden accumulation (garbage pits, other features) is required. If sites are then further investigated, different research designs will be needed, depending on the time and financial resources likely to be expended on the project.

Geological Coring

As will be discussed in detail in Chapters 6, 7, and 8, the analysis of sediment cores taken from archaeological study regions can provide critical information on climate, vegetation, and man–land interactions. This information is available only on a limited and biased basis from archaeological sites because of the human-controlled bias of plant deposition. Geological sediments can be taken from swamps, lakes, or fully terrestrial environments, and a number of coring devices are used depending on the depositional context. The modified Livingston piston sampler is used to extract lake sediments. Hiller borers are appropriate for peats and swamps where sediments are considerably softened, and the vibre core is used in terrestrial settings and some swamps where soils are still fairly compacted. Of the three the vibre core is the only device in which the core barrel is driven into the ground by a motor.

As will be discussed fully in Chapters 6 and 8, all of these depositional environments have proved to be suitable for phytolith analysis, and indeed the time will come in the near future when if phytolith work is not initiated a valuable source of microfossil data will be considered unretrieved. Once removed from the ground, geological core sediments should be sampled much in the same way as for pollen analysis; as we shall see, both can be done at the same time.

In geological sampling, soils are described throughout the length of the cores and stratigraphic horizons are delimited. Sampling intervals within stratigraphic zones depend on how rapidly deposits have accumulated and on what temporal scale we want the analysis to be; a good rule of thumb is not to make intervals more than 5–10 cm unless it is known beforehand that a very rapid deposition of soils has occurred. Samples from adjoining units can always be combined later if contiguous intervals farther apart are desired.

Most Livingston piston samplers are made to use core barrels with diameters of only 1.5 in. or less, so if close sampling intervals are desired there may not be enough sediment for both pollen and phytoliths. We have found it preferable to take replicate cores from lake sediments, one of which is used for pollen and the other for phytoliths. If core tubes with diameters of 3 in. or more are used, a single core should probably suffice for all microfossil analysis. Unlike most soils from archaeological contexts, geological sediments will arrive at the laboratory *in situ*, that is still bound within the core barrel. Depending on the size of the barrel, sediments can be extracted by cutting the pipe open lengthwise with an electric saw or by extruding them. Pollen and phytolith sampling can be done at the same time.

Modern Controls on Sampling

Modern surface soils should always be collected from archaeological sites and coring localities. These serve a number of purposes, including aiding in the recognition and interpretation of fossil phytolith assemblages. Perhaps most importantly in archaeological sites, they provide controlled contexts by which to measure possible mixing between surface and subsurface deposits. This is particularly critical in sites located underneath modern cultivation soils, a frequent occurrence. Here we are concerned with the possibility that phytolith remains of modern cultigens may have been carried below the surface by natural agents of soil movements (earthworms, soil cracking) to contaminate prehistoric levels.

Such a problem can be alleviated if discrete phytolith assemblages are identified in surficial soils. In a Guatemalan site whose upper strata lay in an actively cultivated sugar cane field, Pearsall (1980, 1981a) was able to show that sugar cane phytoliths *(Saccharum officinarum)* bore no resemblance to maize phytoliths, could be differentiated in soils, and were important elements of phytolith assemblages of surface and upper site strata but not of lower archaeological strata. Problems of contamination were thus pinpointed and controlled. In prehistoric fields from the Calima Valley, Colombia (Piperno, 1985d) discrete and distinguishable phytolith assemblages were isolated from modern soil horizons, evidence that old cultivation till stratified below the surface was not infiltrated with more recent material (see also Chapter 7). Obviously, up and down phytolith movement always occurs to some degree in soils; however, the establishment of discrete surficial profiles is a good indication that such movement is insignificant and should not cause erroneous interpretation of data due to the presence of intrusive phytoliths.

Modern Vegetation and Soil Studies

Modern Soil Studies

Gathering some idea of the nature of vegetation surrounding sites is important so that we can start to define modern floral associations and their relationships to climate, soil types, and anthropogenic factors. The collection of surface soils from such contexts is a critical aspect of paleoecological studies because they help define the relationship between standing vegetation and modern phytolith deposition and provide modern analogs for the interpretation of fossil phytolith assemblages. Soils are an integral part of modern sampling strategies, for their phytolith assemblages reveal what a vegetational association "looks like" once phytoliths from all structures in a species are deposited together and less durable phytoliths are removed through dissolution. As with many aspects of phytolith research, the systematic collection of modern soils underneath different types of vegetation has only begun. However, it is useful to review strategies that have been successful, resulting in a close correlation of phytolith assemblages with the standing vegetation.

As part of a project reconstructing prehistoric and early historic agricultural systems in Hawaii, Pearsall and Trimble (1984) sought to define relationships between standing vegetation, climatic zones, and modern phytolith assemblages. Surface soil samples from six different zones of vegetation, such as arid and moist habitats, comprising an area 25 km long and 10 km wide were taken by demarcating 100-m^2 blocks and randomly collecting 60 "pinch samples" (1–2 cm^3/pinch) from each block. Samples were then placed in bags and mixed to form a composite sample for each block. Significant variations among percentages of grass and nongrass phytoliths in the soils corresponded well to actual patterns of grass versus forest cover. Modern phytolith frequencies were then used to reconstruct the nature of plant cover before, during, and after archaeological occupations. In this study, rectangular sampling areas of 100 m^2 from different floristic associations provided good correlations between modern phytolith assemblages and standing vegetation.

In the humid tropics I carried out transect sampling across and within three different vegetation zones comprising old (500 years) and young (100 years) forest and swampy, poorly drained localities. Soil pinch samples were taken every 20 m from 25-m^2 blocks along two transects extending 320 and 180 linear meters. Closely spaced intervals were analyzed because a major goal was to correlate local soil phytolith frequencies with standing vegetation in a mapped and censused 50 hectare area (Hubbell and Foster, 1983). Phytolith assemblages demarcated very clearly the forest and swamp terrain, and microdifferentiation within zones, caused mainly by species distributions on the landscape and distance of the samples from silicon accumulating plants, was apparent. This study will be presented in detail in Special Topic 2, Chapter 6.

Modern Plant Studies

The construction of a modern phytolith collection from archaeological study regions is one of the most important and indispensable parts of phytolith research. At present, for all but a few specific study regions no such collections exist. This aspect of phytolith research will demand considerable time, effort, and funding, and for some years will remain one of the most underdeveloped features of the discipline. The extraction of plant material, description and measurement of phytolith attributes, and interspecific comparisons are, in fact, never-ending tasks, especially in regions where floristic diversity is high.

Specimens for a modern reference collection may come from either plants collected in the field or herbarium folders. Each has their own advantages and limitations. If vegetation in the study area has been greatly disturbed, as is often the case, the extensive use of herbarium specimens may be required to achieve a representable collection of plants that once comprised natural associations. Here, we resort to botanical and climate studies which have indicated the probable nature of the species forming the natural vegetation and then clip small samples of different plant parts from herbaria folders. We can be reasonably certain that in most cases plants have been correctly identified. On the other hand, herbaria coordinators are not particularly fond of having several entire leaves, flowers, and fruits removed from certain folders, so smaller samples often result. However, this seems not to be a problem, as plant specimens weighing only about 0.1–0.2 g have proved to be quite adequate for most analyses.

Plant sampling in the field results in bigger collections of any structure desirable, but one often needs the service of a botanist familiar with the regional flora who can accurately identify many species of plants. Such situations are far more problematic in the tropics where species diversity is high than for temperate regions. In addition to naturally distributed plants, we should have a good collection of introduced domesticates known or surmised to have been grown during prehistoric periods and typical secondary growth plants. The latter are often not hard to find in the field. In modern plant sampling an emphasis should be placed on the recovery of mature specimens and where possible a few mature and immature individuals of each species should be collected.

An efficient sampling strategy, especially in regions such as the tropics that are highly diverse floristically, is to concentrate on silicon accumulating species. Dominants or important ecological markers of vegetational formations that occur outside of these families or genera can then be selectively analyzed.

5

Laboratory Techniques

Introduction

Scientists have been interested in studying deposits of silica found in plants and sediments for a very long time. Over the years since Ehrenberg first observed them in soils in 1841 and Netolitzky described their removal from plants in 1929, methods of isolation and examination have been modified and improved in accordance with technological advancements and increasing significance of phytoliths to the scientific community. Researchers now routinely use scanning electron microscopy and electron probe microanalysis to pinpoint the location of siliceous deposits and study microstructural details.

As Parry and Smithson (1958a) noted in the first modern comprehensive review of study techniques, success in making phytolith preparations demands a knowledge of the properties of plant opal, their chemical stability, density, and refractive index. A variety of techniques for isolation from plants and soils can be used, depending on plant and soil type, available equipment, money, and labor; certain procedures should also be avoided. Phytoliths are very soluble in hydrofluoric acid and strong bases, especially when heated in these solutions. They are resistant to burning, but cannot be heated above the melting point of silica (950° C), when distortion takes place. This chapter presents the various ways by which phytolith preparations are made from plants and soils.

The Phytolith Laboratory

The phytolith laboratory requires certain basic pieces of equipment for extraction of soils and modern plant material. If elaborate techniques and devices to view phytoliths are desired, then more and higher priced equipment is needed, but for routine study the sophistication of equipment need not be great. In terms of equipment the minimum requirements for a phytolith laboratory are a centrifuge (a tabletop centrifuge with heads to hold twelve 16 × 100 mm tubes and four 50-mL tubes is quite sufficient), an acid sink, a fume hood, a biological microscope with low and high power plus an oil immersion lens, a scale (one weighing to at least 2 decimal places), a hotplate, and glassware of various kinds such as test tubes, graduated cylinders, beakers, and stirring rods. This set of laboratory hardware with a total cost of perhaps $4000 is perfectly adequate to carry out efficient and reliable phytolith studies of soils and plants.

A number of different procedures have been used by workers to study phytoliths in plants and soils. Modifications have been developed to suit a particular type of sediment or plant family. Fredlund (1986a, p. 107) succinctly characterizes laboratory procedures: "If you do extraction work in the lab, it is like a kitchen. You get a recipe from someone else, use it in a step-by-step procedure, get it perfected according to their directions, and then start experimenting in an attempt to improve on it." What follows is a review of the various extractive procedures used by different workers.

The Preparation of Soils for Analysis

In proportion to the number of soil particles making up sediments phytoliths often occur in such small amounts that their presence is not immediately obvious when soil particles are microscopically examined. Phytoliths must therefore be separated from soils and concentrated. This is accomplished for the most part by chemical flotation. We first present a detailed explanation of the procedure I have found to be most efficient and reliable, with some discussions of alternative reagents that might be used. Then some permutations of this procedure are outlined, followed by a discussion on soil types that present especially problematic situations for phytolith isolations.

Before chemical flotation a number of steps must be taken to ensure that phytoliths are disaggregated from the soil matrix and unwanted materials such as carbonates and organics are removed, leaving phytoliths free to float in a heavy liquid solution. The first step is deflocculation of soil samples by repeated stirring of 25–50 g of soil in a 5% solution of Calgon or sodium bicarbonate. Sodium hexametaphosphate, the generic equivalent of Calgon, can also be used. The use of an automatic shaker allows an overnight deflocculation, or it can be done by hand over a period of a few days, depending on the looseness of the soil. This step is essential for most types of sediments because phytoliths are in some way bound to soil particles and a very small proportion appears to be free in the interstitial spaces of soils. The same appears to be true for pollen grains (Dimbleby, 1985).

This was brought home to me forcibly when I tried to isolate phytoliths from a few samples quickly without deflocculation, all of which yielded no phytoliths at all. Subsequent reanalysis of the same soils with a proper deflocculation period resulted in very high phytolith yields. A proper deflocculation step will also unbind soil particles to a sufficient degree to allow adequate removal of clays.

The next step is the separation of sands, the soil fraction with particles greater than 50 μm in diameter, which is accomplished easily by wet sieving through a 270 mesh sieve. The sand fraction is set aside for later analysis for this is the soil portion that yields many multicelled aggregates of phytoliths and taxa like *Cucurbita*. Clays, soil particles less than 5 μm in diameter, are now removed. I have found this step also to be essential because light clay particles, if present to any number, will float in the heavy liquid solution and cause soil contamination of the phytolith separations. Clay removal can be accomplished by a number of methods, the most efficient of which seems to be gravity sedimentation, using the technique and sedimentation rates published by Jackson (1956).

Soil samples are placed in large (600–1000 mL) beakers, water is added to a height of 10 cm, the solution is stirred vigorously, and after 1 h the supernatant is carefully poured off, leaving the silt fraction (5–50 μm) behind. Repetition of this step 7–8 times removes most of the clay from soils. There may be a problem with removing particles less than 5 μm in size because a few plants produce identifiable phytoliths that fall in this size range, including *Canna* and Bromeliads. They have been recovered from soils so it seems that some do make it through the clay removal procedure, but probably in a smaller proportion to their actual abundance in sediments. Removal of clay is known to be complete when the supernatant is clear after the limiting time has been met.

Next, the silt fraction can be further divided into fine (5–20 μm) and coarse (20–50 μm) fractions. Many investigators do not carry out this step because it is time consuming and results in a doubling of the number of samples to be extracted. However, I have found it worthwhile and often even necessary to do so, because at times when unfractionated silt is analyzed the assemblage is filled with so many large phytoliths, smaller ones (grass short cells, palms, and other distinctive small bodies) are hard to find and difficult to quantify. The concentration of 5–20-μm phytoliths in fine silt is especially important when one is interested in locating large numbers of cross-shaped phytoliths for maize evaluations. In addition, I have found that more phytoliths seem to be recovered when silt is fractionated.

Silt fractionation can also be accomplished by gravity sedimentation. Samples are placed in 100 mL tall form beakers and water added to a height of 5 cm. The suspension is stirred, allowed to settle for 3 min, and the supernatant, containing the fine silt, is poured off into a 1000 mL beaker. The remaining suspension is diluted again to a height of 5 cm, stirred, and allowed to settle for 2 min and 20 sec after which the supernatant is poured off into the 1000 mL beaker. Repetition of this step 7–8 times usually completes the fractionation of the silt. A silt fractionator, described by Beavers and Jones (1962) can also be used instead of manual methods.

At this stage of the procedure we have three soil fractions to work with—fine silt, coarse silt, and sand. One to 1.5 grams of each fraction is placed in a 16 × 100 mL test tube. If a tabletop centrifuge with a head holding 12 such tubes is available then 12 samples can be processed at one time.

A 10% solution of hydrochloric acid is added to remove carbonates. If carbonates are present, a reaction will be observed in the test tube. Samples are then centrifuged at 500 rpm for 3 min, and the supernatants are decanted. This step should be repeated until no reaction is observed in the tube upon addition of HCl. After washing twice with distilled water, organic material is removed by adding concentrated nitric acid and placing the tubes in a boiling water bath until the reaction, if present, has subsided. This usually takes about 1 hr. A 30% solution of hydrogen peroxide or chromic surfuric acid can also be used. The hydrogen peroxide takes much longer to remove organic material and chromic sulfuric acid cannot be heated, hence tubes must be left overnight for removal of organics to be complete.

Even with the use of nitric acid, a strong oxidizing reagent, many samples often retain high enough levels of organic matter to impede phytolith flotation, and this is not obvious during the oxidation treatment because no reaction occurs in the tubes upon addition of more concentrated nitric acid. However, I have found that in these cases the supernatant in the tube is tinted red or red-orange, a signal that organics are still present in considerable degree and will inhibit phytolith separations. The addition of a few tenths of a gram of potassium chlorate directly to the heated tubes containing soils and nitric acid solves this problem. The appearance of a supernatant with no reddish hue indicates that organics have been oxidized to the extent that successful phytolith extractions are possible.

With carbonates and organics removed, the soils are ready for heavy liquid flotation. The specific gravity of phytoliths ranges from 1.5 to 2.3. That of quartz is 2.65. A heavy liquid solution adjusted to a specific gravity of between 2.3 and 2.4 should therefore cause phytoliths to float while soil particles sink to the bottom of the tube. A number of compounds and solvents can be used to make a heavy liquid solution. They include: cadmium iodide and potassium iodide, tetrabromoethane and absolute ethanol, tetrabromoethane and nitrobenzene, bromoform and nitrobenzene, and zinc bromide and water. I have found the first combination, cadmium and potassium iodide, which was first used by Carbone (1977) to give consistently the cleanest and highest-yielding phytolith separations. It is also much preferable to others that contain organic solvents because it is not carcinogenic and does not give off a noxious odor. It is soluble in water and so precludes the use of alcohol or other reagents as solvents and washing agents.

A 10-mL portion of potassium and cadmium iodide solution at a density of 2.3 (made by adding approximately 470 g of cadmium iodide and 500 g of potassium iodide to 400 mL of water) is added to the soil samples. The density can be checked by simply weighing 1 mL of the solution to three decimal places on a Mettler analytical scale, or specific gravity cubes can be used. The samples are mixed and centrifuged for 5 min at 1000 rpm. After centrifugation, the floating phytolith fraction, which lies at the very top of the tube, is carefully removed with a Pasteur pipette and transferred to another 16 × 100 mL test tube. The samples are then remixed and recentrifuged several more times to remove most of the phytoliths from soils.

After removal of phytoliths from the samples, distilled water is added to the tube containing the phytoliths in a ratio of 2.5 : 1. This lowers their specific gravity to below 1.5 and causes the phytoliths to settle to the bottom of the tube. The tubes are centrifuged at 2500 rpm for 10 min and the supernatant decanted. This step is repeated twice to remove all the heavy liquid from the phytoliths. They are then washed twice in acetone for quick drying and mounted on slides in Permount, which is much less expensive than Canada Fir Balsam and has a proper refractive index for viewing phytoliths.

The refractive index of phytoliths is around 1.42, and they will have a distinctly greater relief than the mounting medium if mounted in agents with refractive indexes of 1.51 to 1.54. These include the aforementioned Permount, Canada Balsam, and Histoclad. Silicone oils, with refractive indexes of about 1.4, are generally to be avoided as phytolith mounting media although Fredlund (1986a) states that phytoliths are viewed with no trouble in silicone oil when phase contrast is used. Either a petrographic or a regular optical microscope can be used for phytolith identification and counting. There are advantages in using a petrographic scope in that extraneous, nonsiliceous matter intrusive to phytolith preparations can be quickly defined as such, a situation especially useful when one is just starting out and is not familiar with the many shapes that phytoliths can take. Phytoliths not mounted on slides can be suspended in absolute ethyl alcohol and transferred to storage vials.

The chemicals and reagents used in making heavy liquid are expensive and many workers prefer to recycle it after every extraction. In practice, this is sometimes more difficult to do than it seems, and since there are usually a few phytoliths left floating in the solution, recoveries are seldom completely pure. At any rate, a number of methods can be used, most of which involve filtering heavy liquid through a fine membrane.

The identification of some plants such as maize requires phytolith rotation and the examination of three-dimensional structure. If slides are read immediately this presents no problems, but if a repeat examination is desired several weeks later, phytoliths must be remounted again because the medium has solidified. There is then a disadvantage in using permanent mounting media such as Permount or Canada Balsam. Rapp's research team at the University of Minnesota has experimented with a number of media that remain liquified. They include benzyl benzoate and styrene. Benzyl benzoate seems most promising because it has a suitable refractive index for viewing phytoliths and a long shelf life (Mulholland, 1982).

Possible Permutations of the Soil Extraction Technique and Some Special Problems Encountered with Different Soil Types

There are a number of variations employed by workers on the procedure presented above that involve shifting some steps ahead or behind in the analysis. For example, Pearsall and Trimble (1982) remove clay during the deflocculation stage and carbonates before samples are wet-sieved. They do not fractionate soils at all, as per many others, but screen soils through a 250 mesh sieve. This seems to rid out pebbles, root hairs and other small, unwanted objects. Carbone (1977) destroyed organics immediately after deflocculation by boiling samples in 30% hydrogen peroxide. Fredlund (1986a) used a swirl technique (Mehringer, 1967) to remove sands, and this also removes some of the clays.

Some soil types pose singular problems and require special processing to allow separation of phytoliths. Peats and other highly organic sediments may fall into this category. They contain humic colloids which cannot be removed with standard phytolith chemicals. The colloids bind to phytoliths and change their specific gravity, thus inhibiting flotation in heavy liquids. The presence of humic colloids in soils can be sometimes detected by the formation of a yellow substance (and no phytoliths) at the top of the tube after addition of the heavy liquid. Humic colloids can be removed by heating samples in a 10% solution of potassium hydroxide for 5–10 min. This step should occur just prior to heavy liquid flotation. Apparently, this short treatment in a very strong alkaline solution is not enough to dissolve phytoliths, as I was able to recover them in substantial amounts from geological peat formations from Panama. However, it may differentially etch or pit phytoliths, so it should not be used unless needed.

Another problem encountered possibly has to do with the nature of the soil examined. Pearsall and Trimble (1982) found that light soils derived from igneous parent material floated in heavy liquid solution, made from tetrabromoethane and absolute ethyl alcohol. I believe that some of the trouble experienced by Pearsall and Trimble has to do with the tetrabromoethane itself, as the same thing happened several times in my work with Panamanian soils, subsequent extractions of which with potassium and cadmium iodide yielded no such problems at all. This is part of the reason why I switched to the use of the latter heavy liquid for chemical flotation. Yet, the fact remains that some sediments may require adjustments in heavy liquid density to ensure that phytoliths and soil particles do not all "come out in the wash."

Extraction of Phytoliths from Modern Plants

There are two basic methods that remove phytoliths from modern plants: wet oxidation which involves the use of chemicals, and dry ashing, in which plants are heated in an oven furnace. The latter procedure has had some controversy surrounding it, which we will see was probably unnecessary, and we will leave it for last. What follows next is a presentation of the wet oxidation procedures, starting first with the one I have found to be most efficient and followed by a number of alternatives.

Plant parts weighing at least 0.1 gram are first soaked in a 1% solution of Alconox for a few hours, placed in 16 × 100 mm test tubes, and washed several times with distilled water. Multiples of 12 samples can be run with tubes of this size; 0.1 g samples are usually sufficient to arrive at representative phytolith collections for most species. A 10-mL portion of Schulze solution, made by adding three parts concentrated nitric acid to one part saturated potassium chlorate (Rovner, 1972) is added to the tubes. They are then placed in a boiling water bath for about 1 h and frequently mixed.

If digestion of all plant material does not occur after about 0.5 h, small amounts of solid potassium chlorate are carefully added directly to the samples until organic material is dissolved. Digestion is known to be complete when no reaction is observed upon addition of more $KClO_3$ or when all material has sedimented to the bottom of the tube. (Organics tend to float in Schulze solution.) The oxidation procedure should take no longer than 1.5 h. If a waxy residue resistant to digestion is present it can be taken into solution with a 1 : 1 mixture of ethyl alcohol and benzene.

The phytolith suspension is washed twice in distilled water, once in a I *N* solution of hydrocloric acid to destroy any calcium present, twice in distilled water, and twice in acetone to quickly dry the suspension. Centrifugation at 1500 rpm for 10 min is sufficient to carry out these steps. The dried phytolith fraction is mounted on slides with Permount or suspended in absolute ethanol and transferred to storage vials.

A number of alternative chemicals can be used in the wet oxidation procedure; these are given in the following list along with their effectiveness and safety.

1. Equal parts 70% nitric acid and 70% percloric acid (Rovner, 1971). Percloric acid is extremely volatile in proximity to organic material and water. Digestion is complete in about 2–4 h in heated water bath. This is probably the least desirable procedure; it tends to blow up in the investigator's hands.
2. Eight parts 70% sulfuric acid and five parts saturated potassium dichromate solution (Moody, 1972). This results in red and green stained residue; i.e., an undesirable method.

3. Concentrated chromic sulfuric acid (Kaplan & Smith, 1980). Room temperature digestion for 24 h. Slower than heating plant material in an acid bath, but it seems to work quite well and is suitable for digestion of very large samples.

4. Concentrated sulfuric acid on a hot plate at 80°C for 2–4 h, followed by addition of 30% H_2O_2 until the liquid is clear and colorless (Geis, 1973). Seems to work quite well and is also suitable for large samples.

5. A ternary acid mixture prepared by combining (in sequence) 10 parts of concentrated nitric acid, 1 part of concentrated sulfuric acid, and 3 parts of 72% $HClO_4$. Samples are predigested for 2 h in concentrated nitric acid. Appears to work well.

All of these procedures are followed by the requisite washes in HCl. Drying of the phytolith suspension can also be accomplished by oven drying for 12 h at 105°C. If digestion of plant samples larger than 0.2 gram is desired, 50-mL tubes or Erlenmeyer flasks should be used.

The Dry Ashing Method of Plant Extraction

In 1929 Netolitsky described many ways of studying silica in modern plants, most of which involved direct ignition or "ashing." The resulting inorganic residues, which contained the silicified microstructures of plants, were known as spodograms (spodos is Greek for ash), and many of the early archaeological references to phytoliths were studies of spodograms found in burnt ceramics and ash heaps. Schellenberg (1908) and Netolitzky (1900, 1914) were able to make detailed comparisons and identifications of these siliceous residues found in archaeological sites. It is ironic that so much confusion and controversy has recently surrounded the dry ashing method, which is actually one of the easiest and most agreeable because it precludes the use of noxious chemicals and for the most part fume hoods. There have been many criticisms [most prominently by Jones and Milne (1963)] that direct ignition of plants changes the physical and chemical characteristics of phytoliths, most notably the index of refraction, surface area, crystalline content, and amounts of residual minerals such as sodium and potassium. Lanning (1965), Lanning *et al.* (1980), and Sterling (1967), among others, have argued the opposite. We need not concern ourselves with this dispute because to build a modern reference collection, we simply want to remove phytoliths from their organic matrix without altering their shapes and sizes.

This can be done most effectively by placing samples in porcelain crucibles and into a muffle furnace and igniting them at 500°C for 1 h. At this temperature opaline silica does not change to other forms of silica and morphology remains unaffected. Samples are allowed to cool and are then washed with 10% HCl, concentrated nitric acid, 10% HCl again, and benzene (Allen *et al.*, 1976). They are dried in acetone. This procedure does not affect morphological characteristics of phytoliths, a fact documented by direct comparisons of wet and dry oxidations carried out by Labouriau (1983) and Andrejko and Cohen (1984). It probably does not affect phytolith size either, although this needs to be proven by similar systematic comparisons. As Andrejko and Cohen (1984) have noted, ashing may result in plant fragments with the phytolith arrangement intact providing more morphological details for comparison and identification. Lanning and Eleuterius (1985) suggest a method whereby the material to be examined is placed between microscope slides and then placed in a muffle furnace. Ashing between glass plates keeps plant fragments with the phytolith arrangement intact so that the overall depositional pattern can be determined. The ash is prepared for light microscopy by removing the upper slide, adding the mounting medium directly to the ash, and covering with a cover glass. This may or may not be desirable, depending on whether size measurements of individual phytoliths must be made and structures observed in three dimensions.

Large or small plant specimens can affectively be analyzed by dry ashing but overall I still prefer wet oxidation because multiple samples of 12 can be run with a single centrifuge, whereas muffle furnaces usually hold only about four crucibles at one time.

Preparation of Archaeological Ash Samples (Spodograms)

It is appropriate to include analysis of archaeological spodograms at this point, having just discussed their preparation from modern plant material. Ash heaps are common in many sites especially from arid zones such as the Near East and Far East. To study phytoliths from these contexts we need simply to mix a small sample in a 10% solution of hydrochloric acid to remove carbonates, wash the sample in distilled water, dry in acetone or a low-temperature oven and mount on slides. At times it need not be necessary to process samples at all, as some can be mounted directly onto slides if phytoliths form a high enough content of the ash. Helbaek (1961, 1969) apparently did this with his samples from Near Eastern sites.

Quantitative Analysis of Silica in Modern Plants

The quantitative determination of silica content is not often of primary interest to paleobotanists, who seek mainly to build reference collections while noting and estimating approximate amounts of phytoliths in different species. At times, however, it is desirable to measure silica content directly, especially when examining new plant structures such as seeds or new families. The procedure followed for quantitative determinations of silica is basically the same as for methods of extracting phytoliths from modern plants.

Specimens are weighed to four decimal places, digested, and at the end of the analysis the resultant dried phytolith residue is weighed to yield silica content expressed as a percentage of dry plant weight. Depending on the accumulator characteristics of particular species it is necessary to start with somewhat more modern material than the 0.1 gram required for simple phytolith extractions. One gram is usually a safe amount. Any of the extraction procedures given earlier can be used.

Quantitative Analysis of Silica in Soils

Quantifying the amount of phytoliths in soils is sometimes desired to measure phytolith distribution with depth or the content of modern soils taken from under different vegetational associations. We carry out this analysis by weighing soils to be extracted to four decimal places, isolating phytoliths by the technique described earlier, and weighing the phytolith residue. The residue is then mounted on slides and scanned for soil particles to derive an estimation of the purity of the extraction and accordingly adjust the percentage silica soil value.

Preparation of Modern Samples for Scanning Electron Microscopy

This procedure again involves isolating phytoliths from their organic matrix by any of the several extraction methods presented earlier. Phytoliths are then suspended in a small amount (1 mL) of 95% ethyl alcohol and stored in glass vials. One or two drops of the suspension is placed with a Pasteur pipette on an aluminum stub which has double sided tape. The suspension is allowed to air dry and then sputter-coated with a 200 Å layer of gold–palladium. Phytoliths isolated from soils are prepared and mounted on stubs in the same fashion.

Phytolith Staining

For a long time phytoliths were considered to be impervious to stains or dyes, but recently Dayanandan, *et al.* (1983) developed a staining process which renders phytoliths reddish-brown, red, or blue. This is primarily used by botanists as another method to detect the location of silica in plants, but it has potential for paleoecological applications as well. For example, stained phytoliths might be used as the exotic type added to soil preparations for absolute (versus percentage frequency) phytolith determinations, or as controls on percentage of recovery studies.

A number of staining solutions can be used—silver ammine chromate (SAC), Methyl red (MR), or crystal violet lactone (CVL). SAC results in reddish-brown colored phytoliths, MR in red colors, and CV in blue colors. SAC is prepared by separately dissolving 34 g $AgNO_3$ and 20 g K_2CrO_4 in 100 mL of water each and mixing these solutions, whereupon a precipitate is obtained. This precipitate is washed in hot water and dissolved in 200 mL of 3% NH_3, prepared by dissolving 20 mL of strong ammonium hydroxide (30% NH_3) in 190 mL of H_2O. The SAC is stored at room temperature and filtered before use when turbid.

Methyl red is used as a saturated solution in benzene. The acid form of MR is preferable to the sodium salt or the hydrochloxide because it stains phytoliths more intensely. Crystal violet lactone is used as a 0.1% solution in benzene. The authors of the staining procedure indicate that a brief acid treatment with 50% H_2SO_4 and mild etching with hydroflouric acid enhances the intensity of the stains.

There are a number of considerations to be made if stained phytoliths are to be used in paleoecological work. It will be desirable to use discrete phytoliths, so they should first be isolated from plants instead of the *in-situ* cuticular procedure developed by Dayanandan, *et al.* on the rice plant. The most appropriate staining times should be explored, and it must be ensured that all phytoliths added as exotics are stained and remain that way after flotation in heavy liquids. It would be easiest to use plants having a single phytolith type, i.e., palms, so it must be determined if phytoliths from such plants stain as effectively as did the rice silica. Some modifications will therefore be needed to prepare stained phytoliths for input into soil extractions.

Other Procedures of Phytolith Analysis

Microscope Techniques

For everyday phytolith identification and counting the standard biological microscope equipped with low (125×), high (400×), and oil immersion (1000×) lenses is used. Many phytolith identifications can be made at magnifications of 315× to 400× with some discrimination of small forms or subtle surface decorations necessary with oil immersion at 1000×. Low powers are useful for scanning after percentage or absolute frequencies have been tabulated, especially for the presence of large (50 μm and greater diameter) phytoliths, such as the *Cucurbita* scalloped types. The scanning electron microscope is desirable for the documentation of morphological variants observed with the optical microscope and for some morphological discrimination.

Proper use of the optical microscope is essential for phytolith identification and requires training and experience. One cannot immediately expect to be able to see three-dimensional structures or subtle features of surface designs without a considerable degree of practice with various kinds of phytoliths.

Documentation of Phytoliths

Phytoliths can be documented by either microphotography or line drawings. Of the first, scanning electron microscopy, of course, offers the finest details and appreciation of three-dimensional characteristics. Line drawings can also present subtle features and the analyst can select critical attributes and supress others to emphasize what he considers most important.

6

The Interpretation of Phytolith Assemblages: Method and Theory

Introduction

How much can be learned from phytolith analysis about the uses of plants, vegetation, and climate of the past? As with any technique that asks questions of once existing ecosystems that cannot be observed directly, phytolith analysis is marked by varying degrees of imprecision in the reconstructions that it offers. Despite the growing body of research designed to improve the accuracy of the technique, this will be an unyielding aspect of the discipline. What paleoecologists must do is elucidate and define the potentials and limitations of phytolith analysis and, in so far as existing methodology and theory permit, interpret the data base to offer sound reconstructions of vegetation and climate. Problems will be continually defined, studied, sometimes resolved, and more often debated for long periods of time, just as palynologists still struggle with such issues as differential production, dispersion, and definition of a source area 70 years after Lennart van Post published the first pollen diagram.

Of fundamental importance always is avoiding gross errors in the presentation of paleobotanical data, such errors usually arising from insufficient understanding of the attributes and behavior of microfossil particles. In this chapter we first use the well-known attributes of phytoliths to posit some fundamental principals of Quaternary phytolith analysis, explore some of the potentials, problems, and limitations as they are now understood, and go on to outline some areas where future basic research might prove profitable.

In the first five chapters we have recounted the research carried out on the characteristics and behavior of phytoliths from modern and fossil plant taxa. Phytoliths possess several fundamental attributes that underlie their applications to Quaternary paleoecology:

1. Phytoliths are not uniformly produced by the plant kingdom. Certain known

taxa are regular accumulators of solid silicon dioxide. When secreted by plants they take on manifold shapes and sizes that are faithfully replicated in species and identifiable in the collection of phytoliths isolated from soils and sediments. Certain known taxa are persistent nonaccumulators of solid silica and evidence for their presence in microfossil assemblages is not to be expected.

2. Plants may contribute several types of phytoliths, any or all of which may be shared by related or unrelated taxa. Different structures of the plant body from one species will often contribute different shapes of phytoliths. Phytolith taxonomy, therefore, involves more complicated systematics than those for other microplant and macroplant remains, which entail a one-to-one correspondence between taxa and associated microbody or macrobody. Many plants do contribute a single phytolith type recognized by distinct shape, sculpturing, and size, and assignable to individual taxa at various levels (family, subfamily, genus, or species).

3. As a result of the decay of organic plant tissue, phytoliths come to occupy various depositional environments in conspicuous numbers. Because they are produced by the vegetation of sites they are a function of that vegetation and, therefore, reflect the plant cover and climate of the landscape at that point in time. They are extremely stable in most depositional environments over long chronological periods, and because they are abundant and small, ranging in size from less than 5 to about 500 μm, they require only small samples, about 50 g or less, for analysis.[1]

Having enumerated the basic attributes of phytoliths, we can now look closely at some theoretical and methodological issues of phytolith interpretation in paleoecology.

The Phytolith Assemblage

In paleoecology, the unit of analysis and interpretation and the basis for intra- and interregional comparisons of the phytolith record is the phytolith assemblage. It is the tabulation and quantification in percentages, absolute numbers, or ratios of all morphological variants observed in a sample. A number of steps are involved in its construction, the first of which involves phytolith identification.

Phytolith Identification

From data presented in Chapter 3 on modern phytolith diversity, it is clear that a large number of morphological variants or "morphologs" will be present in

[1]Points made under number 3 anticipate discussion presented in this chapter and in Chapters 7 and 8, where it is shown that phytolith profiles with depth representing the phytolith deposition over time are achieved and phytolith spectra closely mirror the existing vegetation.

fossil phytolith assemblages. Their assignation to plant taxa is a complex undertaking dependent on a very sound knowledge of morphology in modern plants. Identifications will be made at different taxonomic levels (family, genus, species) depending on the current specificity of forms encountered. Phytolith taxonomies and keys will be in a state of flux for some time as more regional information is compiled on phytolith morphology.

The problems involved in plant identification from collections of fossil phytoliths stem from the characteristics of phytolith morphology outlined earlier. Adequate control of these complex and complicating factors is achieved only by the construction of a large modern phytolith collection comprised of several plant structures from individual species known to be silicon accumulators. Taking the complexity of phytolith production and morphology into account, some general rules can be applied to identification of phytoliths from sediments.

In the case of morphologs found in a large number of species, related or unrelated, they can be subsumed under a general classification heading, for example, polyhedral epidermis or segmented hairs; a list can then be made of the kinds of plants in which they are found. Morphologically redundant phytoliths are not necessarily of little value in interpretation, for they may be restricted to certain growth habits, such as arboreal or herbacious, and shifts in their frequencies over time may help to define concurrent major vegetational shifts. In tropical plants, for example, the sclereid phytolith class is largely confined to arboreal species and very seldom found in herbs or vines. We will discuss in Special Topic 2 the application of this kind of analysis.

The fact that more than one phytolith can be found in a single plant and that some can occur in other plants, is probably more complicating than the redundancy issue taken by itself. One-to-one correlations between a single phytolith and a single taxon will not always be possible. At times, the comparison of modern and fossil morphological type frequencies will permit finer discrimination of taxa, and sometimes it will not. Some plants, like the genus *Zea,* have phytolith assemblages that when compared to the flora of certain study regions seem to uniquely combine several attributes. This situation invites statistical confirmation of the observed and measured differences between plants. This kind of analysis has been demonstrated in Chapter 2 and should be possible for many other plants when the appropriate modern comparative studies have been made.

In many others, however, such as Chloridoid and Bambusoid grasses, phytolith assemblages may not be distinctive below the level of family, and once again, nothing can substitute for a high level of knowledge of the phytolith types to be found in the regional flora and their proportions within species to evaluate the significance of the phytolith record. For many kinds of plants and identifications at broader taxonomic levels, the situation is not as complex since one-to-one correlations between shapes and taxa can regularly be accomplished. Numerous families produce distinctive phytoliths (e.g., nettles, composites, sedges, arrowroots) and genus-level identification is already clearly possible in a significant number of cases (e.g.,

Cucurbita, Musa, bamboos, *Heliconia,* members of the Burseraceae) and probable for many others.

That different structures of a plant may produce different kinds of phytoliths is a feature that at the same time necessitates a considerably greater investment into modern morphology and leads to finer resolution of archaeobotanical data. The archaeobotanical record is a result of selective use of portions of plants, an exploitation pattern to which macrobotanical remains are sensitive but which pollen grains often are not. The finding of fruit, seed, or glume phytoliths in archaeological midden is solid evidence that particular structures of a species were exploited by human populations. Such precision is not always possible in pollen studies where the presence of grains in a site may indicate merely that they were attached to decayed flowers, seeds, or clothing or blown in from local or regional standing vegetation.

Counting Procedures

Counting procedures in phytolith analysis represent a very much open question. As far as I am aware, no published material is available where error estimates have been obtained by counting increasing numbers of phytoliths from the same sample and observing where percentages settle off. Most phytolith analysts have used sample sizes of at least 100–200 particles and assumed on the basis of practical experience and the number of taxa they are dealing with that such numbers result in statistically acceptable counts. Given the lists of taxa published by analysts, these assumptions seem warranted.

Even if experimental data on counting procedures existed, no standard number could be recommended because proper sample sizes, besides depending on the number of taxa found in a preparation, vary according to the information required and the relative proportions of taxa present. If some of the less abundant kinds are ecologically significant, sample sizes of greater than 200 might be necessary to achieve representative percentages. In addition, there will be cases in which it is preferable to set the counts of certain taxa present in excess aside and focus attention on others whose ecological significance is less clear. Counting procedures need to be worked out by phytolith analysts on a site-by-site and region-by-region basis.

The Phytolith Diagram

After the phytolith morphologs in a sample are identified and quantified, data are usually presented in the form of a phytolith diagram, composed of the assemblage from each level or stratum plotted against their stratigraphical depth or age. The diagram consists of two parts, a horizontal line with constituents of the phytolith

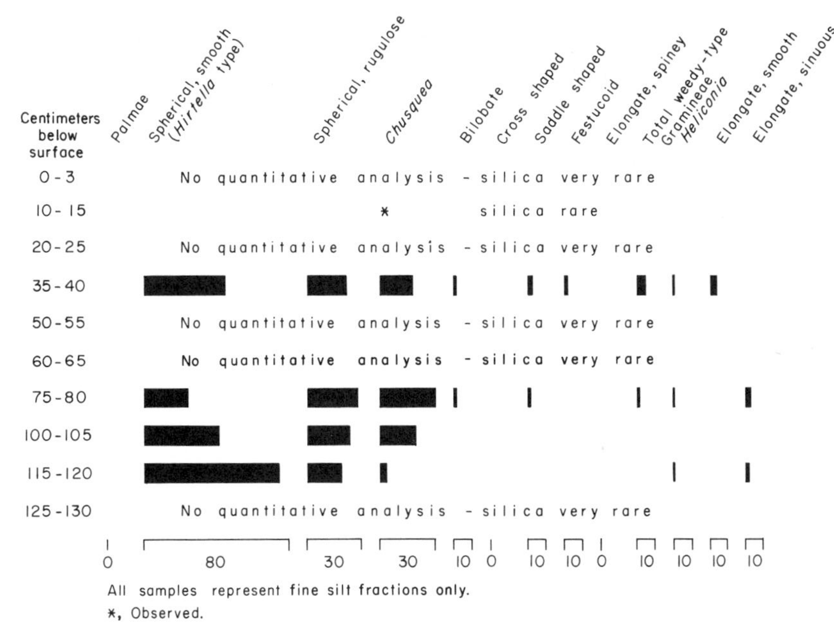

Figure 6.1 Percentage of phytolith frequencies from Casita de Piedra, western Panama.

assemblage plus frequency intervals and a vertical line showing depths and dates of the assemblages. Changes in the percentage of one species are assumed to reflect similar changes in the vegetation composition (due consideration being given to the factors of over- and underrepresentation and problems with percentage frequency sums, which will be discussed later). Phytolith diagrams are particularly informative in cases where sites show a long chronological history, as they may register changes in plant exploitative strategies and their effects on the vegetation cover near sites.

Figures 6.1 and 6.2 show percentage frequency diagrams from two deeply stratified rockshelters located several miles apart in mid-altitudinal elevations (700 m) of western Panama (Ranere, 1980a).[1] Deposits from the bottom to top of Casita de Piedra and 130 to 25 cm at Horatio Gonzalez are associated with preceramic occupations, radiocarbondated from 5000 B.C. to about 300 B.C. Preceramic-phase deposits from both rockshelters are notable for their lack of weedy

[1]In Figures 6.1 and 6.2 and all phytolith diagrams that follow, if a fossil phytolith type is assigned to a given natural taxon, this expresses a high degree of confidence in the identification. If the word *type* is appended to a taxon, this expresses uncertainty that a sufficient number of species have been examined to warrant a positive identification, although a high degree of confidence exists that assignment to growth habit (arboreal, herbacious) and plant family is correct. In all figures, phytolith frequencies are based on sums of 100–200 phytoliths for each silt fraction; some types are not included.

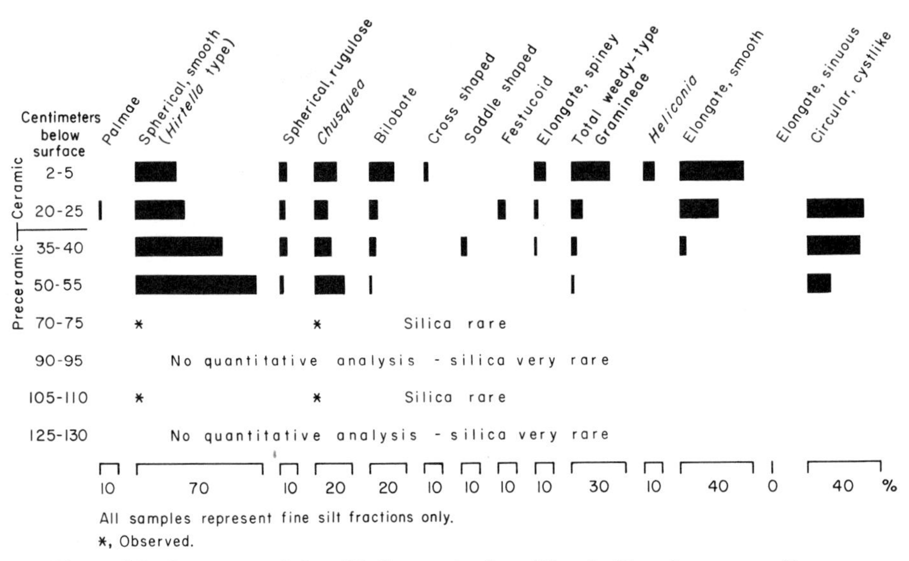

Figure 6.2 Percentage of phytolith frequencies from Horacio Gonzalez; western Panama.

types of grass phytoliths (Chloridoid, Festucoid, and Panicoid shapes), with percentages reaching no higher than 6% of the total sum. Such frequencies are typical of undisturbed and late secondary moist forests in Panama, as will be discussed fully in Special Topic 2 at the end of this chapter. *Chusquea,* a bamboo of semi-evergreen tropical forests, is well represented throughout the sequence, another indication that forest surrounding the site was little disturbed during the whole of the preceramic occupation.

The level 20–25 cm below the surface at Horatio Gonzales marks the introduction of ceramics, ^{14}C dated at about 300 B.C. Weedy grass phytoliths and *Heliconia,* a herbacious plant of secondary growth, show significant increases, indicating a substantial increase in the level of forest modification. The phytolith record correlates well with settlement pattern data from the region, (Linares *et al.,* 1975; Linares and Ranere, 1980) which indicate the presence of small, dispersed horticultural and hunting and gathering populations until 300 B.C., which saw the colonization of the area by maize agriculturalists and rapid environmental degradation.

Documenting the presence or absence of domesticated plants, apart from their importance in past economic systems, is a primary consideration of paleoethnobotany. A major issue in Panamanian archaeology has been the role of maize in preceramic subsistence strategies in different environmental regions of the country. Figure 6.2 shows that cross-shaped phytoliths, which are produced in large numbers by maize, are completely absent in preceramic deposits from western Panama, but appear in the ceramic-phase unit isolated from Horatio Gonzalez along with the

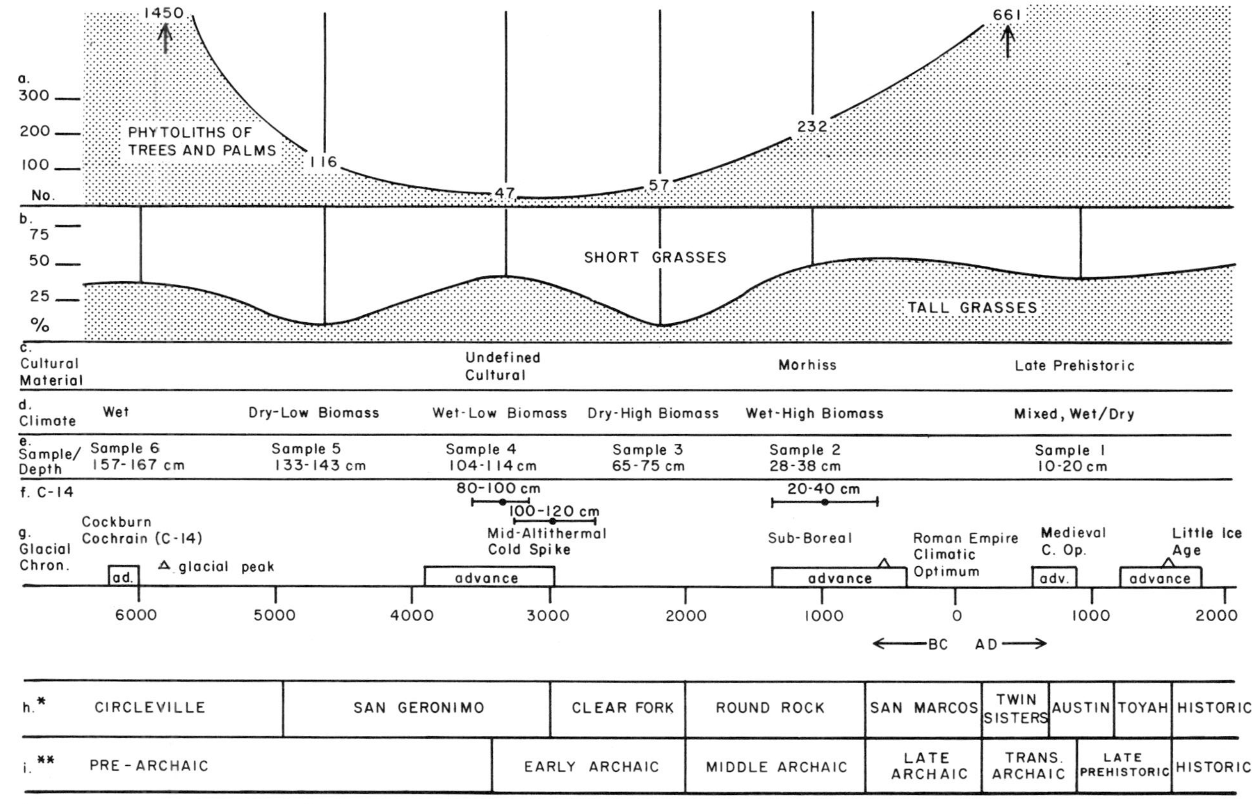

Figure 6.3 Phytolith diagram from an archaeological site in southern Texas. [Courtesy of the Center for Archaeological Research, University of Texas at San Antonio.]

other disturbance indicators. The absence of the ubiquitous cross-shaped type suggests that maize was not part of preceramic subsistence systems in highland western Panama, a pattern which contributed to the maintenance of low population densities and low levels of environmental deterioration over a 4000-year period.

Shifts in frequencies in grass and tree phytoliths, plus other kinds of biogenic silica such as diatoms and sponge spicules, can also help to reconstruct the regional climate of archaeological sites. In archaeological deposits from Texas, Robinson (1979, 1980, 1982) was able to distinguish mesic from more xeric environments in the vicinity of sites based on significant changes in percentage frequencies of short (Chloridoid), tall (Panicoid), and Festucoid grasses, trees such as oak (*Quercus* sp.), elm (*Ulmus* sp.) and hackberry (*Celtis* sp.) and freshwater sponge spicules (Fig. 6.3). Periods of biosilica accumulations characterized by high percentages of short grasses and low percentages of tall and Festucoid grasses, deciduous trees, and sponges probably represent much drier intervals because today short grasses are characteristic of, and dominant in, arid environments with summer rainfall such as the Desert Plains grassland and the short grass prairie. Panicoid and Festucoid grasses and deciduous trees today occupy environments that receive more rainfall or are wetter and cooler. These data will be fully discussed in Chapter 7.

Absolute versus Percentage Frequency Diagrams

In the last few years, there has been a great deal of discussion by palynologists surrounding the strengths and limitations of percentage frequency versus absolute diagrams. In the former, numbers are expressed as percentages or proportions of the sum of all taxa. This sometimes presents problems because the curve for each plant is partially dependent on the composition of the rest of the diagram. If one pollen type, especially a poor producer, decreases in relative frequency, other type(s), especially prolific producers, may artificially increase even though the quantity of those species has not changed at all because the sum of all frequencies must add up to 1.

Various pollen studies [e.g., Davis (1969) and Ritchie and Cwynar (1982)] have shown that percentage frequency diagrams give misleading pictures of vegetational composition when drastic changes in vegetation and, hence, in pollen input to the site have occurred. Davis (1969) demonstrated this factor quite convincingly at Rogers Lake, Connecticut. In Late Glacial period sediments (14,300–11,700 B.P.) percentage frequencies of such taxa as *Artemisia, Rumex, Pinus* and *Picea* suggested that both herbs and trees contributed significant amounts of pollen into the regional landscape. The significance of long-distance tree pollen transport was unclear. Some of the problems of percentage frequencies can be overcome with the use of absolute frequencies, which estimate the numbers of individual pollen taxa being deposited on a surface over a given time period. Independence between taxa in a pollen sum is achieved either indirectly by comparing them to a known amount of exotic pollen added to preparations or directly by counting all pollen in a subsample of known volume or weight.

If the chronology of a sequence can be determined by radiocarbon, then sedimentation rates can be evaluated and the absolute numbers of grains falling onto a surface per unit time are calibrated. Absolute frequencies constructed by Davis (1969) demonstrated that tree cover indeed was sparse to absent in the Rogers Lake region during the Late Glacial period. Their prominence in the percentage diagrams resulted from negligible pollen input from herbacious plants, which rendered the effect of long-distance transport indecipherable.

Other changes in pre- and early Holocene vegetation at Rogers Lake such as the initial repopulation by various genera of trees, which took place over several thousand years, were made much clearer with the use of absolute diagrams. For most of the post-Glacial period at Rogers Lake, when deposition rates for total pollen were relatively uniform, i.e., when species compositional change was not nearly on the order of magnitude as during the late Glacial, the percentage diagram conveyed basically the same information as did the absolute diagram.

In addition to the individual representation of the influx of taxa per year made possible with absolute diagrams, the total influx of all types constituting a vegetation association per unit time (variously called pollen influx, pollen deposition rate, or pollen accumulation rate) is of considerable ecological significance. A number of studies from modern contexts have shown that different vegetational communities such as tundra, mixed coniferous–deciduous forest, and deciduous forest contribute significantly different amounts of pollen per unit area on a yearly basis. (e.g., Lamb, 1984; Davis, 1969; Ritchie and Cwynor, 1982). Tundra and other open, cold-adapted communities exhibit characteristically low concentrations while coniferous and deciduous forests often have concentrations at least twenty-fold greater than those of open landscapes. In fossil diagrams where radiocarbon chronology is sufficiently dense to permit estimations of sedimentation rates, and when pollen deposition rates are then calculated, vegetational formations with different degrees of openness or tree cover may be identified solely on the basis of total pollen influx. This is often particularly important when species composition and frequencies of fossil assemblages have no modern analog.

Absolute frequencies are not without their interpretative problems. Birks and Gordon (1985), who are not strong proponents of absolute pollen studies, have summarized them. Little is known about factors controlling sediment input, focusing, and redeposition in lakes, where absolute diagrams are usually computed. These, at any rate, probably vary with the size of the basin and prevailing climatic conditions. One can also question whether a few cores taken from the center of lakes are really representative of sediment and pollen input into the entire lake. Studies by Davis *et al.* (1973) have shown that modern and fossil pollen accumulation rates commonly vary from two- to five-fold within and between lakes despite relatively constant sediment accumulation rates.

Despite these difficulties, which seem to be most serious when formal, numerical analyses of pollen assemblages are undertaken (Birks and Gordon, 1985; Maher, 1972), and the increase in counting and laboratory time, the advantages of using absolute pollen diagrams can be substantial, especially in more intuitive interpretations of past vegetation. Some time has been spent discussing these advantages because it is believed that phytolith studies of lake and other geological sediments will soon become a potent method of paleoecological analysis; this will be explained later. Both absolute and percentage frequency counts will become desirable in certain situations. Some techniques that might be attempted in constructing absolute phytolith frequencies are described in Chapter 5.

It seems quite possible that in using absolute frequencies the phytolith analyst will derive the same benefits as does the pollen worker and generate data and insights not possible with either pollen analysis or phytolith percentage frequencies. In many regions of the world transitions from grassland and other open formations to woodland formations, and vice versa, probably entail considerable changes in phytolith inputs to sites. In contrast to the situation for pollen, phytolith influx in herbacious communities may be substantially higher than in mixed forest–open, boreal, or deciduous forest. Silicon-accumulating plants belonging to the Gramineae, Cyperaceae, the genus *Artemesia*, and other Compositae are important constituents of treeless, cold-adapted biomes of the northern and southern hemispheres. Other herbacious taxa poorly represented in the pollen record and not yet tested for the presence of silica may be found to have distinctive phytoliths. Tundra herb vegetation, which is very difficult to interpret with pollen data because of low pollen productivity and redundancies in pollen types, may prove much less infractile to resolution using a phytolith record.

There are major questions concerning the composition of Late Glacial herbaceous cover in regions, such as Beringia, that were near the ice sheets or were greatly affected by cold, dry conditions, and subsequent early Holocene transitions (Ritchie and Cwynar, 1982; Matthews, 1982). These questions, in turn, have a considerable bearing on determining the productivity of the landscapes for supporting large and diverse faunal populations and, by implication, early humans (Colinvaux and West, 1984). Such issues should become more amenable to resolution with absolute and percentage frequency phytolith data.

In the humid tropics, pollen sequences from Guatemala and Venezuela have shown that the Late Glacial plant cover was comprised of open, herbacious vegetation which was replaced between 11,000 and 10,000 B.P. by deciduous (dry) forest (Bradbury *et al.*, 1981; Leyden, 1984). However, it is unclear whether Late Pleistocene plant cover was a savanna in the present-day sense of the word or a combination of grasses, other herbacious plants, and thorn-scrub species which are often "silent" in pollen assemblages due to poor production. Again, the composition and prevalence of taxa in these herbacious zones are major issues in elucidating early man–land relationships in the humid tropics (e.g., Ranere, 1980b; Lynch,

1983), issues which should become better understood with the generation of phytolith data.

Problems in the Interpretation of Phytolith Assemblages: Some Considerations of the Factors Affecting Phytolith Representation in Soils

The relationship between plants composing the existing vegetation and plants utilized by site residents on the one hand, and plants represented by phytoliths recovered in sediments on the other, is governed by variation between species in the following four factors: (1) the magnitude of phytolith production, (2) the taxonomic significance of phytoliths, (3) the mode of dispersal of phytoliths to the site under investigation, and (4) the resistance of phytoliths to destruction. The interactions of these factors and how they relate to reconstructions of past plant usage, vegetation, and climate are extremely complex and very poorly understood. We will deal with them by first considering separately each of their possible impacts on interpretations of fossil phytolith sequences, and then judging their final collective effects by exploring what we know about modern phytolith assemblages.

Phytolith Production and Morphology

In Chapters 2 and 3 the most up-to-date information on phytolith production and morphology has been presented. The great variance in phytolith production in the plant kingdom will lead to overrepresentation, underrepresentation, and the complete absence of many species in the phytolith record. Plants where underproduction is a problem can be found in Tables 2.1–2.4. In addition, little can be said for those plants which contribute either minute numbers of recognizable phytoliths, as some of the Euphorbiaceae, or only forms that are repeated often across the plant kingdom. The probabilities that such plants will be incorporated into sediments and detected by phytolith analysts, or identified if recovered, are virtually nil. Therefore, phytolith assemblages from sediments in many cases will not adequately reflect the diversity of species exploited by man or which formed the existing vegetation but will be skewed in favor of that smaller percentage of plants that produce distinctive phytoliths.

The results of the analysis of domesticated plants to date have indicated that few will leave behind identifiable siliceous remains in fossil records. Many, such as beans (*Phaseolus* and *Canavalia*, spp.), cotton (*Gossypium*, spp.), manioc (*Manihot esculenta*) and sweet potato (*Ipomoea batatas*) are severely limited by either the absence of phytoliths or the presence of siliceous forms that are commonly repeated in other plants. However, a number of major Old and New World domesticates, such as rice (*Oryza sativa*) wheat (*Triticum* spp.), barley (*Hordeum distichum*), bananas (*Musa* spp.), and squash (*Cucurbita* spp.) offer a considerable amount of diagnostic potential in terms both of production and morphological specificity.

It is important also to remember that although phytolith production can occur throughout the plant body, seeds, tubers, and fruits may be devoid of silicified cells. Many wild and domesticated plants are exploited for precisely these structures and so may leave behind an inadequate phytolith record even if they were used and deposited into occupation middens. Maize and bananas are cases in point here. The production of the diagnostic cross-shaped silica bodies and phytoliths with troughs appears to be highest in leaves. Maize husks and banana fruits and rinds, which presumably would have decayed commonly in living areas of sites, have been shown to be depauperate in silica (Chapter 2). This factor has not negatively biased the representation of maize in many of the tropical sediments that have been studied, apparently because maize leaves were used in economic activities, e.g., for mats or to line pits. Still, it must be considered that the phytolith record may be negatively biased in documenting domestication even when plants in question have distinctive phytoliths in some part of the plant body. The analysis of old fields rather than habitation areas may be required to document the presence of some crop plants.

The nature of the cropping system will also affect phytolith representation in soils in a manner independent of the magnitude of phytolith production. Primitive and incidentally used crop plants often leave behind only meager evidence whereas food staples grown in a full-blown agricultural system would contribute substantially more biomass into sites and have a far greater degree of probability of recovery.

The Mode of Phytolith Dispersion and Deposition

With respect to the dispersion factor, phytoliths, at first glance, do not seem to present the theoretical problems that attend pollen transport. Much attention has been given in the palynological literature to differences in pollen modes of dispersal, efficiencies of dispersal, and the relationship between pollen source strength and dispersal distance (for example, Bonny, 1978; Anderson, 1970; Bradshaw, 1981). These factors determine the source area of fossil pollen grains and, hence, the spatial scale of vegetational reconstruction. Because phytoliths are often liberated from plant matter upon its death and decay and are released into the soil, not into the air, we can expect a very large proportion of the phytolith record to be an *in-situ* product of plant deposition.

The local character of phytolith deposition and transport makes the archaeological phytolith record a very good indicator of on-site plant usage. Ruderals and other plants growing very near occupation areas may also be incorporated into middens by a small-scale natural movement of plant matter or soil. The nature of phytolith deposition will, of course, vary according to the type of site. Partially enclosed rock shelters offer further protection from possible wind-blown elements, but sites in floodplains deserve special scrutiny because of the possibility of sediment redeposition. Archaeological sites comprising the remnants of prehistoric fields will benefit from phytolith analysis since interpretations will not be confounded by the probability of intrusive, wind-blown particles.

Like the archaeological record, the geological phytolith record may often be localized and provide inadequate representations of upland and regional vegetation. This at once offers promise and problems of interpretation. Long-distance transport will not present the complicating situation that palynologists must try to resolve, and paleovegetation and climate reconstructions will be extremely sensitive to detecting changes within small tracts on the landscape. In tandem, pollen and phytoliths would offer powerful data on two spatial scales, the local and regional, and allow comparisons between the two in assessing the magnitude of vegetational change. However, if the phytolith record of a single site does not adequately portray the regional plant cover, data from a suite of sites would be necessary for adequate control. Anderson (1970) and Bradshaw (1981) showed that within closed forests most pollen does not travel beyond 20–30 m from its source. In Special Topic 2 we present evidence that on level terrain in a tropical rain forest, phytolith movement is of the same small magnitude and even less.

All of this said, we now proceed to show how it would probably prove misleading to assume that phytolith sequences always predominantly represent local vegetation. It has been demonstrated that under certain conditions of climate and resulting plant cover, phytoliths are susceptible to long-distance movement and may be blown or carried well beyond the geographical distribution of the source vegetation. The silt fraction of top soils often contains a substantial phytolith content sometimes reaching values of over 50% (Folger *et al.,* 1967), which may then become a major component of any wind-blown sediment.

Remember that phytoliths were first classified by Ehrenberg (1854), who found them in dust samples collected by Darwin from the sails of the H.M.S. Beagle many miles at sea. The origin of the dust was northwestern Africa, where the strong northeast trade winds and desiccated vegetation create a considerable windborne soil load. A number of studies over the last 30 years have documented the conspicuous presence of phytoliths in aerosols and deep-sea cores of the North Atlantic (Folger *et al.,* 1967; Parmenter and Folger, 1974; Melia, 1980). Downwind transport distance can be as far as 6000 km; however, the majority of phytoliths are deposited within 500 km from the shore. Transport distance and phytolith content covary with the velocity and direction of the wind and degree of dessication of the vegetation (Melia, 1980; Folger *et al.,* 1967; Twiss, 1983).

Two primary conditions under which considerable horizontal phytolith movement may occur are, therefore, strong winds and a marked degree of vegetation-poor, open terrain. These may be considered to have synergistic effects; that is, the magnitude and distance of transport is most marked when the two conditions are operating together. Examples of studies that have documented late Quaternary climatic change on the basis of airborne phytoliths found in deep sea sediments are presented in Chapter 8.

River and ocean currents may also carry phytoliths considerable distances (Melia, 1980), possibly to be redeposited in lakes or on shores and terraces where archaeological and geoarchaeological studies are being carried out. In addition, a considerable amount of movement can take place by means of soil erosion and surface run-off after rains. Large numbers of phytoliths may also be injected into the atmosphere from plant tissues during major fires.

In geological phytolith research, recognizing the origin of phytoliths will be especially important in lake sediments. Phytoliths from lacustrine sediments have not been used thus far as a tool of paleobotanical reconstruction. However, they offer a considerable promise for providing an independent source of paleoecological data and, in combination with pollen analysis, for achieving finer-scale resolutions of past vegetation and climate. The significance of phytolith production and taxonomy in many plants suggests that fossil assemblages characteristic of a number of vegetational associations should be present in stratified lake muds. The many advantages of a phytolith record over a pollen record, in terms of taxonomic precision for many major plants (sedges, nettles, grasses, ferns, composites) and production of distinctive phytoliths in many species that are palynologically "silent" (rain forest trees and herbs), indicate that in tandem phytolith and pollen analysis will generate powerful paleoecological data.

There is no reason why, especially in the case of lakes whose deposits contain a considerable allocthonous input of soil, phytolith content should not be quite high. In contrast, lakes with sediments comprised mainly of black, organic gyttja (autocthonous material derived from the decomposition of lake-life such as diatoms and sponges) may have depauperate phytolith assemblages. It stands to reason that much of the phytolith content of lake sediments will derive from the influx of soils from the catchment basin, either via slope wash or inflowing streams, with deposition from aerial transport occurring to a lesser degree and as a function of the prevailing climate and land cover. This said, we can go on to discuss in more detail some aspects of phytolith recruitment and sedimentation in lakes, which are very poorly studied and no doubt complex.

A critical and difficult area of paleobotanical lake studies is recognizing the source area of fossils found in deposits. Are they derived from local (within 20 m), extra-local (between 20 and several hundred meters), or regional (at greater distances) vegetation [using Jacobsen's and Bradshaw's (1981) distance scales] and what proportion of the botanical record is represented by input from spatially disparate areas? Palynologists have struggled with these questions for years (e.g., Oldfield, 1970; Tauber, 1965, 1977; Jacobson and Bradshaw, 1981; Birks and Gordon, 1985), for they determine how much of a particular study region's vegetation and climate in areal extent can be reconstructed from lake data. Many palynologists consider the problem to be among the most vexing and intractable (for an older but still highly pertinent review see Oldfield, 1970).

Recognizing the source area of phytoliths will be as important as recognizing it for pollen. As noted previously, phytolith input should derive predominantly from the sediment load into the lake via either slopewash or inflowing streams. In lakes with no inflowing streams it can be argued that deposition should largely reflect local and extra-local sources of vegetation, depending on the size of the basin and topographic characteristics of the watershed it drains. Phytoliths from lakes will be somewhat analogous to macrofossils in representing a nonregional environmental record. In situations where past climate and vegetation have combined to create optimal conditions for long distance transport (as discussed previously), more of the phytolith record will represent extra-local and regional vegetation.

In cases where there are inflowing streams, the situation becomes more complicated. For example, Bonny (1978) showed that in a particular lake in England only 15% of the total pollen influx to the sediments of the open water came from airborne and local sources; an inflowing stream contributed most of the pollen load. Depending on the rate of flow of incoming streams and the size of the watershed feeding the stream, a lake may receive phytoliths from a source area much larger than the immediate lake environment might suggest. Human factors such as land clearance as well as climatic and vegetational conditions affecting the nature and degree of cover of the topsoil will all affect the streamborne and other soilborne fraction of phytoliths into a lake.

In summary, factors relating to phytolith dispersion will vary depending on the choice and location of a site (rock shelter versus floodplain occupation; terrestrial versus lake deposits) as well as past environmental conditions. As Jacobson and Bradshaw (1981) have emphasized, the limitations and problems of each kind of site should be considered beforehand. The question of whether phytoliths in a paleoecological study represent local, extra-local, or regional vegetation can then be better evaluated, and in many cases the answers are not likely to be simple ones.

This brings us to the subject of how long-distance phytolith transport and phytolith source areas can be recognized by the paleoecologist. It should be emphasized that phytoliths, like other macro- and microfossils, are seldom studied in isolation. They are interpreted within a cultural or natural context, which entails their association with macrobotanical remains, pollen, nonbiotic cultural debris, stratigraphic zones, and spatial divisions of sites. Correlations of the phytolith record with other archaeological and paleoecological data can often help sort out potential problems of origin and association. For example, in moderate to large-sized lakes that act as catchment basins for the regional pollen rain, changes in phytolith frequencies that are independent of the pollen stratigraphy imply *local* changes. Changes in pollen frequencies that do not correspond to the phytolith stratigraphy also suggest a local phytolith record. In archaeological or geological sites the presence of pine pollen but not pine phytoliths suggests aerial transport of the pollen with no local decay, whereas presence of both phytoliths and pollen indicates near-site growth of pine trees.

The appearance of phytoliths and other biogenic silica found in sites, such as diatoms and sponge spicules, may suggest how far they have traveled. Phytoliths recovered from Panamanian rock shelter deposits as old as 9000 years and almost certainly representing *in-situ* deposition typically have clear, relatively unabraded surfaces. In contrast, phytoliths isolated from sections of geological sediment cores possibly derived from riverbank deposits (see Chapter 8) were noticeably more abraded and fragmented. The presence of phytoliths from the Podostemaceae, a family of aquatic plants that live in rapidly flowing water, provided independent data that soils were partially fluvial in origin. In the same fashion, sponge spicules that have blown into sites will often be fragmented and eroded whereas in primary swamp deposits sponge spicules are often complete with unetched surfaces (Piperno, 1985c).

The Resistance of Phytoliths to Destruction

The utility of phytoliths as a paleoecological tool also depends on its stability in soil environments. In Chapter 2 some factors were discussed that influence the rate of solid silica dissolution and, hence, phytolith stability (Wilding *et al.,* 1977). Some of these factors are: (1) soil pH (values approaching pH 9 and above tend to rapidly accelerate dissolution), (2) iron and aluminum absorbed to silica surfaces (these seem to protect phytoliths from dissolution), (3) other characteristics of the soil environment (contexts protected from weathering processes such as rock shelters and lake sediments may create more favorable preservation conditions), (4) phytolith surface area, a function of particle size and three-dimensional structure (generally speaking, the greater the surface area, the more rapid the dissolution), (5) the presence of occluded carbon (greatly retards dissolution), and (6) the particular taxon that is silicified.

Differential preservation of plant taxa will be a significant factor affecting the interpretation of phytolith assemblages. Wilding and Drees (1974) and Bartoli and Wilding (1980) have provided experimental verification for differential solubility of temperate zone tree opal and opal of grass origin by means of cold-water and hot-water dissolution in the laboratory. What this seems to indicate for natural situations is that some kinds of tree phytoliths will be less well represented in fossil soils (in proportion to their true past abundance) than will, say, grass phytoliths. Based on my experience with tropical tree litter and A horizon soils, it should *not* indicate that tree phytoliths will be rarely isolated from fossil soils or that forested vegetational associations should not present identifiable phytolith assemblages. The subject of phytolith representation in modern soils underneath tropical rain forest will be discussed fully in Special Topic 2 at the end of this chapter.

Using information compiled from studies of modern and fossil phytoliths we can at this time offer some broad assessments of phytolith durability in various taxa. On one side of the spectrum we find phytoliths formed by a complete or almost complete silicification of cells. Some of these include grasses, sedges, some ferns, palms, members of the Marantaceae, Podostemaceae, Bromeliaceae, Moraceae, Urticaceae, Acanthaceae, Musaceae, Burseraceae, Dilleniaceae, and Annonaceae, among many others. We can expect that such phytoliths will not be subject to much dissolution. At the other and less durable side of the spectrum, we may call into question the ability of many dicotyledon epidermal and hair cell phytoliths to survive in a proportion representative of their true abundance in vegetation, given that very often they are formed as incrustations of cell walls and, hence, are much more soluble. On the other hand, the particular soil environment of sites may ameliorate (or exacerbate) the degree of dissolution and, therefore, this problem must be resolved on a site-by-site and region-by-region basis.

Other Factors Affecting Phytolith Representation in Depositional Environments

Vertical Displacement of Phytoliths

There have been a few commentaries by nonphytolith specialists suggesting that phytolith illuviation or downward movement in stable soils may be a serious, phytolith-specific problem (Dunn, 1983; Starna and Kane, 1983). The authors of these papers unfortunately have not benefitted from sufficient practical experience with phytolith extractions from soils and correlations with other site data to offer judgments on phytolith movement. Just as seriously, they are not aware of or have not carefully evaluated the very substantial body of evidence accumulated over the last 30 years showing that phytolith illuviation is not a problematic factor.

Numerous temperate-zone pedological profiles have been quantitatively analyzed for phytolith content by soil scientists and botanists (e.g., Beavers and Stephen, 1958; Jones and Beavers, 1964a,b; Norgren, 1973). Results and their implications have also been thoroughly reviewed by Rovner (1986b). In all of the soil studies that measured phytolith distribution with depth, phytoliths were concentrated in surface (A) horizons and phytolith quantities decreased dramatically with depth, so that in uppermost B horizons opal content was virtually nil. In the very few cases where phytolith content was substantial at significant depths below the surface (to 30 in.) (cited by Starna and Kane, 1983 as demonstrating that illuviation is a problem) the nature of loess accumulation at the site, not illuviation, was cited as the probable cause (Jones and Beavers, 1964a).

The stability of phytoliths in vertical profiles is further indicated by their use as an "index mineral" for the presence and location of buried A horizons in paleosols. A major criterion used by pedologists for identifying buried A horizons is the abundance of phytoliths found in them, whereas layers immediately above or below display a depauperate phytolith content (Beavers and Stephen, 1958; Dormar and Lutwick, 1969), a phenomenon hardly possible if phytoliths were to any significant degree moving up and down in soils.

The vertical stability of phytoliths has been amply demonstrated in archaeological contexts, especially sites from the humid tropics whose soils were subjected to intensive weathering and leaching. In numerous deposits from a range of site types, occupations, and time periods extending back 23,000 years, phytolith distributions in soils that showed no visible sign of disturbance or mixing consistently displayed the following characteristics (Piperno, 1983, 1984, 1985b–d):

1. Culturally sterile contexts stratified immediately underneath artifact-bearing deposits were devoid or virtually devoid of silica. Cultural levels just above yielded considerable quantities of phytoliths.
2. In deeply stratified rock shelters, there were marked shifts in phytolith assemblages corresponding to changes in the nonphytolith cultural inventory, such as the introduction of ceramics or the appearance of different types of ceramics. In one rock shelter whose 60-cm deep deposits were extremely compacted, preceramic levels showed a paucity of cross-shaped phytoliths whose sizes were very small, while ceramic-bearing levels just above were associated with a seven-fold increase in cross-shaped phytolith frequency, their sizes being significantly larger and characteristic of maize.
3. The absolute phytolith quantity in site deposits peaks when cultural materials such as ceramics and stone tool debitage are found at their highest numbers; therefore, phytolith abundance is correlated with the intensity of human activity at sites.
4. Correlations of phytolith with pollen and macrofossil assemblages demonstrated close agreement in the types and frequencies of taxa and inferred vegetational associations.
5. Phytolith assemblages were discrete across stratigraphic boundaries, consistent within stratigraphic boundaries, and showed little sign of scattered haphazard movement characteristic of intrusive or mixed particles.

Under what circumstances could phytolith movement result in displaced intrusive fossil assemblages? Substantial phytolith movement will take place in the same situations that paleobotanists have long been aware of; bulk soil movement or mixing either by natural processes or human agents, and other serious postdepositional disturbances of sites by root systems, burrowing biota, etc. The probability that such postdepositional events have occurred can be evaluated in the usual ways by looking for signals of mixed assemblages, such as intrusive cultural materials or suspect C14 dates. Rovner (1986b) succintly sums up the case for phytolith stability: "Vertical movement cannot be ignored, but it is a non-issue warranting no special attention. It is certainly no invalidation of phytolith analysis in archaeology."

Modern Phytolith Assemblages

The processes that underlie the production, taxonomy, dispersion, and preservation of phytoliths interact in extremely complex and yet very poorly understood ways to create what we know as fossil phytolith assemblages. Often we cannot even begin to evaluate many of the factors in the past that may have contributed to variability in plant species and plant population production, dispersion, and solubility. However, it is possible to ameliorate considerably the difficulties of interpreting past phytolith records by circumventing the intermediate effects of these processes and studying only their end product. We achieve this goal by constructing modern phytolith spectra from the surficial or "modern" soils underneath different kinds of plant communities. Birks and Gordon (1985, p. 142) state: "Studies of modern pollen preserved in modern soils provide, in many ways, the simplest and certainly the soundest and most repeatable approach to the reconstruction of past plant populations and communities as well as paleo-environments." We consider that such studies relating modern phytolith assemblages to their parent plant communities will equally improve the accuracy and reliability of phytolith-based paleoecological reconstructions.

Modern or surface samples can be taken from a variety of depositional contexts; terrestrial soils, lake muds, river and streamborne sediments, peats, and swamp mucks. Studies of modern phytolith assemblages in relation to contemporary vegetation can be carried out at several spatial scales, from local collections within one small tract of forest or grassland to regional situations in which vegetation covering thousands of square kilometers is systematically sampled. Sampling technique may vary and comprise small-volume (but perhaps regionally significant) sediments taken from the mud–water interface of a large lake or a long series of terrestrial soils taken along transects through several contiguous, clearly defined, and widely spaced vegetation communities. As Caseldine and Gordon (1978) have noted, transect data is roughly comparable to fossil data obtained from a core sequence, relating the spatial variation in communities as expressed by the transect data to the temporal variation as seen by stratigraphic changes in the fossil record of cores.

A minimum goal of modern phytolith studies would be to isolate and define modern phytolith spectra that correspond with known, contemporary patterns of vegetation and climate. If close relationships exist between extant vegetation and the soil phytolith assemblages that result from it, then comparisons can be drawn between modern phytolith assemblages from known vegetation types and fossil assemblages, and the latter identified on the basis of overall similarity to modern profiles. This is called the modern comparative or analog approach to the interpretation of fossil phytolith data (*sensu* the palynological usage of modern pollen spectra cf. Wright, 1967; Birks and Gordon, 1985). The comparative approach utilizes the entire phytolith assemblage including minor plants that do not form dominants of the vegetation but whose presence may be ecologically significant. The major plant formations (and by inference climatic conditions) that occupied the region surrounding sites are then more accurately reconstructed.

Modern phytolith studies can also help define the geographical and vegetational scales at which phytolith data most closely correspond with and, therefore, characterize vegetation. In addition, more specific relationships between phytolith assemblages and vegetation that refine the accuracy of fossil phytolith data can be elicited from modern phytolith spectra. The durability of phytoliths from individual taxa can be accurately measured. Relative representation of phytolith types in soils in proportion to the abundance of parent taxa in standing vegetation, a function of both production and solubility, can be evaluated. The significance of overrepresentation and underrepresentation of plant taxa in phytolith records is of considerable importance in estimating the true abundance of particular plants in past vegetation.

In cases where adequate forest or other plant inventory data is available, it would be possible to generate *R* (representation) factors, also called correction factors (*sensu* the palynological terminology), to estimate more directly the abundance and population size of taxa in past vegetation. Here, the percentage frequency of plants in phytolith records are compared to actual frequency in vegetation, and correction factors for modern and fossil assemblages are computed accordingly. Some critical limitations affect the use of correction factors in comparing modern and fossil assemblages. It must be assumed that an assemblage is derived entirely from a known source area whose spatial and vegetational characteristics are invariant across time. Obviously, these assumptions are rarely if ever satisfied except on a very local scale. We have already discussed some of the problems involved in defining source areas, and furthermore, plants are seldom evenly distributed throughout an arbitrarily selected source area in the present let alone in the past.

The interpretation of past vegetation and climate from modern analogs relies on the principle of uniformitarianism, which holds that the processes that operate in the present also operated in the past. However, it is well established that during certain periods of the past the vegetation and climate conditions in many parts of the world were unlike any known today, a situation that seriously impairs the use of the comparative method. In such cases systematic and detailed modern studies can still provide invaluable information on the distribution of indicator species, plants that individually or collectively have known ecological requirements or habitat preferences, and whose presence in fossil assemblages indicates certain characteristics of plant cover and/or climate (Ritchie and Cwynor, 1982; Lamb, 1984).

In many regions of the world, the application of phytolith analysis to paleoenvironmental reconstruction is still not very much past an embryonic level of development. Thus, it is hardly surprising that detailed descriptions of modern phytoliths and calculations of their frequencies in various vegetational formations and climate zones are rarely encountered. In archaeology such studies have been mainly limited to the upper horizons of soils above primary occupation zones of archaeological sites or located at sufficient distances from occupied zones to offer a noncultural natural vegetational picture. Pearsall (1981a), Pearsall and Trimble (1984), Carbone (1977), and Piperno (1985b,d) have used phytolith assemblages from these contexts to evaluate the significance of vertical phytolith movement and establish modern, baseline phytolith spectra from which to assess the significance of fossil data and identify past vegetation.

Evaluations of modern soils by soil scientists and botanists engaged in paleoecological phytolith research have been made primarily on a gross quantitative basis (Jones and Beavers, 1964a,b; Verma and Rust, 1969). Phytoliths isolated from A horizons underneath present-day forest or grassland were grouped and expressed quantitatively as a percentage of the total soil weight, which then indicated whether grassland or forest cover had dominated the soil's history. In many cases, fine silt soil fractions (5–20 μm), which contain substantial numbers of distinctive silica bodies, were not analyzed and no attempt was or could be made to provide percentage frequency diagrams and, therefore, phytolith assemblages characteristic of various environmental zones.

What follows is such an analysis of modern soils taken underneath mature tropical vegetation from the Republic of Panama where detailed information on species composition, distribution, and abundance is available.

Special Topic 2: Phytolith Representation in Tropical Forests: Comparison of Modern Phytolith Spectra with Mapped Vegetation

Barro Colorado Island (BCI), Panama, is a 1600-hectare wildlife preserve located in Gatun lake, a freshwater reservoir formed by the construction of the Panama Canal in 1914. The island has been a biological preserve since 1923; no logging or agricultural activity has occurred since that time. The Smithsonian Tropical Research Institute maintains a research station on BCI, and studies of tropical forest dynamics and community ecology have greatly benefitted from the long-term intensive research conducted by Smithsonian and visiting scientists (Lee, *et al.*, 1982).

The potential vegetation of the area is classified as tropical moist forest. Annual precipitation averages about 2500 mm, and there is a long and marked dry season from January to April when less than 500 mm of rain is received. During this period, many species in the forest shed their leaves. Therefore, using the classification of Beard (1944) the area can be considered semi-evergreen forest.

Forests of different age occur contiguously on the same soil type on BCI. The northeastern part of the island was used for agriculture (sugar cane, banana, and cacao) shortly before its establishment as a preserve. Forests in this area average between 65 and 130 years old, with some parts as little as 35 years old (Foster and Brokow, 1982; R. Foster, personal communication, 1985). The southwestern end of the island has much older forest which has been the subject of intensive research by botanists (Lee *et al.*, 1982) and archaeologists (Piperno, 1986a). It is a mature-looking, species-rich association which Foster and Brokow (1982) and other botanists felt had been little disturbed since the Spanish Contact, except for the selected felling of trees such as mahagony (*Swietenia macrophylla*) and tropical cedar (*Cedrela odorata*). Archaeological investigations within a 50-hectare mapped and censused area of old forest, the focus of the modern phytolith studies, uncovered abundant prehistoric materials but no evidence of historic-phase occupation. ^{14}C dates from the archaeological middens indicated that prehistoric groups abandoned the area about 500 years ago (Piperno, 1986a). The forest indeed was very ancient and provided an excellent context in which to evaluate the phytolith record of a humid lowland tropical forest.

The Sampling Site and Strategy

The sampling site lies in a 50-hectare area called the Hubbell–Foster plot, on the top of BCI, on level terrain at an elevation of 163 m above sea level. Every tree and shrub in the plot with diameter at breast height (dbh) of at least 1 cm was identified, mapped, and censused, with recensusing occurring every 5 years (Hubbell and Foster, 1983, 1987). The species composition, approximate abundance, and relationships of other woody (liana) and herbacious vegetation are also well known (Croat, 1978; Lee *et al.*, 1982). Detailed information on the plant cover permitted very accurate comparisons between the phytoliths in the standing vegetation and in the soil beneath. In addition to the more general goals of constructing modern phytolith assemblages from tropical moist forests, it was possible to determine which kinds of silica were or were not remaining in soils because of differential solubility, which taxa were over- and underrepresented, and by precisely how much in relation to standing cover and biomass.

Samples for phytolith analysis were taken at 20 m intervals along two transects which covered three distinct vegetational units: mature forest on fairly well-drained silty clays, a swamp formation on clays seasonally inundated by a stream, and a small tongue of younger, perhaps 100-year-old forest (R. Foster, personal communication, 1985) protruding into the old forest in the northeastern end of the plot (Fig. 6.4). Samples for phytolith analysis were taken as soils, because they more accurately predict the nature of fossil phytolith assemblages after liberation of siliceous particles from plant matrix and dissolution have occurred.

Details of the vegetation of these units can be found in Croat (1978), Foster and Brokaw (1982), and Hubbell and Foster (1983, 1987). Briefly, in species number, composition, and structure, the forest is typically a semideciduous association. The flora is composed of species characteristic of a wide array of climatic zones, wetter to drier, while most, including the most abundant, are largely restricted to the intermediate zones neither wholly wet nor wholly dry that BCI typifies. In the Hubbell–Foster plot there are approximately 318 species of trees and shrubs, with an additional 200 species of lianas, vines, and herbs. There are nearly 60 species of trees over 20 cm dbh in a single hectare. Canopy height is mostly between

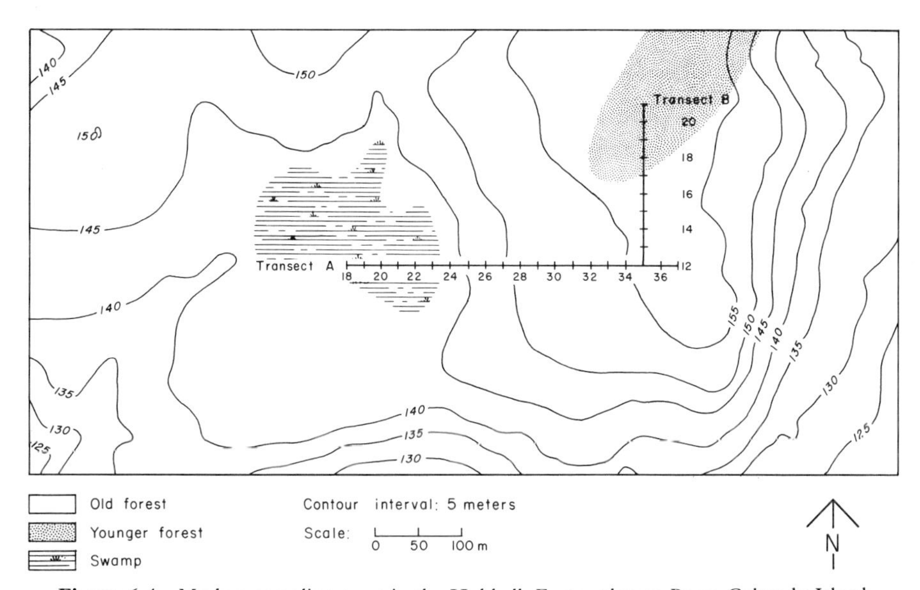

Figure 6.4 Modern sampling areas in the Hubbell–Foster plot on Barro Colorado Island.

30 and 40 m with occasional individuals reaching 50 m. Herbacious undercover is sparse and comprised of species not well adapted to high temperatures, intense solar radiation, and moisture stress, such as the bamboos *Chusquea* and *Pharus*, the forest grasses *Streptochaeta* and *Rhipidocladum*, the Marantaceae, Zingiberaceae, and various ferns. Weedy-type Panicoid, Chloridoid, and Festucoid grasses that inhabit open and disturbed cover are virtually nonexistent, even in the tongue of younger forest.

The upper 1 cm of soil was taken every 20 m along 320 linear m of Transect A and 180 linear m of Transect B, each sample being comprised of 5–7 trough-fills from a 5 × 5 meter area located due west (Transect A) or due north (Transect B) of each sampling stake (Figure 6.4). The strategy was designed to recognize, in so far as possible, the vegetational variation occurring across significant amounts of space, as represented by the three distinct units, as well as microdifferentiation within each unit caused by differential occurrence and/or clusters of species created by tree fall gaps in the older and younger forest.

Phytolith identifications were made by comparison to a modern phytolith collection comprising over 1000 species from 70 different families. For many species of silicon-accumulating plants, seeds, flowers, nonleaf vegetative structures, and replicate samples from different individuals were analyzed; 250 or 79% of the tree and shrub species and approximately 35% of the over 200 species of lianas, vines, and herbs occurring in the plot were extracted. Trees not included in the phytolith collection were mostly rare species, being represented by less than 5 individuals in the 50-hectare area. Many of the lianas, vines, and herbacious taxa not examined can be identified on the basis of taxonomic affinity as silicon nonaccumulators and are expected to leave no phytolith record, although future analysis of these taxa is planned.

Constructing the modern collection entailed long-term intensive research carried out over 7 years. Results describing production and taxonomy of phytoliths have been discussed in part in Chapters 2 and 3. Some additional comments are included here. The sheer number of phytolith shapes contributed by tropical vegetation will tax the memory of the phytolith analyst no less than they do the palynologist; 109 or 44% of tree and shrub species can be characterized as silicon accumulators. Of these, 36 or 14% produce distinctive phytoliths, defined as being characteristic of plants at the family level or below. In almost every case discrimination below the family level seems possible. The future identification of several peculiarly shaped unknown phytoliths found in modern soils but not yet in modern plants will add to this inventory (Plate 93). About 50% of rainforest arboreal species, therefore, will be silent in phytolith assemblages, while some not producing taxa-specific phytoliths may be assignable to either growth habit types or limited groups of taxa. Approximately 36% of the liana, vine, and herbacious species analyzed are silicon accumulators with over half of these producing distinctive silica. The herbacious phytolith record is highly diagnostic of a moist, dense forest.

General Characteristics of the Soil Phytolith Record

Tropical forests contribute an abundant amount of durable phytoliths to soils. Though many individual species are silent, the tropical forest formation certainly is not. In Figs. 6.5 and 6.6 percentage frequencies for phytoliths isolated from all transect soil samples are given. Figure 6.7 is a composite phytolith diagram of the tropical forest achieved by averaging values from all samples. Such a profile averaging phytolith deposition over a broad but still mainly extra-local source area is probably more representative of phytolith input into lakes and other sites of fossil studies.

The phytolith assemblages are characteristic of and define a tropical forest and may even be associated with *moist* versus dry or deciduous forest. Seven different taxa of tropical trees comprising 5 genera of arboreal dicots (*Guatteria, Licania, Unonopsis,* etc) and 2 subfamilies of palms are identifiable by their phytoliths. The composition of the herbacious phytolith record also typifies a forest environment. Grasses are predominantly represented by the remains of *Chusquea, Pharus,* and *Streptochaeta,* all shade-tolerant taxa. The percentage of grass phytoliths from open-ground weedy kinds of Gramineae that produce the familiar saddle, dumbbell, cross-shaped, and hat-shaped forms is extremely low. The dumbbells and saddles that do occur are probably from *Rhipocladum* and *Streptogyne,* two forest grasses that produce these kinds of silica bodies, respectively. The Marantaceae, which are present as herbacious plants in the understory, also contribute significant inputs into phytolith spectra, and *Trichomanes,* a fern confined to moist forest habitats, is present in low frequencies. Such taxa as *Heliconia* and members of the Compositae, plants much more characteristic of disturbed, open environments, are present in very low frequencies, as they are in the vegetation.

A few more general phytolith categories were also defined which indicate not individual taxa but the prevalence of arboreal, nonherbacious cover. These categories are sclereids and celtlike edges. Out of 46 species in the comparative collection producing silicified sclereids, 37 or 80% are trees and the rest are lianas (4) and vines (5). Out of 8 species producing phytoliths that look like celts, four were trees, three lianas, and one a vine. These two categories, especially the sclereids, contributed a significant input to the phytolith record and can be taken as additional indicators of closed vegetation.

Phytoliths of several other categories were frequently present in soils. Polyhedral and anticlinal epidermis can be derived from many dicotyledons, arboreal and herbacious, and a few ferns. Presumably, many of these in the BCI soils were a product of arboreal vegetation, but equal amounts may conceivably be contributed in situations where secondary growth herbacious dicots such as composites have invaded an area. Therefore, in situations where vegetation is being reconstructed from fossil profiles, they cannot be interpreted as independent markers of arboreal taxa.

Solid chunks of square to rectangular and bulliform cell-like phytoliths were

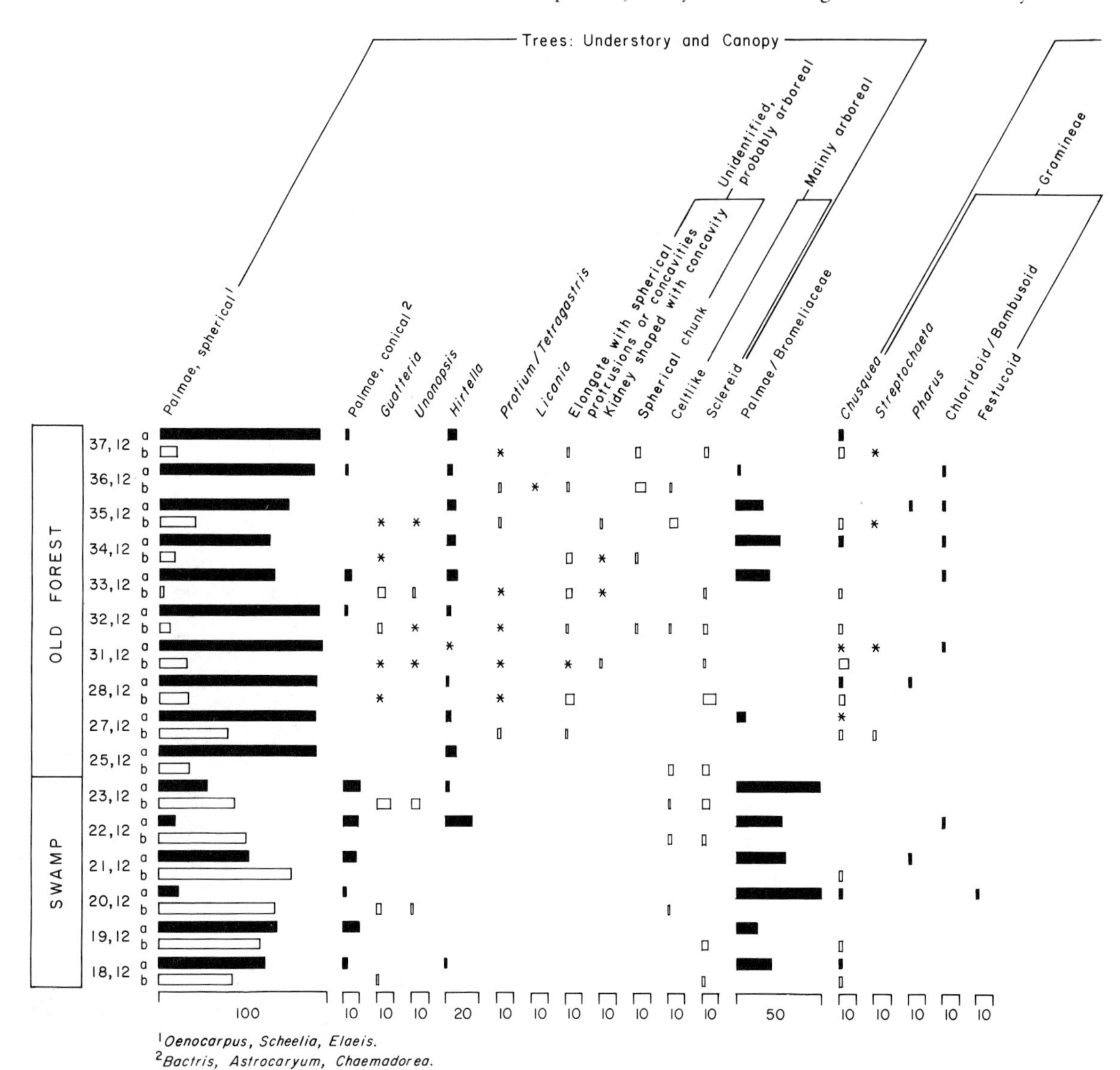

Figure 6.5 Percentage of phytolith frequencies from the modern soil transects in the Hubbell–Foster plot, transect A. a, Fine silt fraction; b, coarse silt fraction; * = <1%.

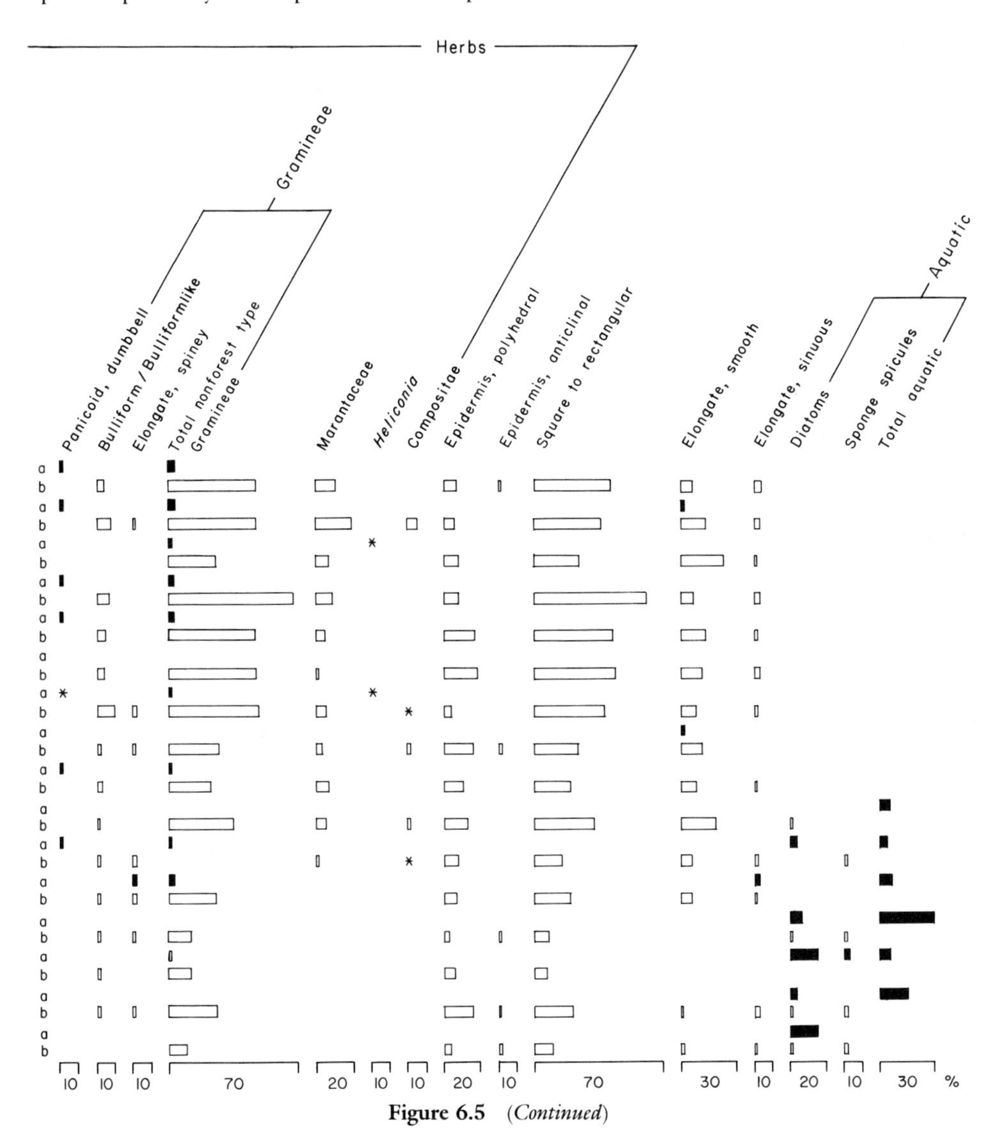

Figure 6.5 (*Continued*)

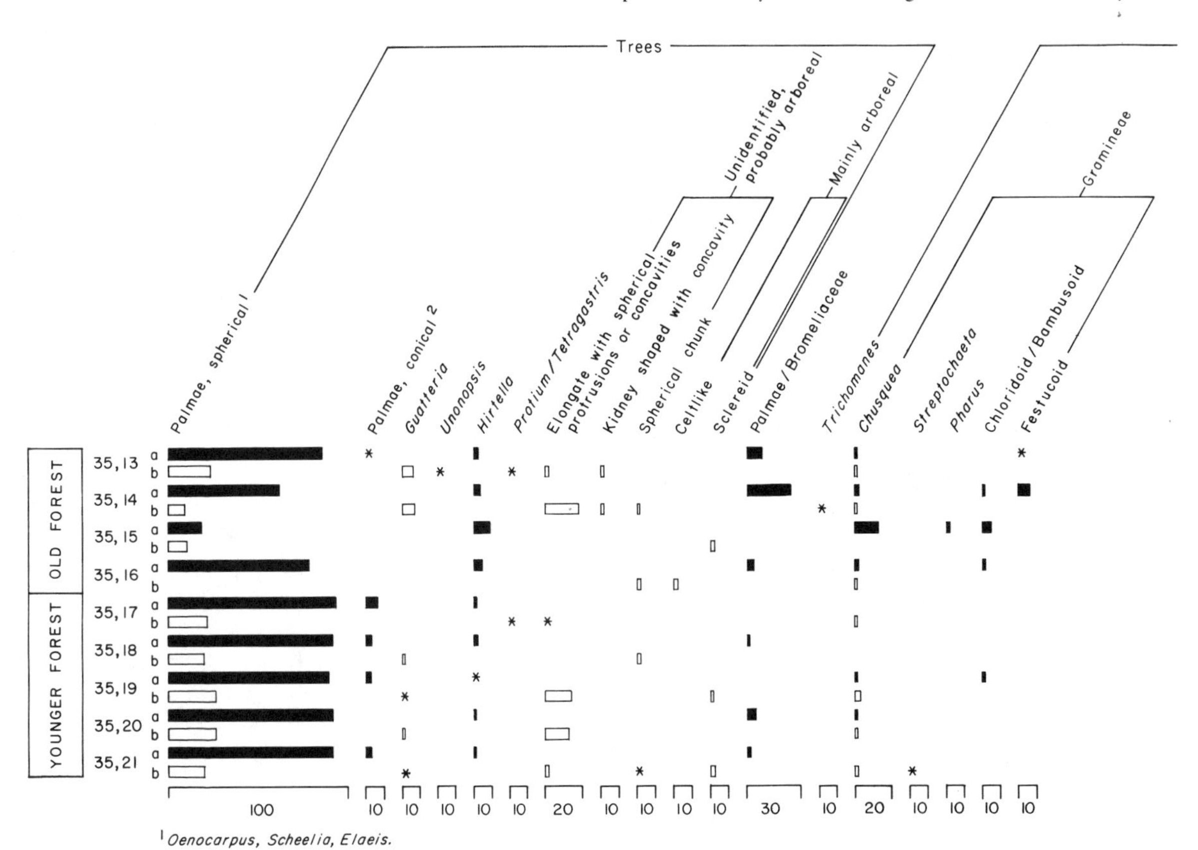

Figure 6.6 Percentage of phytolith frequencies from the modern soil transects in the Hubbell–Foster plot, transect B. a, Fine silt fraction; b, coarse silt fraction; * = <1%.

commonly isolated from soils. It is important to point out that although these types are often associated solely with the Gramineae, a small number of woody and herbacious dicotyledons, including the Piperaceae and Chrysobalanaceae, produce them in fairly substantial amounts. Because grass frequency is so low in the BCI vegetation, a significant number probably came from woody vegetation. In fact, values of less than 50% characteristic of BCI soils may mark the presence of forested or closed plant cover in fossil profiles. If the square to rectangular and bulliformlike phytoliths are not included in the Gramineae sum, the total percentage of phytoliths attributable to arboreal species in the fine silt soil fraction averages

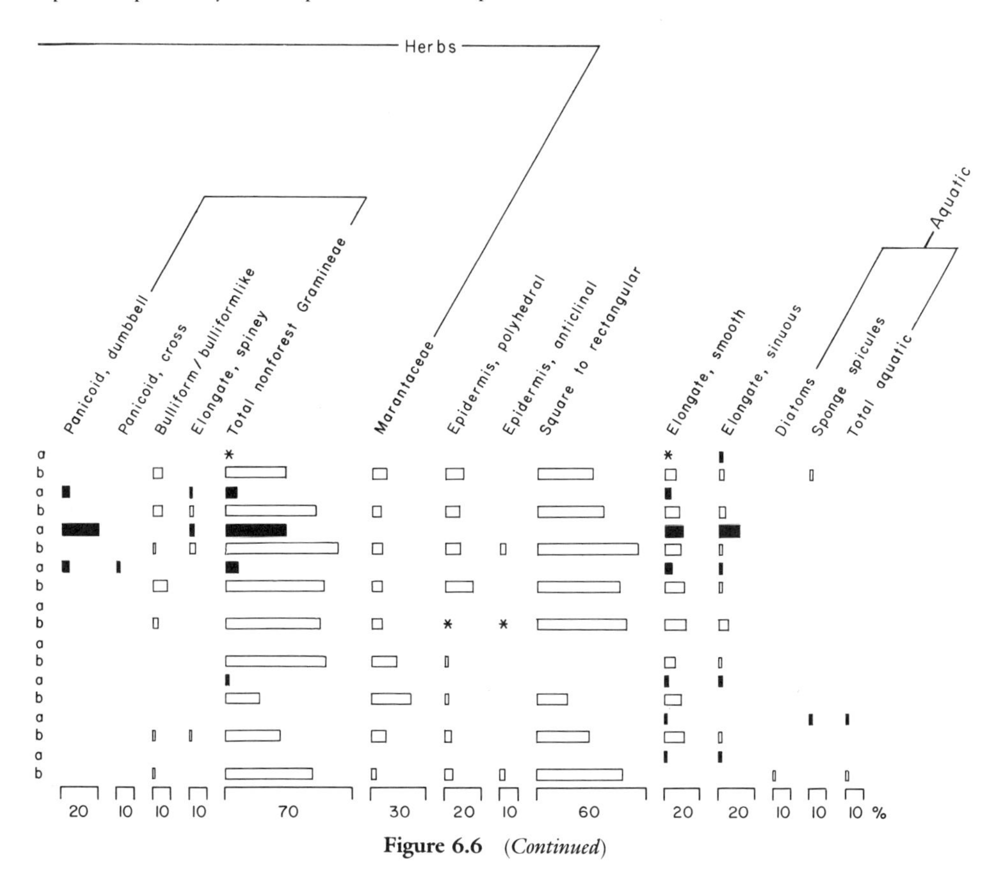

Figure 6.6 (*Continued*)

to 78%, and in the coarse silt it averages to 30%. In contrast, herbacious plants comprise only 2.5% of the fine silt and 7.4% of the coarse silt phytolith assemblages.

In summary, the phytolith assemblages can indicate little else but dense tropical forest. Similar analyses of soils underneath mapped deciduous forest in Guanacaste province Costa Rica (Hubbell, 1979) are underway. When comparisons between moist and dry forest phytolith assemblages are made, it may be possible to differentiate between the two types, and therefore general climatic as well as vegetational conditions, on the following basis: (1) the presence of different kinds of forest grass phytoliths (*Chusquea* and *Pharus* were not observed growing in the dry forest; *Streptochaeta* and *Olyra*, another distinctive phytolith producer, were), (2) different percentage frequencies of weedy-type grass phytoliths (during the dry season in

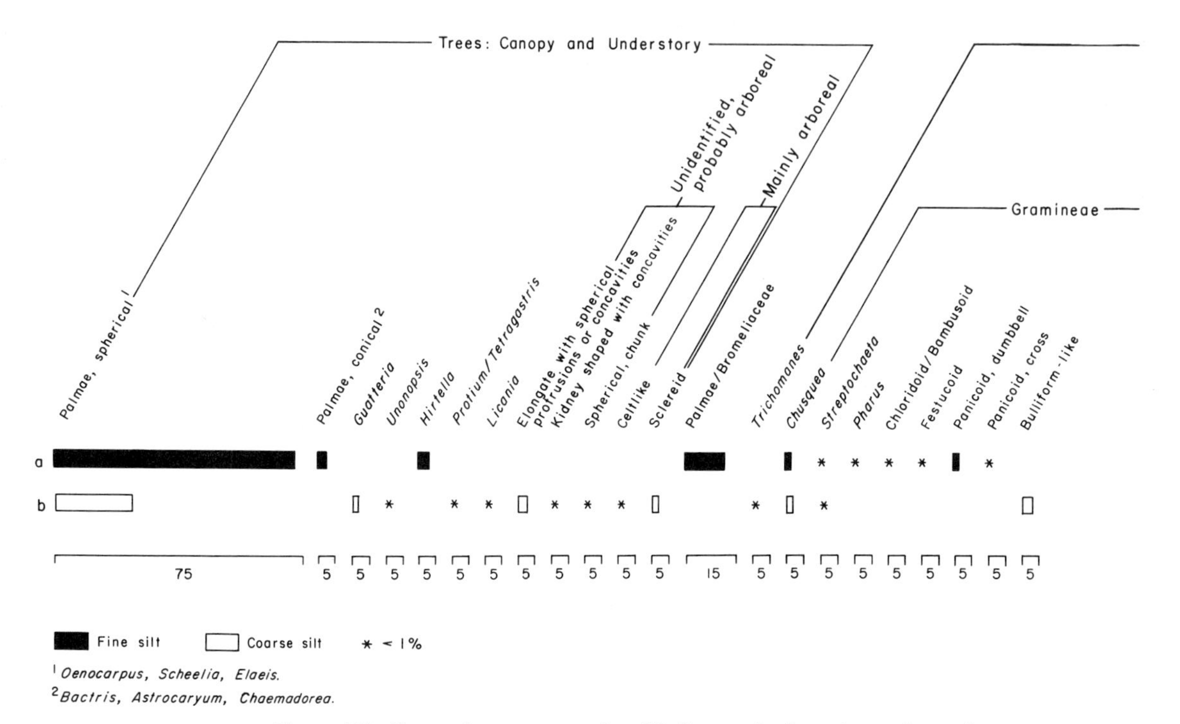

Figure 6.7 Composite percentage phytolith frequencies from the modern soil transects.

deciduous forests significant areas underneath the canopy may be covered by such Gramineae), (3) significantly lower percentages or the absence of palm phytoliths (palms are rare in Costa Rican dry forests), and (4) different kinds of phytoliths from trees (*Tetragastris*, *Guatteria*, etc. are largely restricted to intermediate climates producing semi-evergreen associations; dry forests may yield their own distinctive tree silica or assemblages).

Phytolith Assemblages across Boundaries

Major discontinuities are seen in the phytolith assemblages taken from the three vegetational units. The swamp formation is demarcated by points along the coordinates from 23,12 to 18,12. In this zone only plants such as the terrestrial Bromeliad *Aechmea magdalenae* and the oil palm *Elaeis oleifera* grow and small understory palms such as *Bactris major* are found in dense clumps. Many trees

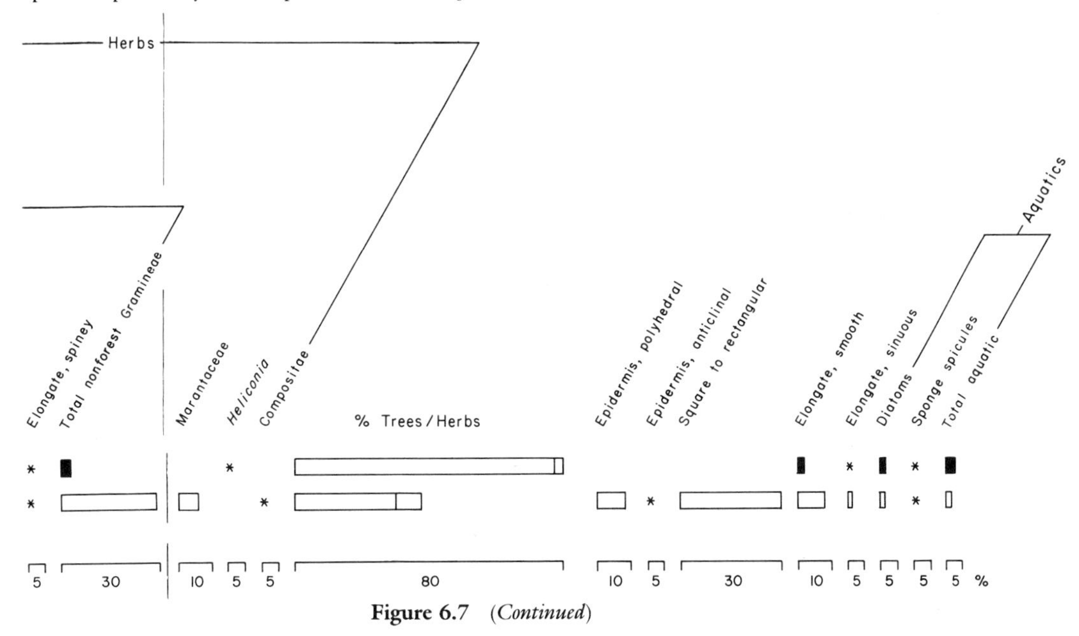

Figure 6.7 (*Continued*)

common in the upland forest decline in frequency or are absent. These changes are amply reflected in the phytolith record (Fig. 6.5). Percentages of Palmae/Bromeliaceae-type phytoliths increase greatly, representing the terrestrial Bromeliad *Aechmea,* a dominant in the swamp. *Elaeis* (oil palm) phytoliths, recognizable by their larger sizes and markedly asperical shapes, appear here also. Dense stands of *Bactris major* are reflected by the large increase in conical-shaped palm phytoliths. The trees *Protium* and *Tetragastris* disappear from the phytolith record as they do in the vegetation of the transect area. Finally, the appearance of biogenic silica from aquatic species marks the poorly drained soils of the swamp vegetation.

Discontinuities between the old and young forest are also apparent, though much more subtle. The tongue of younger forest protruding into the northeast corner of the Hubbell–Foster plot (Fig. 6.4) is, in species composition, very similar to the old forest, but is still marked by dense stands of successional trees such as *Gustavia superba,* tangles of lianas, and lower numbers of old forest trees. Its phytolith record, from coordinates 35,17 to 35,21, is marked by an apparent reduction in *Guatteria, Tetragastris,* and *Protium* phytoliths, corresponding to the decreased densities of these trees in the vegetation. Marantaceae phytoliths also appear to be better represented. Though higher sample counts for these less abundant taxa would be required to verify these conclusions, it appears that the younger forest displays subtle differences in its phytolith record. Also, though no quantitative analysis was carried out, phytolith content generally seemed to be lower in the coarse silt fraction of younger forest soils, reflecting decreased frequencies of mature forest taxa contributing large phytoliths.

Microdifferentiation within Boundaries and Phytolith Movement

Microdifferentiation within vegetation boundaries is visible in the phytolith record and much of it appears to relate to the distance of silicon-accumulating taxa from sampling points along the transects. It is therefore possible to measure the extent of horizontal phytolith movement across the forest landscape. Four samples showed a drastically reduced absolute number of spherical palm phytoliths in the fine silt fraction, and because they dominate many fine silt assemblages, a markedly reduced fine silt phytolith content. In three out of four cases, the nearest contributing palm, *Oenocarpus,* to the sampling point was 25–30 m away. In one case an individual of *Oenocarpus* was only 3 m away from the sampling quadrat but still did not influence the phytolith content of fine silts. All other samples produced considerable amounts of spherical palm phytoliths and here individuals of *Oenocarpus, Scheelia* or *Elaeis* were located anywhere from a few meters to up to 20 meters from sampling locations.

The soil sample of coordinate 35,15 exhibited far greater proportions of *Chusquea* and grass short cell phytoliths than others. This is attributable to its position in the middle of a gap created by a tree fall. In addition, soils from an area only 1 × 1 square were collected because of interference from the fallen tree. Soils taken at localities 20 m from either side of 35,15 exhibited the typical tropical forest phytolith spectra and the Festucoid-type and *Pharus* silica bodies present in sample 35,15 disappear from the assemblages. Sample 35,15 perhaps more than any other demonstrates how localized the phytolith record can be.

It appears that on a level surface in a tropical forest setting, most phytoliths do not move more than 20 m from their primary depositional loci. Such evidence underscores the localized nature of the phytolith record and for the first time includes quantitative data on the relationship between modern phytolith spectra and standing vegetation. With regard to fossil phytolith spectra from geological sediments, however, the BCI results are probably very much of an idealized situation. For reasons discussed earlier in the chapter, catchment areas of the past would probably derive a considerable phytolith input from source areas further than 20 m.

Phytolith Solubility

The extent and nature of phytolith solubility can be accurately judged by comparisons of phytolith production in standing vegetation to the kinds of silicified forms recovered from soils. There are two major classes of phytoliths produced in copious amounts by species common in the vegetation that are never or rarely isolated from soils. Therefore, they must be extremely susceptible to dissolution in tropical forest soils. The two classes of nondurable phytoliths are hair cell and hair base phytoliths. Not a single hair cell and only one hair base phytolith were recovered despite prolific production of these types in the leaves of several commonly occurring species such as *Poulsenia armata, Sorocea affinis,* and *Cordia* spp. The major barrier to preservation seems to be a silicification process involving only cell walls, which leaves the interior of cells hollow. Completely silicified hair cells and bases (with silica filling up both cell wall and lumen) are rarely encountered in modern plant specimens, and unfortunately it appears that this requirement must be met for some plants to leave durable and well-represented silicified remains in tropical soils.

This problem is further exemplified by the fact that although a plethora of tropical forest species were found to produce polyhedral and anticlinal epidermis, virtually the only such phytoliths isolated from soils were solidly silicified versions. In contrast, solidly silicified plugs and other chunks of silica from palms, grasses, sclereids, and many other plants are well preserved in soils. There is some evidence to indicate that tropical forests may be particularly harsh depositional environments adversely affecting the solubility of more fragile kinds of phytoliths. Year-round high soil temperatures, heavy rainfall, and intensive soil leaching are all contributing factors, but perhaps most importantly, tropical soils contain high amounts of tannins, secondary compounds derived from the breakdown of vegetal matter that are known to dissolve silica. We may surmise, therefore, that soils underneath forests of different climatic zones may prove more hospitable to the long-term survival of epidermal, hair cell and hair base phytoliths; and even in tropical regions, lake and archaeological deposits may offer more favorable conditions of preservation.

Some other kinds of phytoliths found in modern specimens were not encountered in soils, though it is difficult at this point to ascertain whether poor preservation or limited occurrence of phytoliths and/or species in the vegetation is the problem. For example, cystoliths were isolated in small numbers from several species of *Ficus,* not one of which attained significant numbers in standing vegetation. Here, uncommon representation of both the phytolith type and the producer species may explain the blind spot in soils. A few species of uncommon trees not occurring in transect areas (*Licania hypoleuca, Sloanea terniflora*) produced abundant numbers of highly distinctive forms which were either rarely or not found in soils.

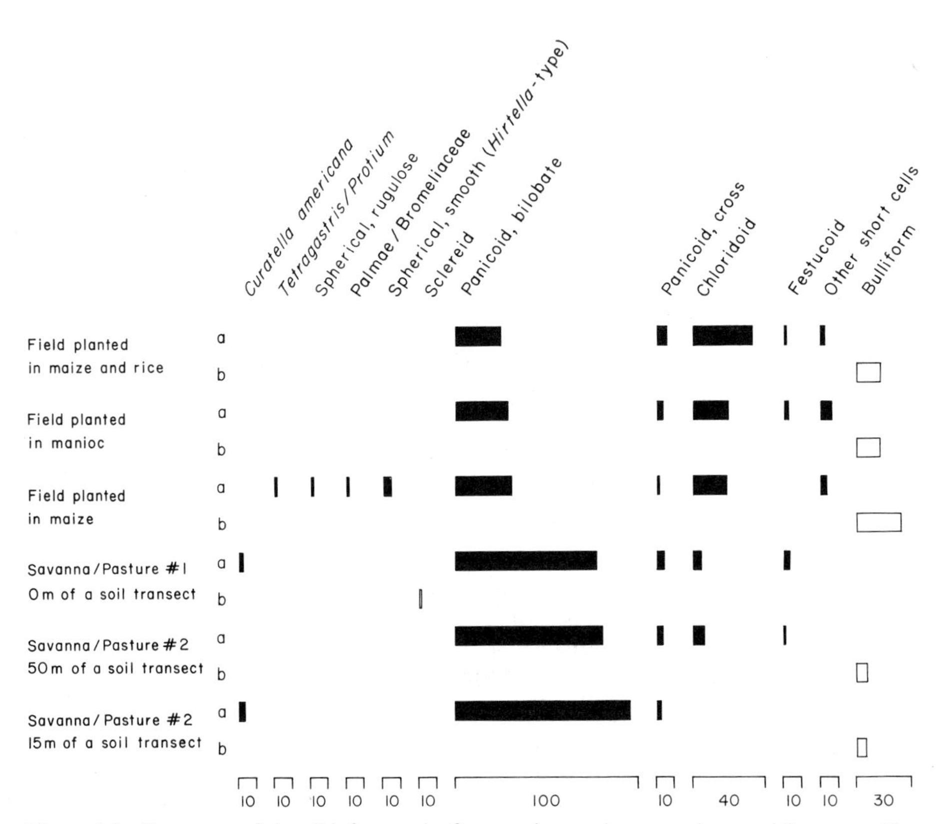

Figure 6.8 Percentage of phytolith frequencies from modern environments in central Panama. a, Fine silt; b, coarse silt; *, observed. Compositae categories: A, abundant in sand fraction; R, rare in sand fraction.

In these cases, limited horizontal movement of a phytolith from a tree poorly distributed across the landscape, not preservation, may be the problem. Soils taken from immediately underneath such trees will be analyzed to evaluate these possibilities.

Phytolith Representation

There are many species in which a considerable degree of over- and under-representation quite obviously occurs, and effects could be seen even if detailed

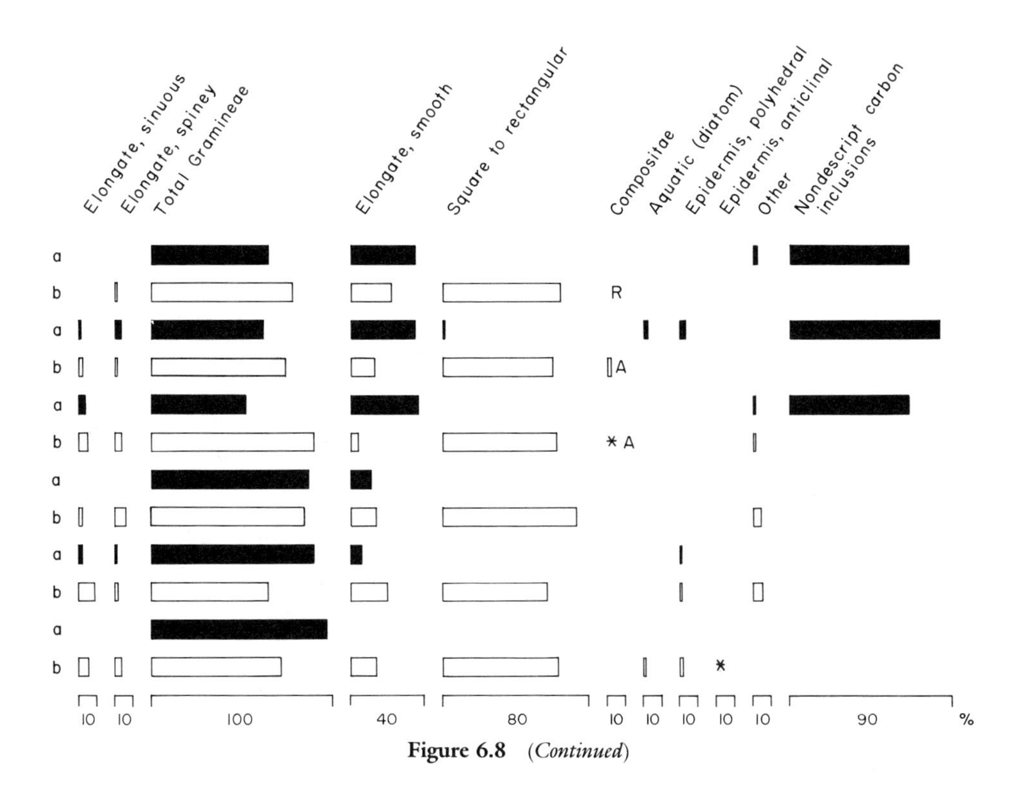

Figure 6.8 (*Continued*)

forest inventory data were lacking. Spherical phytoliths from two species of palms contributed an average of 72% of the fine silt and 23% of the coarse silt content in a flora containing over 600 species of vascular plants. In contrast to the abundant evidence left by palm trees, the three most common arboreal species (*Hybanthus prunifolius, Faramea occidentalis,* and *Trichilia tuberculata*), constituting 33% of the tree cover, are nonsilicon accumulators and therefore completely silent.

Results of the forest censusing project allow some quantitative estimation for the degree of overrepresentation of some palms. The two species contributing spherical phytoliths in old and young forest, *Oenocarpus panamensis* and *Scheelia zonensis,* have a total of 1858 individuals (over 1 cm dbh) in standing vegetation, amounting to only 0.8% of the tree and shrub cover. Their representation in modern (and, by implication, fossil phytolith spectra) is a gross exaggeration of their true abundance in vegetation. Other phytolith-producing taxa, such as the trees *Guatteria dumetorum* and *Unonopsis pittieri,* leave silicified remains in numbers that more closely approximate their true proportions in vegetation. They account for 1.2% and less than 1%, respectively, of the coarse silt record, while comprising 0.7% and 0.32% of the tree cover.

In conclusion, phytolith assemblages from soils of tropical moist forest exhibited close relationships to the standing vegetation, hence corresponding patterns (Webb, 1974) between modern phytolith spectra and modern plant cover have been established. The entire assemblages, with their dominance of forest taxa and paucity of weedy, open, or seasonally open-cover herbacious phytoliths, characterized a dense, tropical forest environment. The comparative approach to the identification of fossil assemblages should therefore be of great value in phytolith analysis. In cases where past assemblages have no modern analogs, the diversity of phytolith morphology should lead to the identification of various indicator species, those with known ecological preferences and habitats. The presence of several shade-loving tropical grasses, for example, in fossil assemblages presupposes the past presence of forested environments even if the entire assemblage cannot be utilized by comparison to modern spectra. These kinds of analyses need to be carried out on many vegetational formations occupying different climatic zones of the world. Phytolith analysis as a paleoecological discipline will advance significantly in maturity when the requisite modern studies have been made.

A few additional examples of modern phytolith assemblages from Panama can be offered. Figure 6.8 shows phytolith percentage frequencies from three agricultural fields planted in (1) maize and rice, (2) manioc, and (3) maize. Weedy Gramineae, Compositae, and nondescript phytoliths with occluded carbon (a product of plant burning) are abundant, while many other phytolith classes typical of forest or non-field herbacious plant cover are absent or present in low frequencies. Figure 6.8 also shows phytolith frequencies from soils in Central Panama beneath a mosaic of pasture and anthropogenic savanna. The vegetation cover is dominated by an introduced African Panicoid grass, *Hyparrhenia rufa,* with significant proportions also of fire resistant dicotyledons such as *Curatella americana, Byrsonima crassifolia,* and *Xylopia frutescens.* Three samples were analyzed, two taken 35 m apart along a vegetation transect.

Phytolith spectra are dominated by Panicoid phytoliths from *Hyparrhenia,* with small proportions of other taxa, including *Curatella,* a major indicator species for this kind of severely affected plant cover in the seasonal tropics. *Curatella* phytoliths should then be represented in fossil profiles and serve as markers of severe anthropogenic disturbance. The two samples taken 35 m apart show very similar phytolith profiles. *Curatella* is not represented at the 50-m mark, presumably because it occurred along the transect only from 0 to 30 m. Once again, we have a situation where on level terrain (and in this case in an extremely wind-blown open environment) the extent of phytolith movement beyond 20 m was insufficient to influence the composition of phytolith diagrams.

In Panama, three distinct vegetation units, forest, agricultural fields, and savanna/pasture, all yielded three distinct, identifiable phytolith assemblages. The comparative approach to paleoecological reconstruction holds much promise in the interpretation of fossil phytolith assemblages.

7

The Role of Phytoliths in Archaeological Reconstruction

Introduction

Archaeobotanists have accumulated considerable insight into the potentials, problems, and significance of phytoliths recovered from archaeological strata. An impressive number of research goals have been addressed and many specific, important issues in prehistoric human behavior and ecology are understood to be very amenable to resolution, partially elucible, or infractile to study through phytolith analysis. This chapter recounts this research and evaluates, so far as can be evaluated now, the role that a mature phytolith discipline, one attended to archaeological problems on a worldwide scope with years of experience and very large accumulations of data, may come to assume.

Research carried out in the Old World, North America, and the New World tropics has shown that phytoliths are capable of providing an independent avenue of data and interpretation for four major areas of archaeobotanical studies: (1) the origins and dispersals of domesticated plants and development of agricultural systems; (2) the availability, economic usage, and non-economic roles of wild plants; (3) the nature of environments and environmental modification associated with past human occupation of sites; and (4) the relationship between technology, economy, and social organization.

The phytolith analyst will be able to (and at any rate should endeavor to) go beyond mere identifications and catalogs of plants isolated from sites and offer sound reconstructions and insights into the dynamic interactions between prehistoric peoples and their plant world. The manipulation and transfer of plants beyond natural ranges, culturally induced changes in the genetic composition of plants, patterns of plant husbandry and agriculture, and changes in species availability in natural and anthropogenic plant communities may all be discernible in phytolith records. Chronological depth will be great because of the stability of plant opal in soils and detailed site-by-site comparisons of a regional sequence are made possible for the same reason. The following major areas of research are sufficiently investigated to offer research summaries and evaluations.

Prehistoric Agriculture and Agricultural Technology

The New World

It is fitting that when archaeobotanists reopened investigations of archaeological phytoliths during the 1970s (after a hiatus of over 50 years) they did so by examining silica bodies from the leaves and floral bracts of major cereal crops. It was precisely this area that had attracted the attention of German botanists in the first third of this century during the botanical period of phytolith research. Working with ash samples and ceramic vessels from the Near East and China sent to them by archaeologists, they were able to isolate and identify the silicified remains of wheat, barley, and rice.

In nonarid parts of the world ash heaps are not conspicuous components of archaeological middens; therefore, large articulated pieces of silicified plant tissue are not as easily recovered. Here, the main body of the phytolith record will be isolated from soils as discrete, disarticulated shapes. Pearsall (1978, 1979) first showed that disarticulated phytoliths from maize were identifiable in archaeological soils from the New World tropics. She focused her studies on a single phytolith type, the cross-shaped form, and demonstrated that maize cross shapes were significantly larger than those of the native Panicoid grass cover of southwest Ecuador. She then isolated cross-shaped phytoliths from soils associated with the Valdivia-phase occupations of southwest Ecuador, radiocarbondated from 3000 B.C. to 2300 B.C., and found that their size distributions indicated the presence of maize.

Pearsall's original study (1978) used a four-tiered ranking system of cross-shaped phytoliths to identify maize (see Special Topic 1, Chapter 3). In sites from the American tropics, the maize identification procedure has been refined and expanded to include additional morphological and size criteria (Piperno, 1984; Piperno and Starczak, 1985, Special Topic 1, Chapter 3). Three-dimensional structures for cross-shaped phytoliths are determined, based on eight categories called variants; short-axis measurements are tabulated in micrometers, and mean size values for each three-dimensional Variant in samples are determined. Mean percentages of cross-shaped phytoliths, out of all dumbbells and cross shapes combined, are tabulated (see Special Topic 1 for details of this analysis). The result is numerical information on cross-shaped phytoliths from an unknown archaeological population, and the object is to determine if they can be identified as deriving from maize or wild grasses.

When modern maize and tropical Panicoid grasses were compared as two groups, comprising all maize races combined versus all wild species combined, it was found that mean values for each character were quite different, but their distributions overlapped, sometimes considerably. On the basis of one character we could not identify unknown cross-shapes with any degree of accuracy from archaeological samples as belonging to one or the other of the two populations. On the basis of intuitive observations and comparisons we could judge that no wild grass combined attributes present in many maize races, and so when cross shapes bearing all of these attributes were isolated from Panamanian archaeological soils they were identified as maize (Piperno, 1984; Piperno *et al.*, 1985). A more formal statistical analysis was obviously desirable and multivariate analyses were undertaken. A discriminant function showed that modern populations of maize and wild grasses could be separated into two groups based on the measurements made. The details of this analysis were presented in Chapter 3, Special Topic 1.

Archaeological unknown cross-shaped phytoliths can be classified in the same manner. Tables 7.1 and 7.2 contain measurements and percentages of cross-shaped phytoliths isolated from various tropical sites in central Panama and southwest Ecuador comprising preceramic (7000 B.C.–3000 B.C.), Early Formative (3000 B.C.–1000 B.C.), and Later Formative (300 B.C.–A.D. 500) occupations (Piperno and Starczak, 1985; Piperno, 1986b; Pearsall and Piperno, 1986). Cross shapes are commonly found in soils where contributing grasses have decayed, making large sample sizes easily achievable.

The formula for calculating discriminant function values for archaeological cross-shapes, derived from a three-variable analysis of modern specimens, is

$$\text{D.F. value} = 0.8082\ (\overline{X}\ \text{size Var 1}) + 0.1025\ (\overline{X}\ \text{Size Var 6}) + 0.0215\ (\%\ \text{Var 1}).$$

Figure 7.1 shows the results of this analysis. It can be seen that many samples dating from the fifth millennium B.C. onward contribute cross-shaped phytoliths whose discriminant function values fall within the 95% confidence intervals about the means of modern maize, well outside of the area of maize/wild overlap. We can state with a high degree of confidence that such strata contained the remains of maize. Several D.F. values, including deposits ^{14}C dated between 7600 B.C. and 6500 B.C., fall clearly in the cluster for wild grasses, while many fall into the area of maize/wild overlap. In the latter case, it becomes difficult to place an identification on archaeological specimens, and indeed in some of the situations in Fig. 7.1 such as the Euadorean Valdivia I and Machalilla occupations, maize was probably present. Substantial contribution of wild cross shapes via decay of roof thatch and matting could easily mask maize occurrence. However, if we are going to err, it is important to do so on the conservative side in these matters by sometimes not detecting the presence of maize rather than identifying decayed wild grasses as maize.

Table 7.1
Characteristics of Cross-Shaped Phytoliths from Panamanian Archaeological Sites

	% Variant		$\overline{X}$ Size: Variant			*n*	Excavation
	1	6	1	6	% Cross shapes[a]	Cross shapes	leaders
Cueva de los Vampiros preceramic deposits; 6610 B.C.	95	5	10.7	10.3	12	50	Cooke and Ranere (1984)
Cueva de los Ladrones preceramic deposits; 4910 B.C.	70	25	13.5	11.7	20	100	Bird and Cooke (1978)
SE-189 preceramic deposits; 5125 B.C.	82	15	12.4	11.9	19	67	R. Cooke (personal communication, 1986)
Aguadulce early ceramic deposits; 2000 B.C.–1000 B.C.	80	15	13.8	12.4	18	83	Ranere and Hansell (1978)
Sitio Sierra 300 B.C.–A.D. 500	82	18	14.0	13.2	16	50	Cooke (1984)

[a]Cross and dumbbell sum.

Table 7.2
Characteristics of Cross-Shaped Phytoliths from Ecuadorean Archaeological Sites

	% Variant			*n*
	1	6	$\bar{X}$ Size: Variant 1	Cross shapes
Vegas, Site OGSE-80, preceramic deposits[a] 7600 B.C.–6220 B.C.				
Sample 1	44	54	13.3	50
2	44	54	12.8	50
3	40	54	13.5	50
4	44	56	12.2	50
6200 B.C.–4800 B.C.				
5	64	36	13.3	50
6	52	44	13.4	50
7	60	34	13.3	100
Real Alto - Early ceramic deposits				
Valdivia I; 3250 B.C.	48	34	12.4	50
Valdivia I; 3250 B.C.	73	23	12.8	30
Valdivia II	56	26	13.3	30
Valdivia III; 2350 B.C.	83	17	13.2	50
Later ceramic deposits				
Machalilla; 1500 B.C.	33	50	11.9	50

[a]Sites excavated by Stothert (1985).

The identification of archaeological maize phytoliths by multivariate statistics reduces dependence on intuitive inspections of archaeological assemblages in plant identification. (It does *not* mean that qualitative phytolith characteristics become less important, as we shall see shortly.) Phytolith evidence indicates an early introduction of maize into Lower Central America and South America. Discovery of maize pollen in preceramic deposits from the Cueva de los Ladrones, Panama (Piperno *et al.,* 1985) corroborated the identification of maize through phytolith analysis. Wild phytolith spectra obtained from early preceramic (7600 B.C.–6100 B.C.) deposits, when maize was most probably not present, show that wild cross-shaped producing grasses leave their own, singular profiles in fossil records that are differentiable from maize, and maize phytolith profiles obtained from sites such as Sitio Sierra, Panama, where macroscopic remains of maize were abundant, points to a close correspondence between these two methods of archaeobotany.

The percentage of cross-shaped phytoliths (of the sum of dumbbells and crosses) may also be calculated from archaeological samples and used in a five-variable discriminant analysis, whose results for modern grasses have been presented in Special Topic 1, Chapter 3. Figure 7.2 presents such an analysis from the Panamanian sites. We can see that although values from several sites, including late first millennium B.C. deposits from Sitio Sierra where macrobotanical remains of maize

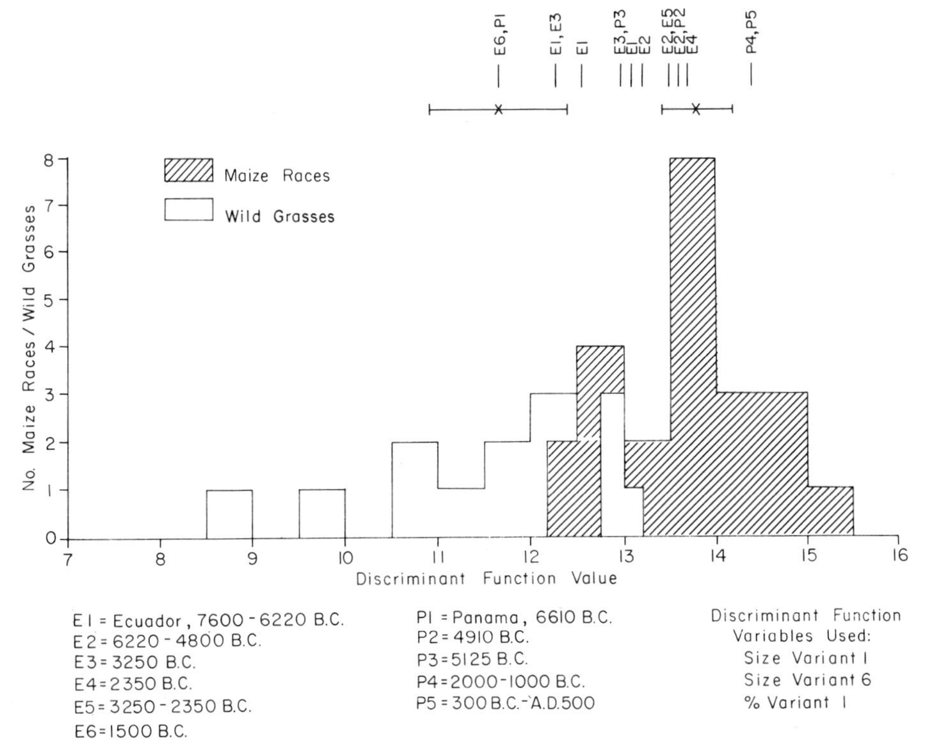

Figure 7.1 A three-variable discriminant function analysis of cross-shaped phytoliths from archaeological soils of Panama and Ecuador.

were recovered, do not overlap wild grass distributions, they are slightly too low to fall into the maize cluster. Cross shapes from 2000 B.C. deposits that were clearly classified as maize with the three-variable analysis, are classified as wild when the percentage of cross-shapes is included. Although the frequency of cross-shaped phytoliths in these deposits is quite high and definitely maizelike, there is apparently enough contribution from wild dumbbells to skew measurements and create an ambiguous picture. The three-variable analysis, which does not include the percentage of cross shapes, appears more sensitive to maize presence in archaeological soils.

This brings us to an important point relevant both to three- and five-variable analysis. In a significant number of cases maize, though present in archaeological assemblages, might leave a record statistically indistinguishable from wild grasses due to (1) substantial decay of wild cross-shaped producing species, (2) contribution by maize races with smaller-sized cross shapes and/or lower percentages of Variant

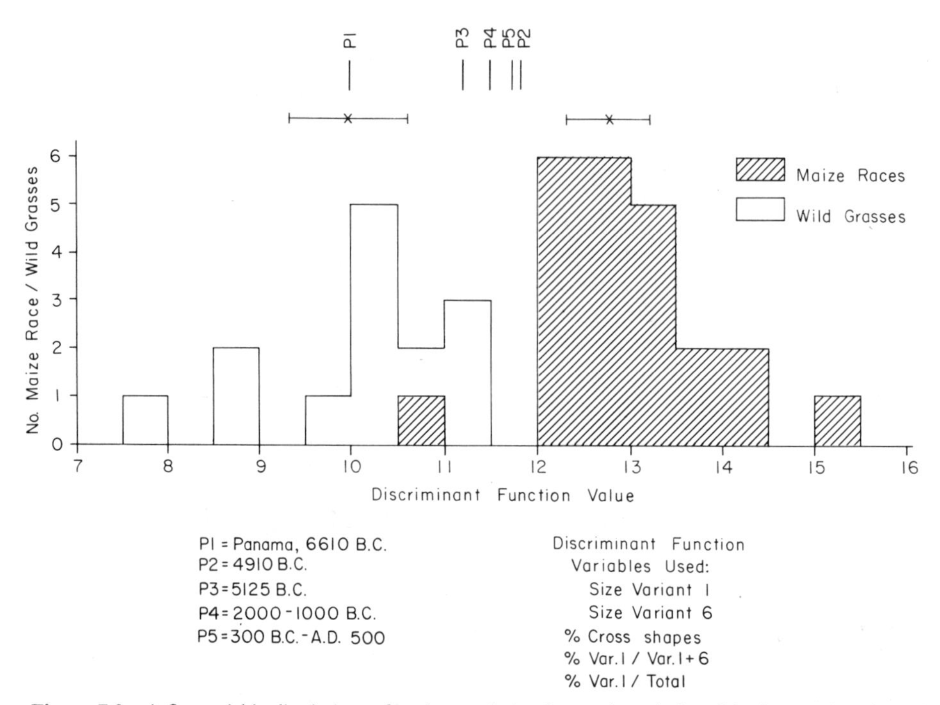

Figure 7.2 A five-variable discriminant function analysis of cross-shaped phytoliths from archaeological soils of Panama.

1 type crosses, and (3) contribution of maize largely through husk, tassel, or ear decay. In relation to the last possibility, we discussed in Chapter 3 how tassels, and especially husks, contribute significantly fewer cross-shape numbers, plus sizes and morphologies which may skew the spectrum to look more like wild grasses. Cobs and kernels produce no cross shapes at all. Indeed, the patterns found so far point to maize leaves as the contributing structures of archaeological cross shapes. Not a single husk-type cross shape was observed in soils, and sizes plus three-dimensional structures of phytoliths correspond to leaf origin and not husk or tassel origin; such a pattern argues for on-site cultivation of maize. It appears that a significant number of prehistoric Central and South American maize races present identifiable phytolith profiles, but relevant data from other regions of the New World are yet to be accumulated.

As corollary to the statistical tests of phytolith attributes, qualitative phytolith characteristics should be assessed to determine the taxonomy of archaeological cross shapes. As we have discussed in Chapter 3, certain cross-shaped producing tropical wild grasses also produce high frequencies of durable, apparently genus-specific phytoliths or forms found in a limited number of genera not including maize. They are *Oplismenus, Cenchrus, Andropogon,* and *Paspalum plicatulum*. In Panamanian deposits none of the distinctive forms isolated from these grasses were recovered. Grasses like *Olyra, Cryptochloa, Lithachne,* and *Tripsacum* can be ruled out on the basis of cross-shaped, three-dimensional considerations. Hence, it is reasonable to conclude that 7 out of 10 major cross-shaped-producing wild taxa *(Oplismenus, Cenchrus, Paspalum plicatulum, Olyra, Cryptochloa, Lithachne,* and *Tripsacum)* and the taxa producing large Variant 1 types *(Cenchrus, Oplismenus)* did not influence the composition of the archaeological assemblages. With archaeological representation from *Oplismenus* and *Cenchrus* excluded, the Panamanian site assemblages bearing discriminant function values between 12.8 and 13.1 may also be identified as maize, since these are the wild grasses creating overlap with modern maize in this region of the graph. Many of the dumbbells in the Panamanian deposits were Variant 1 in three-dimensional structure; and Variant 6 cross shapes were thin (3–6 μm), as in maize. Thus, the phytolith pattern is entirely consistent with the presence of maize.

However, the same cannot be said for Ecuadorean assemblages, where many Variant 5/6 bilobates characteristic of the genus *Cenchrus* were observed (Plate 33). In fact, *Cenchrus echinatus* may be implicated as the wild grass contributing large-sized Variant 1 cross shapes into pre-6000 B.C. deposits, whose D. F. values sometimes fell into the area of maize/wild overlap. In post-6000 B.C. deposits Variant 5/6 bilobates were also present; however, a drastic increase in percentages and continued large size of Variant 1 crosses (Table 7.2) indicated that maize was also contributing cross-shaped phytoliths. We thus have examples of two contrasting cases; one where the presence of a single cross-shaped producing grass—maize—is suggested, and the other where it appears that both a wild grass, probably *Cenchrus,* and maize left a phytolith record.

The productivity and economic importance of crop plants recovered from sites are usually difficult to evaluate. They are no less so in phytolith studies. In the case of maize, primitive popcorns, flint, and flour corn may all contribute high numbers of large, cross-shaped phytoliths, and so it is difficult to interpret the role of maize in prehistoric economies with phytolith evidence. In Panamanian sites, there appears to be a gradual increase in maize phytolith size over time (Table 7.1), which may be a result of changing maize races or the quantity of maize used. Much comparative and quantitative work must be undertaken before relationships between phytolith size, maize type, and maize productivity can be postulated.

However, the morphology of maize cross-shaped phytoliths isolated from early fifth millennilum B.C. deposits in Panama may offer some intriguing hints into the antiquity of maize in Mexico. We discussed in Chapter 3 how phytoliths from many teosinte populations, including the race Balsas, are predominantly non-Variant 1 in three-dimensional structure and small when compared to maize. By at least 5000 B.C., maize had developed to the point where cross shapes were predominantly

Variant 1 and phytolith size was large. It is by no means clear how long it would have taken for the much better understood genetic and macromorphological changes under domestication that resulted in the transformation of the teosinte spike into a primitive ear of maize let alone for postulated changes in phytolith morphology and size. However, these are issues that after proper study may provide valuable information on the antiquity of maize domestication. The analysis of phytoliths, macromorphology, and genetics of modern maize and teosinte in tandem should be initiated to generate information on the relationship between phytolith characteristics, the genetics of *Zea,* and the antiquity of maize.

North America has seen some extension of the maize identification procedure with Pearsall (1982) and Piperno (n.d.) examining deposits from Illinois and Delaware, respectively. The native grass cover of North America has not yet benefitted from extensive comparative studies with maize, so that accumulated evidence was interpreted in way of either ruling out the presence of maize in sites due to low cross-shaped content or suggesting possible maize presence by reference to phytolith size in related wild tropical taxa. Pearsall (1982) found that Archaic period strata from a multicomponent site in Illinois yielded cross shapes in very rare numbers whose sizes were quite small, whereas Late Bluff period samples contributed far more crosses with sizes characteristic of maize. Interestingly enough, several features from the later Mississippian occupation yielded extremely few cross shapes with medium size distributions. As Pearsall notes, the lack of evidence in the Mississippian structures, when maize is known to have been cultivated, may be because food storage was not a function of the features analyzed. It is also possible that contribution was by husk and ear rather then leaf decay, accounting for the paucity of the cross-shaped type.

Deposits from the Delaware Park site, Delaware (Piperno, 1986c), ^{14}C dated from 1850 B.C. to A.D. 640, contributed phytoliths in substantial numbers, but again, cross shapes were extremely rare and their sizes small. If deposits from north Central and eastern North America dated before the time of Christ continue to be characterized by such assemblages, phytolith evidence will indicate a late arrival of maize, supporting evidence derived from macrofossil analysis.

Squash (*Cucurbita* spp.) is another major domesticate whose presence is detectable in archaeological phytolith assemblages. The distinctive phytoliths from squash rinds first observed by Bozarth (1986) have been described in Chapter 3 (Plate 14). They have so far been isolated in large numbers from late first millennium B.C. deposits from Panama (Piperno, 1985b) and late eighth to early fifth millennium B.C. deposits from southwest Ecuador, where they are present in number from the bottom to the top of the preceramic Vegas culture typesite OGSE-80 (Piperno and Stothert, 1987). Studies contrasting wild and domesticated *Cucurbita* have not yet been initiated and wild *Cucurbita* has been described from around Guayaquil, Ecuador (Cutler and Whitaker, 1968); thus, it is not possible to determine if Ecuadorian preceramic squash represents cultivation activity. However, the conspicuous presence of scalloped phytoliths throughout the Vegas occupation suggests that *Cucurbita* was regularly exploited over several millennia by preceramic populations. The Vegas-phase *Cucurbita* remains represent the oldest evidence for the plant in South America.

Cucurbita is not naturally distributed in Panama; thus, fossil squash here indicates domestication. It is interesting that squash phytoliths first appear so late in central Panama, when maize was introduced so early, by the fifth millennium B.C. We have referred in Chapter 3 to the apparent vagaries in squash rind production of spherical scalloped forms, not being found so far in several modern varieties of *C. pepo* from Panama and exhibiting a similar notable capriciousness in southwest *Cucurbita* (Bozarth, 1986). The late appearance of *Cucurbita* in Panama may not indicate a late introduction of the genus itself but perhaps of different species moving as a complex with productive races of maize from South America. Before this suggestion can be properly evaluated, considerable work must be invested in comparing inter- and intraspecific phytolith production of *Cucurbita* rinds. Not a single squash-type hair cell phytolith (Plates 15 and 16) was recovered from rind-bearing deposits, indicating either nondeposition of squash leaves or, more probably, poor preservation of hair cell phytoliths.

Achira *(Canna edulis)* may be a third New World domesticate evidenced with phytolith analysis. Pearsall (1979) recovered chains of spiny phytoliths similar to those in modern achira leaves from Early Formative deposits in Ecuador but notes that further work on related taxa and ecological associates is required before the identification can be considered secure.

The New World has also seen research into phytolith distributions in prehistoric ridged and raised field systems. Remnants of ancient wetland agriculture taking the form of channalized and/or raised fields are widespread in the American tropics (Denevan, 1982). Their discovery has led to revision of traditional viewpoints on agricultural technology, carrying capacity, and population numbers (Hammond, 1978; Turner and Harrison, 1981, 1983). Varying opinions exist as to the function of fields with regard to what crops were grown, or if fields were used mainly for subsistence or cash crop production.

Soils of prehistoric fields analyzed from Vera Cruz, Mexico and the Calima Valley, Colombia (Piperno, 1985d, 1987b) demonstrated a very high phytolith content, and maize phytoliths were recovered in great numbers. Table 7.3 shows characteristics of cross-shaped phytoliths that exhibit the typical maize spectrum. Percentages of cross-shaped phytoliths from Vera Cruz are the highest yet reported in fossil assemblages, indicating that maize may have been a major crop grown in the field systems.

In both sites, bulk transport of soils from outside of the field system by prehistoric cultivators was ruled out (A. Siemens, personal communication, 1986; Bray *et al.,* 1985) and, hence, it could be stated with confidence that remains of crops in cultivation strata represented plants actually grown in fields and not blown

Table 7.3

Characteristics of Cross-Shaped Phytoliths from Vestiges of Prehistoric Fields in Mexico and Colombia

Mata de Chile, Vera Cruz, Mexico

Depth (centimeters below surface)	% Variant 1	% Variant 6	$\overline{X}$ Size: Variant 1	$\overline{X}$ Size: Variant 6	% Cross shapes[a]	% Cross shapes[b]	n Cross shapes
90–110	92	6	12.9	13.8	10	31	50

Calima Valley, Colombia

Depth (centimeters below surface)	Stratum	% Variant 1	% Variant 6	Cross-shape size: Small	Medium	Large	X-Large	Dumbbell to cross shape ratio	n Cross shapes
5–10	S1	88	5	32%	64%	4%	0	14:1	50
20–25	S2[c]	75	25	50	50	0	0	6:1	12
30–35	S2, maize	80	18	26	58	12	2	1.5:1	50
35–40	S2, maize	97	3	14	73	13	0	4.5:1	30
62–72	S4/1	83	17	38	52	10	0	4.5:1	40
72–82	S4/1, maize	75	25	28	53	17	2	2.4:1	40
92–100	S4/2	None observed out of 10,000 phytoliths scanned.							
100–108	S4/2	None observed out of 5,000 phytoliths scanned.							
120–131	S5	None observed out of 5,000 phytoliths scanned.							
140–149	S7	92	8	52	42	0	0	1:1	40
156–162	S7	None observed out of 5,000 phytoliths scanned.							
170–180	S7	None observed out of 2,000 phytoliths scanned.							

[a] Sum of all phytoliths.
[b] Sum of crosses and dumbbells.
[c] Stratum 2 is the prehistoric cultivation surface.

into or artificially incorporated into cultivation soil during field construction and maintenance. Such an origin of the cultivation till was thought to be a real possibility in the Pulltrouser field system of Belize and prevented Wiseman (1983) and Miksicek (1983) from firmly stating that their maize pollen and macroremains represented raised field crops.[1] Where such possibilities are considered unlikely, taphonomic characteristics of phytoliths make it possible to affirm that remains of plant domesticates represent the in-place decay of field crops. Phytoliths will also be of considerable value in locating and identifying remnants of prehistoric fields when other markers like channelization, soil mounding, and terracing are absent. Even in the Vera Cruz raised fields, a concentration of cross-shaped phytoliths at 90–100 cm below the surface helped to place the agricultural horizon in a soil profile where stratigraphy was difficult to discern.

At the Calima Valley site, column samples were analyzed from a 2 m deep cut made into a small group of ditched fields (Bray *et al.*, 1985) (Figure 7.3). Seven different strata were identified that provided details of the vegetation of the Valley over the last 23,000 years. It is pertinent here to discuss this sequence in some detail because it clearly (1) demonstrates characteristics of phytolith distribution in a soil profile spanning a demonstrably very long chronological period, (2) shows that phytolith profiles can be obtained with depth that are a function and reflection of vegetational changes in the study region, and (3) suggests that anthropogenic influence on vegetation can be distinguished from natural processes affecting the composition of floral communities. In the excavations, apparent prehistoric cultivation surfaces, defined by a soil darker gray in color with higher clay content, Stratum (S) 2, were underlain by soils resulting from a number of depositional and environmental events, including possible slopewash from the side of the valley (S4/1). Beneath this stratum, and considerably below the present day surface and water table, soils were uncovered containing large tree trunks and sizeable branches lying horizontally and vertically (S7). The upper surface of Stratum 5 yielded a radiocarbon determination of more than 23,000 years before present, thus placing Stratum 7 firmly within the late Pleistocene.

Figure 7.3 shows that phytolith assemblages from each strata reflect the different cultural and environmental episodes in the Calima Valley over the period spanned by the soil profile. Stratum 1, postabandonment soil, is marked by an abundance of phytoliths from the Panicoid grass subfamily, but the sizes of cross shapes are small and the dumbbell to cross-shaped phytolith ratio high (Table 7.3); thus, the presence of maize is not indicated.[2] The valley bottom pasture land is

[1]I believe that the water lily *(Nymphaea)* phytoliths reported by Wiseman (1983) from Pulltrouser swamp are calcium oxylate crystals and not secretions of silicon dioxide. *Nymphaea* species extracted by the author have yielded no phytoliths, in the sense that they are defined in this volume.

[2]The Calima Valley samples were analyzed before the variable percentage of cross shapes was substituted for the variable dumbbell to cross-shape ratio. Use of the latter is discussed in detail in Piperno (1984).

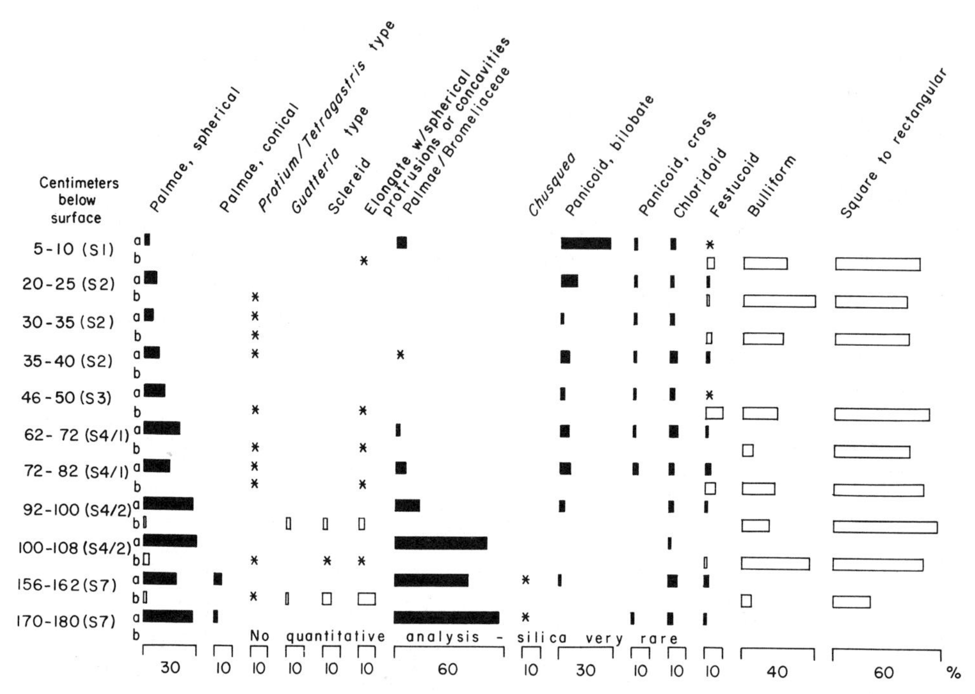

Figure 7.3 Percentage phytolith frequencies from the Calima Valley, Colombia. a, Fine silt fraction; b, coarse silt fraction; *, observed.

today planted in *Panicum purpurascens,* an introduced African species, which no doubt contributed many of the Panicoid grass phytoliths. Stratum 2, thought to be the prehistoric cultivation soil, indeed contained abundant phytoliths of maize. Cross shapes from the top of S2 at 20–25 cm were too small to identify as maize and this, coupled with the still high (6.3: 1) dumbbell to cross-shaped phytolith ratio, indicated that superficial levels of S2 were a mixture of S1 and S2 soils. Low dumbbell-to-cross ratios in the remainder of S2 soils are clear evidence that old cultivation strata were not infiltrated with more recent material.

It can be seen also in S2 that Chloridoid, Festucoid, and Panicoid bilobate types contribute a very small input to the phytolith record, far smaller than values from contemporary maize and manioc plots in Panama, where percentages of Chloridoid grasses range from 18 to 32%, and of dumbbells from 24 to 35% (Fig. 6.8). It appears that Calima agriculturalists allowed only a very small weed component to exist in the fields, an activity that probably required a high labor investment.

Shifts in frequencies of biogenic silica from diatom and sponge spicules also provide a picture of environmental conditions, especially as they relate to soil

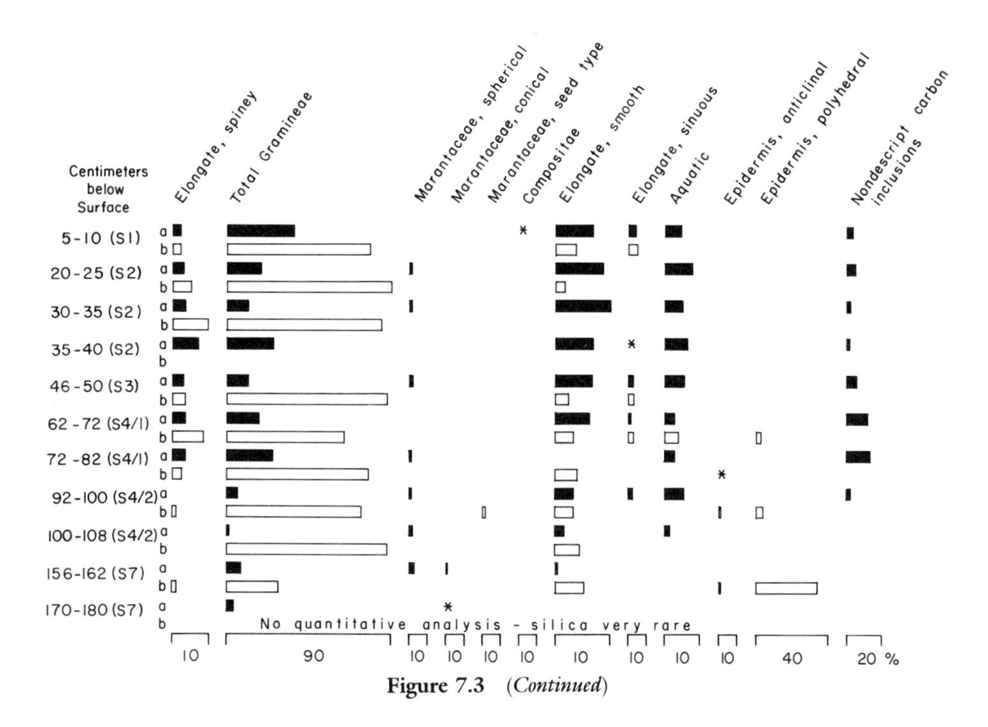

Figure 7.3 (*Continued*)

drainage. Levels of aquatic biogenic silica in S2 levels are higher than in post-abandonment and modern S1 strata, revealing the poorly drained nature of the valley bottom when it was converted into fields. Stratum 4/1, deposited before the valley bottom was artificially drained, was thought by pedologists to represent slopewash from the side of the valley, a conclusion also pointed to by the phytolith record, which registered (1) a drop in levels of aquatic silica, suggesting that soils were in part derived from better-drained upland contexts, (2) the presence of maize phytoliths after a brief hiatus in S3, and (3) a sizeable increase in silica having occluded carbon, believed to be the remains of burnt vegetation and unlikely to have been derived from the swampy valley bottom. In combination, these features support a hillslope origin of S4/1 strata and suggest that hillsides were cleared, burned, and planted in maize before the valley bottom was converted into fields.

Another abrupt change in the phytolith record occurred in stratum 7, whose deposits were marked by the presence of large tree trunks and big branches lying horizontally and vertically throughout the unit, and whose age is greater than 23,000 years. Soils below the top of S7, at 156–180 cm, indeed presented phytolith spectra entirely typical of undisturbed humid tropical forest. They are very similar to patterns found in modern soils underneath mature forest on Barro Colorado Island, Panama (Fig. 6.7) and can be identified by reference to these modern assemblages in accordance with the modern analog approach to fossil vegetation studies. Weedy-type grass levels are low, palm levels high, and phytoliths from *Chusquea,* the forest bamboo, and other forest taxa appear for the first time. Yet another indication of the emergence of a well-drained, fully terrestrial environment is the disappearance of aquatic biogenic silica from diatoms and sponge spicules. It is significant that the fossil vegetation of stratum 7 provides no hint of climatic aridity in the Calima Valley because paleobotanical records from elsewhere in Colombia also indicate that Late Pleistocene climates started to dry out after 20,000 years ago (Van der Hammen, 1974). It is demonstrated from this sequence that, in isolation, phytolith data provide reliable and highly informative reconstructions of plant usage and vegetation and can serve as time–stratigraphic markers in soil profiles whose chronological history is long.

The Old World

The Old World has seen fewer systematic studies of archaeological phytoliths oriented toward prehistoric agricultural development. Miller (1980) and Miller-Rosen's (1985) studies with wheat and barley in the Near East have been benchmark works and can be used as the standard references for future investigations. She initially carried out comparative analyses on modern emmer and einkorn two-rowed wheat and barley and developed a preliminary classification system for the identification of glume phytoliths from these plants. The taxonomic criteria have been illustrated and discussed in detail in Chapter 3. In way of a brief summary, they rest on the configuration of the glume epidermal long cells, which are generally squared or pointed in barley and rounder or smooth in wheat (Fig. 3.4), as well as the glume microhair and short cell spacing.

Because phytoliths from Near Eastern sites are often multicelled sections of silicified epidermal tissue virtual casts preserving the surface features of once living plant structures are formed, permitting identification of wheat and barley using the characteristics just described. The frequent formation of multiple silicified cells of epidermal tissue in some regions of the world may itself be a character of paleoecological significance, as we shall see later. In two contemporary Chalcolithic sites from Israel dating to the late fifth millennium B.C. Miller-Rosen (1985) identified wheat and barley glume phytoliths. Macrobotanical remains of the two crops had already been identified at one site, Horvat Beter, where wheat seeds predominated.

Wheat phytoliths also dominated at Horvat Beter, while at the other site, Shimrat, phytoliths of barley clearly prevailed. Miller-Rosen proposed that edaphic conditions determined the different crop mixtures present; moister, less-saline conditions characteristic of habitats surrounding Harvat Beter are more favorable to wheat, while the drier habitats near Shimrat would favor the growth of barley.

Miller-Rosen cautions that more modern comparative material must be analyzed before wheat and barley phytoliths are independently identified in sites; however, she has demonstrated that three important issues in Old World agricultural development are amenable to resolution with phytolith data; the domestication and spread of wheat and barley, the determination of crop mixtures at sites, and the relation of crop mixtures to the nature of environments in the immediate vicinity of sites.

Miller (1980) also drew attention to the frequent occurrence of multicelled sections of silicified leaf epidermis in sites from certain regions of the world, such as the Near East and American tropics. Such phytoliths are formed when the "atypical" grass cells (long and stomatal cells) are commonly silicified, (see the discussion in Chapter 2, pp. 17–18.) It has been discussed in Chapter 2 how the nature of the growing environment may affect the total content of solid silica in those plants that accumulate it. This appears to result from either soil conditions that raise the level of soluble silica available for plant uptake, or increased levels of plant transpiration that intensify the silicon precipitating mechanisms. Experimental studies have shown that the uptake of silica by plants increases with increasing water content and temperature of soils (McKeague and Cline, 1963a). Thus, it is expected that warm, wet environmental regimes would favor increased silica content in plants, possibly accounting for the heavy amounts of multicelled sections of phytoliths found by Miller (1980) in archaeological sites in Belize. She found the same to be true in sites from the Nile Valley, Egypt, where transpiration promoted by the hot, arid environment may be the implicating factor. It may be possible, then, to detect the nature of soil hydration and rainfall or drainage patterns prevalent during site occupation by assessing the numbers of multicelled sections of silicified cells present in archaeological soils. As Miller has suggested, if natural environmental conditions in regions can be held constant, it may be possible to distinguish irrigation from dry-farming agriculture.

Negative phytolith evidence has been shown to be important in assessing the age of macrobotanical findings. Wendorf *et al.* (1979) reported the recovery of barley and wheat cereal grains from a buried hearth associated with Late Paleolithic contexts in Egypt, dated between 18,300 and 17,000 radiocarbon years B.P. If true, this would challenge many major assumptions about agricultural origins in the Near East. Wendorf then sent Tack and Kaplan (1986) soil samples from the site, from which they isolated phytoliths in large quantities, but none having characteristics of wheat or barley. Subsequent accelerator dating of the grains showed them to be about 4800 years old (Wendorf and Schild, 1984) hence, the negative phytolith record accurately registered the absence of the crops from Late Paleolithic contexts.

Preliminary phytolith studies have been carried out on an early agricultural site in the highlands of Papua, New Guinea (Wilson, 1982). Wilson sought to investigate agricultural origins at the Kuk site, proposed on the basis of evidence for human drainage of the swamplike terrain to have housed food productive economic strategies for at least 9000 years (Golson, 1977). Of the three major domesticates thought to have been introduced into the region, taro *(Colocasia esculenta)*, yams (*Dioscorea* spp.) and banana (*Musa* spp.), only the latter produced phytoliths, in the form of the *Musa*-specific bodies with troughs (Plate 22). Wilson examined soils from remnant ditches cut into swamps by prehistoric occupants and isolated *Musa* phytoliths from several strata. However, with multivariate statistical techniques Wilson could not confidently separate modern forms of introduced *Musa* from the native wild species indigenious to the study region, and therefore could not say unequivocally that fossil *Musa* phytoliths represented a domesticated plant and one not merely a colonizer of disturbed areas. He has demonstrated, though, their durability in soils over long periods of time.

Pearsall and Trimble (1984) carried out the first archaeological phytolith studies in the Hawaiian Islands. They studied known agricultural sites comprising terraced hill slopes, fields with retaining wall boundaries, and other field complexes. Phytolith data provided information on the nature of premodification vegetation (open–herbacious versus woody–forested) and identified cropping, fallow, and other agricultural activities. Pearsall and Trimble isolated some limitations of the phytolith approach and information desirable from other disciplines before phytolith analysis was initiated, in that (1) the identification of a site as an agricultural field had to be done by independent means (surface features, vegetation patterning) since many Hawaiian domesticates were not silicon accumulators, (2) studies of soil profiles by pedologists proved useful in evaluating erosion episodes and helping to explain nonoccurrence of phytoliths in certain horizons and, (3) controlled studies of modern surface soils and vegetation provided a necessary and reliable index by which to judge the prehistoric record.

Phytoliths and the Paleoecology of Sites

Three attributes of phytoliths promise to solidify their role as highly important contributors to vegetational reconstructions in the vicinity of archaeological sites. They are produced in high numbers by many wild plants, in distinctive morphologies by a significant number of these, and exhibit longevity and constancy of preservation, a factor that enables accurate comparisons of within-site and between-site assemblages.

The first investigations of the relationship between archaeological phytolith records and the paleoecology of sites, with correlations of inferences derived from phytolith data to those from other microfossil data sets, were carried out by Robinson (1979, 1980, 1982). He extracted and classified phytoliths from several sites in Texas with long chronological records and was able to independently generate phytolith profiles for vegetation and climate change that agreed with paleoecological sequences proposed on the basis of pollen evidence derived from other regions. The subtropical Texas climate and acid soils quickly destroy pollen grains. Robinson employed six silica accumulator biota as indicators of vegetation and climate; grasses, palms (Palmae), elms and hackberries (Ulmaceae), oaks (Fagaceae), diatoms (Bacillariophyceae), and freshwater sponges (Spongillidae). These have silicified cells that are both identifiable and durable in the acid and sometimes alkaline Texas soils.

Through careful study of present paleoenvironments in Texas and probable past vegetation associations before human interference, Robinson made the following generalizations of environmental preferences of the indicator taxa. The Chloridoideae (short grasses) are characteristic of and dominant in arid environments with summer rainfall, such as the Desert Plains grassland and Shortgrass prairie found presently in western and central Texas. Tall grasses (Panicoid-type phytoliths) are characteristic of the moister eastern parts of Texas and were also closely associated (along with palms) with the climax forest vegetation of east Texas (oaks, elms) before it was destroyed by agricultural activity. Festucoid (Pooideae) grasses prefer cool climates or growing seasons and winter rainfall, with moisture being the secondary limiting factor. Other taxa producing biogenic silica are important in reconstructing past landscapes. Diatoms and Chrysophytes (both algae), along with sponge spicules, live in ponds and streams and reflect a well-watered environment. There exist in Texas, then, several silica accumulator taxa whose shifts over time would suggest concurrent changes in vegetation and climate.

Figure 6.3 has shown the phytolith sequence from a site of the south Texas coastal plain and its correlation with Holocene climatic episodes defined on the basis of paleoecological data from other parts of the world. Biosilica frequencies and types vary considerably from horizon to horizon. Sample 6, with an age around 6,000 B.C., exhibits very high numbers of tree (oak, elm, hackberry and others) and palm phytoliths, plus high frequencies of tall grasses, indications of a wet climate. Tree phytoliths are properly identified on the basis of hair cells and hair bases, not polyhedral epidermis, and in Texas soils survived for long periods of time. Robinson chose to quantify trees and palms on an absolute basis separately from grasses because in a percentage frequency diagram with all taxa combined, grass phytoliths would be present in disproportionate numbers and mask the true abundance of other taxa whose distinctive cells are less durable or less frequently silicified than grass short cells. Absolute tree and palm frequencies were arrived at by counting all phytoliths in a known volume of sediment.

In Sample 5, tree and palm phytoliths decline drastically in frequency and short grasses predominate, indications of a dry period. Robinson also estimates biomass from the numbers of biogenic silica found in soils, and because Sample 5 showed the least amount, he suggests a period of relatively low vegetation biomass, including grasses, in response to dry conditions. Samples 4 to 2, dating from 3000

B.C. to 2000 B.C., demonstrate a continuing paucity of tree and palm phytoliths with varying mixtures of tall and short grasses, perhaps suggesting relatively moister and drier intervals. At the end of the sequence tree phytoliths and tall grasses peak again in frequency, signaling a return to moister conditions in south central Texas.

As Robinson notes, closer interval sampling would better reveal details of climatic change, and all aspects of climate change cannot be studied with phytoliths, but the phytolith record appears to have documented broad periods of changes in rainfall and vegetation over the landscape. The mid-Holocene Hypsothermal interval was revealed and mesic periods seemed to correspond with Holocene glacial advances, while xeric periods corresponded to glacial retreats as postulated by paleoecologists for North American regions, including Texas.

It is worthwhile to review some other attributes of biogenic silica taken by Robinson to signal shifts in moisture regimes and associated landscape cover, for they are quite simple to identify and corroborate the general trends independently evidenced by the indicator taxa. In Sample 5, a dry episode, a freshwater sponge *Trochospongilla leidyi* which is tolerant to alkaline water (itself correlated with periods of less water) was isolated. During this period sponge spicules were infrequent and broken into small pieces, suggesting aerosol transport in a dry, sparsely vegetated environment, and the only examples of grasses adapted to eroded soils were found, again suggesting an extremely dry environment. Conversely, during wet periods sponge spicules were more often complete and diatoms were present in considerable frequency. Cool, humid Festucoid grasses, which dominate in the northern Texas plains, were represented in the prehistoric soil samples by only one phytolith; hence, there is little reason to believe that conditions were cool enough for these grasses to have become important elements in prehistoric vegetation.

It would be useful to have pedologic data on the nature of soil deposition and possible erosion at sites to ensure that rapid aggradation of soils does not account for low phytolith biomass, or that the opposite, implying longer stable surfaces for periods of suggested high biomass, has not occurred. It can also be questioned whether the phytolith record of a few sites may be taken to indicate regional paleoecology. One possible control over this problem is to examine sequences from many sites located in present day ecotones where shifts would be most visible, studies which Robinson is carrying out.

Some caveats are also pertinent with regard to the use of biological assemblages from archaeological sites to reconstruct vegetation and climate. The kinds of phytoliths deposited into sites may have been heavily biased by the factor of human choice; indeed, the degree of human preference is unknown. In response, it can be argued that the phytolith spectra reflect the species most readily accessible for use and that since major indicator taxa are grouped by habitat preference, the broadscale shifts in taxa clearly documented in the phytolith record do reflect vegetational shifts in response to climatic episodes. This conclusion, in turn, rests on the assumption that humans have not significantly influenced the composition of the vegetation through fire, forest clearance, and agricultural activity. In Texas, this appears to have been true for the time period and region under study. Otherwise, as we will see shortly anthropogenic influence on the vegetation confounds climatic interpretations for changes in plant species composition.

In classifying grass phytoliths, Robinson used the scheme of Twiss *et al.* (1969), by which Gramineae silica bodies are placed in three divisions, Panicoid, Chloridoid, and Festucoid. As has been discussed fully in Chapter 3, changes have been made in the higher level taxonomy of grasses to include more subfamilies, and shape distributions do not always conform neatly to taxonomic affinity. For example, *Phragmites,* a tall grass in the subfamily Arundinoideae produces saddles, and *Danthonia,* of the same subfamily, produces bilobates. Nevertheless, good knowledge of the grass cover of Texas and its ecological implications enabled Robinson to rule out potential confusers, and he was even able to identify *Bouteloua,* a dominant of the Texas short grass prairies, which were abundant in intervals identified as dry. In addition, distinct shifts in the fossil grass cover that correlated to shifts in other fossil taxa can most probably be taken to signal a change in moisture regimes, not the occurrence of the small number of possible confuser grass taxa.

The region in which Robinson is working is in many ways ideal for the reconstruction of climatic impact on ancient landscapes, for the confounding influence of human perturbation on environments is not nearly as great as in parts of the world where prehistoric population densities were higher and food-producing economies were developed during the mid-Holocene or earlier. In Panama, for example, the phytolith records of several sites show marked changes over time in the frequencies and types of biogenic silica of the native plant cover, but here food-producing populations were in place by 7000 years ago (Piperno *et al.,* 1985) and the changes are interpreted as reflecting mainly anthropogenic disturbance of the vegetation in proximity to sites, not climatic change. Human modification of vegetation, of course, is also an extremely important issue, one that appears to be amenable to resolution with phytolith data from archaeological sites.

In Figs. 6.1 and 6.2 percentage frequency diagrams from two western Panama rock shelters occupied from 5000 B.C. to about A.D. 500 were presented. They suggest no significant alteration of the tropical forest vegetation during the preceramic period (5000 B.C.–300 B.C.) since weedy types of phytoliths do not appear in any quantity until ceramics are introduced. Phytolith data correlate well with settlement pattern and economic data from the region (Linares *et al.,* 1975; Linares and Ranere, 1980; Ranere, 1980a), which indicate the presence of small, dispersed horticultural and hunting and gathering populations until 300 B.C., which saw the colonization of the area by maize agriculturalists and rapid environmental degradation.

Phytolith records from occupations located in Central Pacific Panama tell a different story. Figures 7.4 and 7.5 show percentage phytolith frequencies from two rock shelters, Aguadulce and Cueva de los Ladrones, bearing preceramic and

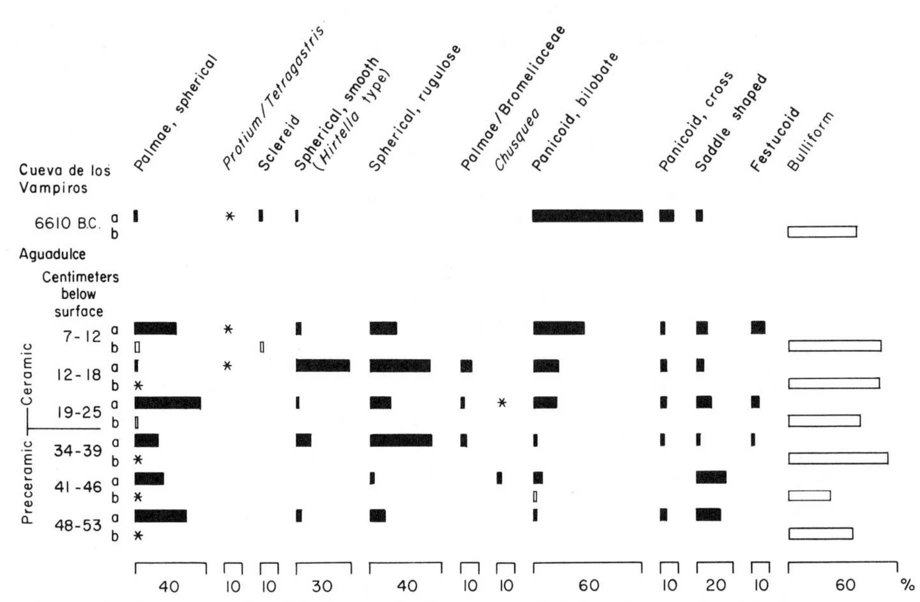

Figure 7.4 Percentage phytolith frequencies from two rock shelters in central Pacific Panama. a, Fine silt fraction; b, coarse silt fraction; *, observed.

early ceramic deposits that are radiocarbon dated from about 5000 to 1000 B.C. (Ranere and Hansell, 1978; Bird and Cooke, 1978; Cooke, 1984). Both sites demonstrate notable frequencies of weed-type grass and *Heliconia* phytoliths in preceramic levels, suggesting that a significant degree of local environmental modification is occurring. Many of these may in fact be garden or field weeds associated with the seed cropping practices documented to have commenced by 5000 B.C. (Piperno *et al.,* 1985). At the Aguadulce Shelter, saddle shapes are present in high numbers from the bottom to the top of the deposits, while at Ladrones they do not appear in any number until ceramics are introduced. This may relate to different kinds and intensities of food producing systems at the sites (preceramic maize is present at Ladrones but not at Aguadulce) associated with different kinds of weeds.

However, caution is required in interpreting the Panicoid and Chloridoid phytoliths as signalling significant, local cultural manipulation of vegetation. Aguadulce and Ladrones are located in a low-lying region that supported deciduous (dry) forest and experiences a long and marked dry season from January to April. Annual precipitation averages only about 1200–1600 mm. Therefore, Panicoid and Chloridoid grasses may simply have been present as a natural, seasonal component of the dry forest, establishing themselves when the canopy opened, and appearing in sites as economic-activity grasses collected from relatively undisturbed contexts. That this possibility is real is made clear by the extraordinary numbers

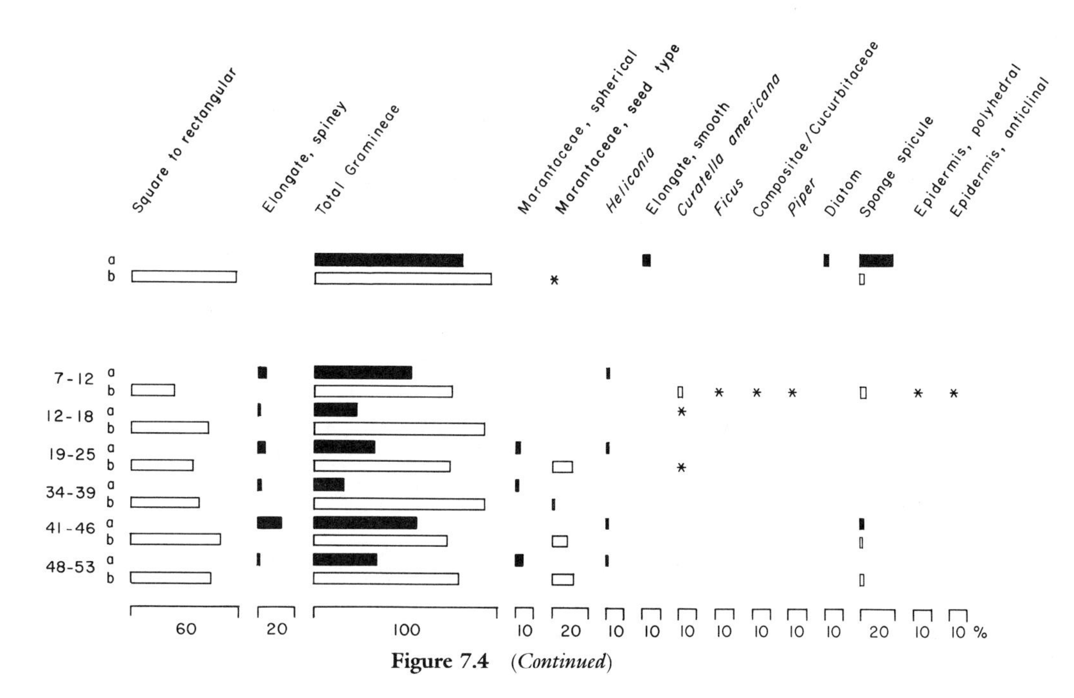

Figure 7.4 (*Continued*)

of bilobates present in preceramic levels of Cueva de las Vampiros, dated to 6610 B.C. (Fig. 7.4), a time when significant local vegetation perturbation from horticultural activity would seem unlikely. In addition, saddle-shaped phytoliths in preceramic levels may derive from selectively used forest taxa *(Bambusa)* and not from Chloridoid grasses. These possibilities are unlikely because food production in association with sizable numbers of people was impacting on the vegetation of Central Panama by 7000 years ago (Weiland, 1987); however, they cannot be discounted.

Phytolith evidence further indicates that the vegetation immediately surrounding several sites in the Central Pacific watershed was not experiencing the same trajectory or level of intensity of anthropogenic alteration. Figure 7.6 shows percentage phytolith frequencies from two rock shelters located at higher elevations than Aguadulce or Ladrones; at 600 m (Vaca de Monte) and 1000 m (Cobre), which places this site very near the Continental Divide. Rainfall increases as one moves from the coast closer to the Divide (the actual linear distance covered is only 45 miles) and the natural vegetation becomes evergreen montane forest in contrast to lowland, deciduous forest on the coastal plain.

Phytoliths are abundant both in preceramic and ceramic-phase levels, the former radiocarbondated to begin at 3600 B.C. at Vaca de Monte (R. Cooke, personal communication, 1987). The ceramic-phase occupation at Cobre is radiocarbon

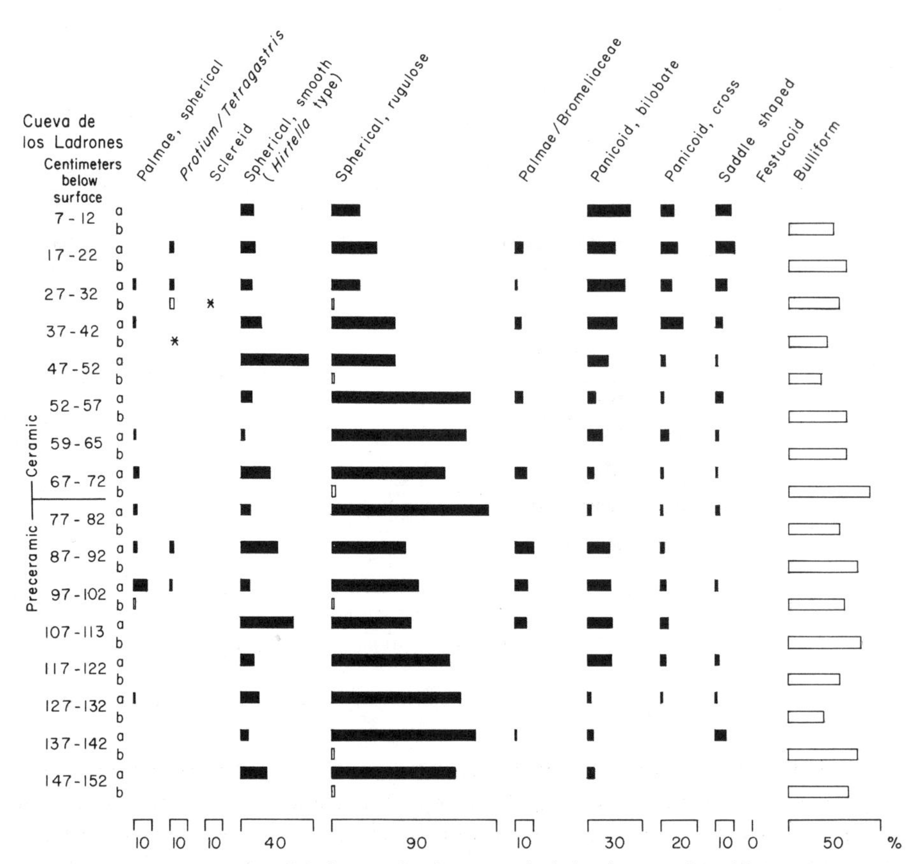

Figure 7.5 Percentage phytolith frequencies from a rock shelter in central Pacific Panama. a, Fine silt fraction; b, coarse silt fraction; *, observed.

dated to 2000 B.C. (R. Cooke, personal communication, 1987), and preceramic stone tool inventories are characteristic of assemblages radiocarbon dated elsewhere from ca. 5000 B.C. to 2500 B.C. However, weedy–grass and *Heliconia* frequencies are extremely low from the bottom to the top of the deposits and there is no sign of maize, except perhaps in the uppermost levels of Vaca de Monte where cross-shaped phytoliths first appear. Assemblages are instead dominated by the remains of *Hirtella*-type fruit phytoliths, with significant frequencies also of *Trichomanes*, ferns of deep, moist forest, *Tetragastris*/*Protium*, *Chusquea*, and other forest-type taxa not yet identified [elongate phytoliths with spherical protrusions or concavities, (Plate 93)]. The phytolith record from Cobre is particularly idiosyncratic, in that it contains very high numbers of unknown phytoliths, presumably because the modern comparative collection is oriented toward plants of lowland deciduous and

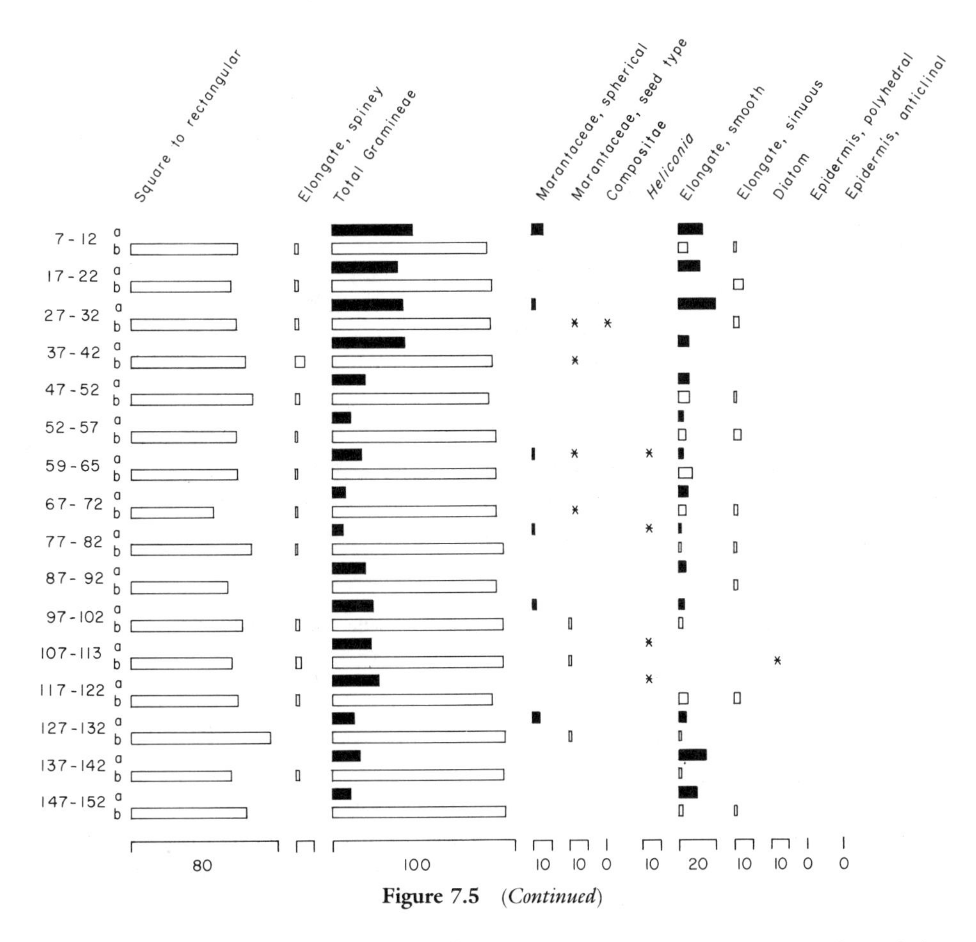

Figure 7.5 (*Continued*)

semi-evergreen forest and not montane formations receiving year-round precipitation. The picture that emerges from Vaca de Monte and Cobre is a prehistoric vegetation little modified over a very long period of time. The presence of moist forest elements not recovered at Aguadulce, Vampiros, or Ladrones, documents the nature of wetter, denser forest that surrounded Cobre and Vaca de Monte. Such environments posed a greater difficulty for seed cropping and habitat modification and may explain why these activities are not evidenced here but are reflected from an early time in the Aguadulce and Ladrones phytolith records.

A clear case for anthropogenic impact on the vegetation of Panama, one that we envision occurred on a dramatic scale, is seen in the level from the Aguadulce shelter containing Period IV ceramics (7–12 cm b. s.), dated elsewhere in central Panama from 300 B.C. to A.D. 500 (Cooke, 1984) (Fig. 7.4). Here, phytoliths from *Curatella americana* (Plate 94) at least double in frequency; and though percentages are still low, they are probably greatly underrepresented because of

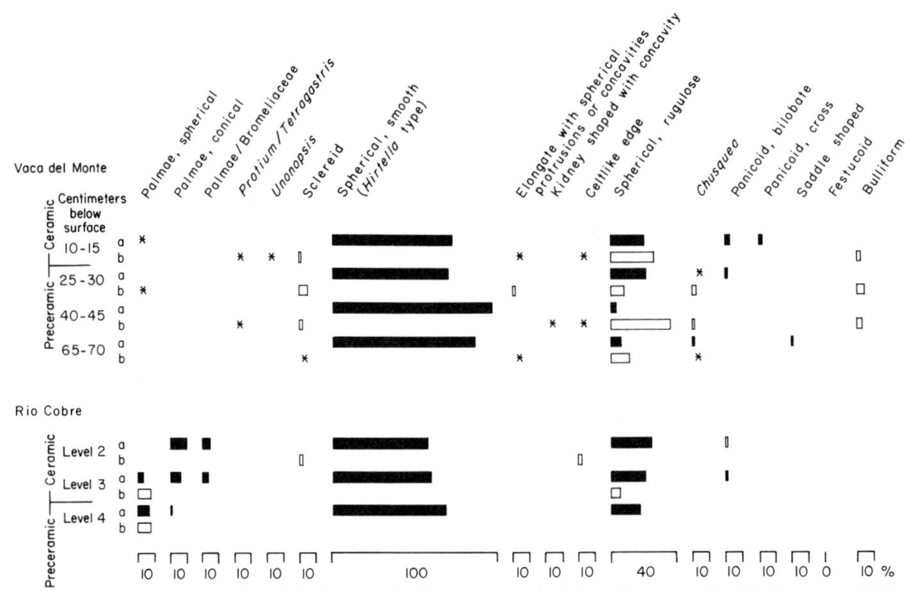

Figure 7.6 Percentage phytolith frequencies from two rock shelters in central Pacific Panama. a, Fine silt fraction; b, coarse silt fraction; *, observed.

swamping by phytoliths from square to rectangular forms from grasses. *Curatella* phytoliths are absent entirely from preceramic deposits. *Curatella americana*, a fire resistant shrub, is a major indicator species for fire disclimax vegetation in central Panama, where it grows commonly on the man-made savannas. The large increase in Period IV contexts probably indicates severe deforestation and habitat destruction near the shelter during the first millennium B.C., when sedentary villages dependent on productive crop plants are first evidenced in the region. This compliments Cooke's (1984) faunal information from nearby Sitio Sierra where almost all of the sample consists of aquatic, ditch, savanna, and disturbed land forms, implying that forested habitats were also scarce around this site. There are increasing indications that by the end of the first millennium B.C. significant areas of land had been repeatedly and severely denuded of forest.

Another record for the nature of environments surrounding a site comes from the preceramic phytolith spectra of Cueva de los Vampiros (Fig. 7.4), where 6610 B.C. levels exhibited high frequencies of biogenic silica from sponge spicules and the presence of marine diatoms. This spectrum is very similar to that from the surface of the site, where sponge spicules blown upward from the surrounding *alvinas* (mud flats landward of mangroves) occur in large quantities. Spicules from both contexts are fragmented as a result of transport by wind, whereas in primary

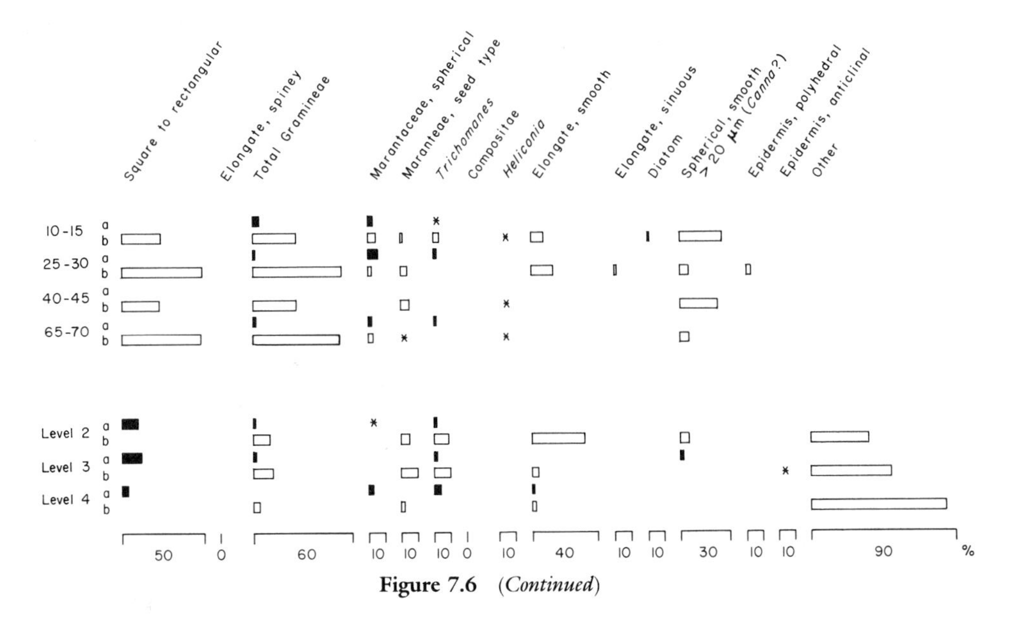

Figure 7.6 (*Continued*)

swamp deposits from geological sediments (Piperno, 1985c) sponge spicules were often complete with unetched surfaces. The site is presently only 2½ km from the coast, and from the similarity of modern and fossil deposits we can infer that at the time of its initial occupation almost 9000 years ago it was also very close to the sea, and perhaps within an open, alvinalike setting.

In Panama, the study of five sites located in different but contiguous environmental zones of the same region resulted in phytolith records that showed significant differences in the nature of regional site ecology and ecosystem modification over several millennia. In terms of evaluating inter- and intraregional patterns of ecology and subsistence, such records are probably more informative and sensitive than, say, a geological lake record, which might reflect the vegetation of one small area or perhaps average and thus confound the changes occurring over a larger area.

Pearsall (1981b) has contributed an important body of evidence on the vegetation near North American sites through studies at Natural Trap Cave, Wyoming. Here, the history of the deposits was very long, beginning at about 110,000 B.P. and ending shortly after 12,000 B.P. Pearsall, in much the same manner as Robinson, examined the grass composition of the site as reflecting the relative abundance of grasses of the three major subfamilies: Festucoid, Chloridoid, and Panicoid. She recognized that another source of phytoliths besides those brought in with soil falling or blowing into the cave would have been the decay of the viscera of grazing animals entombed in the cave. Consequently, several levels from the upper strata were sampled in two locations, one in an area rich in faunal material, the other in an area low in faunal remains. Her results are presented in Table 7.4. From the bottom to the top of the deposits, Festucoid grasses dominate the assemblages, followed in abundance by Panicoid grasses, with Chloridoid grasses always the lowest. It appears that Gramineae cover near the Cave was consistently dominated by Festucoid grasses. As discussed earlier grasses of the Festucoideae favor cool to cold growing seasons with moisture preferred, but moisture is secondary to temperature as a limiting factor.

Today the vegetation around the cave is a short grass (Chloridoid)/sagebrush *(Artemisia)* cover. Therefore, phytolith assemblages indicate that cooler, and perhaps moister, growing conditions prevailed before 12,000 years ago with a concurent shift in the composition of the grass cover. The low occurrence of Chloridoid and Panicoid phytoliths argues against any short grass or tall grass-dominated prairie formations during this time. There may also be another climatic shift evidenced in the phytolith record, in the slight decline of Festucoid grasses after 20,000 B.P. and increase in Panicoids and Chloridoids. A change involving the warming of growing season temperature may have occurred, leading to a more mixed assemblage of grasses.

This study further indicates that the Festucoid, Chloridoid, Panicoid classification can be applied to paleoecology and that the small number of grasses within

Table 7.4
Grass Phytolith Frequencies from Natural Trap Cave, Wyoming[a]

Years B.P.	Level	% Festucoid	% Panicoid	% Chloridoid	Total short cell
	1	77	15	8	136
12,000					
	2	77	15	7	142
14,000					
	3	75	18	7	60
17,000					
	4	81	16	3	172
	5	74	20	6	148
20,000					
	6	80	16	4	161
	7	94	2	4	100
	8	95	5	0	100
	9	95	4	1	100
	10	96	3	1	100
	11	90	10	0	100
110,000	12	93	7	0	100

[a]From Pearsall (1981b).

and outside of these groups contributing confuser silica bodies do not often confound interpretations. Brown (1984) and Piperno (1987a) have found that no Festucoids produce bilobates or saddles whereas some Choridoids and a few Panicoids contribute Festucoid-type bodies, but all except the tribe Aeluropodeae contribute as well a plethora of the diagnostic bilobates or saddles. Therefore, as Brown (1984) notes, "Paucity of these two phytolith shapes in earth materials, coupled with the abundance of trapezoids, probably indicates a dominance of C3 (Festucoid) grasses."

Lewis (1978, 1981) conducted phytolith studies at several bison kill sites or processing stations and rockshelters in the high plains of Wyoming, Nebraska, and Colorado. At the Hudson–Meng site, Nebraska, a bison processing station dated to 9800 B.P., the abundance of Chloridoid phytoliths suggested a shortgrass prairie environment, while persistently high levels of Festucoid silica bodies probably derived from a nearby spring or stream. Other sites generally displayed few forms of taxonomic significance, except in bone beds, where Chloridoid phytoliths were interpreted as deriving from the visceral contents of slaughtered animals.

Lewis' and Pearsall's studies, which dealt with the changing proportions of Panicoid, Festucoid, and Chloridoid phytoliths, demonstrate how the Gramineae phytolith record can be used to indicate the nature of vegetation and, by implication, of climate surrounding sites. If enough sites from a region are investigated, the record then becomes a regional one and sensitive interpretations of changes occurring across a major landscape can be made.

At the site of Alta Toquima Village, Nevada, a high-altitude seasonal camp dating to circa A.D. 1000, Rovner (1985) recovered abundant numbers of pine phytoliths, which largely occurred outside of prehistoric structures. This suggested that the site, located in an open meadow today, was in a pine grove at the time of occupation. Conspicuous presence of phytoliths from pine is, of itself, very important as it indicates that certain species of this plant will leave readily recoverable remains in fossil soils.

Phytoliths and Wild Plant Usage

The same characteristics that make phytoliths important contributors to studies of vegetation around sites make them important in assessing wild plant usage as well. Such studies have not been carried out as yet on any significant intra- or interregional scale; hence, case studies from Panama will first be presented in detail. We will refer again to Figs. 7.4, 7.5, and 7.6 which present phytolith frequencies from five rock shelters in the Central Pacific watershed of Panama. As explained earlier, their locations form a transect from the coast to the Continental Divide, with Vampiros, Aguadulce, and Ladrones occupying highly seasonal, lowland environments, Vaca de Monte intermediate elevations with substantially higher rainfall, and Cobre situated near the Divide in a montane forest receiving year-round rainfall.

A total of 22 different wild or nondemonstrably cultivated taxa were identified in the phytolith assemblages of the five sites (plants like palms were probably domesticated or tended but at present there is no way to prove this with phytolith attributes). The identification of at least 15 more recurrent phytolith shapes not yet in the modern collection will add to this inventory. In contrast, only seven pollen taxa and carbonized remains of palms and the Sapotaceae have been identified thus far. The phytolith record is certainly a very small representation of the totality of plant exploitation, but it is a considerable improvement over the situation for pollen and macrofossils. The phytolith record invites as well an effective, systematic, intra- and intersite comparison, since the preservation aspect is held constant. This is another feature not possible with either pollen or macrofossils, both of which exhibited spotty and often poor preservation.

There is a considerable degree of variability between the phytolith records from the rock shelters. On-site cultivation of maize and squash is not indicated at two, Cobre and Vaca de Monte, during either the preceramic or ceramic periods, but bear in mind that major root crops like manioc and otoy *(Xanthosoma)* are exceedingly difficult to document, as they leave no identifiable siliceous remains. Palms are poorly represented in every site except the Aguadulce shelter where percentages reach 37% of the phytolith sum (Plate 95), and carbonized palm fragments are also abundant here only.

Phytoliths from the fruit exocarp and seeds of tropical trees like *Hirtella* and *Tetragastris/Protium* are present at several sites, providing evidence for the exploitation of tropical fruits. These genera are still eaten in some quantity by indigenous groups today (Duke, 1968, 1975; Gordon, 1982). In fact, at Cobre and Vaca de Monte between 70% and 80% of the fine silt record is contributed by the remains of *Hirtella* spp.; and the frequencies of grass, Compositae, and *Heliconia* phytoliths offer little suggestion of habitat disturbance near the sites. A long-term and stable orientation toward the exploitation of tree products is suggested, bearing in mind that we have little idea of what roles root crops may have played. It would surely be misleading to assume that all residential groups from 5000 B.C. to 300 B.C. were cultivating crops, the same crops, or the same crop mixtures. Rock shelters located in different microhabitats probably served as seasonal occupations for groups who moved to exploit the ecosystem as it produced its goods during various times of the year. The tremendous intersite variability seen in the phytolith record is evidence of such activity.

The remains of shoots from a number of herbacious monocotyledons and ferns evidenced in the phytolith record probably functioned as nondietary economic items. For example, Duke (1968) reports that the leaves of the Marantaceae are used by Panamanian Indians for wrapping food and making mattresses and baskets. *Heliconia* leaves are similarly used today for bedding by tropical forest groups. The fern *Trichomanes* may have been used to treat snake bites (Duke, 1968). And of course, phytoliths from such plants as the Compositae may simply have found their way into middens as ruderals growing in the disturbed habitats near occupations, although the possibility exists that they were gathered from gardens and fields and consumed as greens. In sum, phytolith frequencies suggest major differences in plant usage and cultivation in Panamanian rock shelters. It is suggested that phytolith records are amply reflecting a number of interrelated causal factors: ecological setting of sites, variation in subsistence practices, seasonality of occupation, and functional specialization.

Before we leave the Panamanian phytolith record, another finding is worthy of mention. How the sampling of house floors and other areas of roof collapse might yield evidence for materials used in construction was discussed in Chapter 4. At the first millennium B.C. agricultural village Sitio Sierra, Cooke (1984) took soils from a house floor that were in association with wasp nets bearing impressions of palms and other monocotyledons. This sample was dominated by palm phytoliths (all spherical shapes, with a frequency of 66%), which occurred rarely in other samples (frequency below 1%) from the site. The sample also contained much higher remains of *Heliconia* than did other contexts. The palm, and possibly *Heliconia* phytoliths, must then represent decayed thatch from the house. Many Panamanians still use the palm *Scheelia,* a producer of spherical phytoliths, as material for thatching.

Phytoliths, Dental Microwear, and Diet

Silicon dioxide is an extremely hard and durable substance. Baker *et al.* (1959) reported that plant opal is harder than enamel and may therefore cause wear and detectable abrasions on teeth of animals who masticate plants high in phytolith content. Covert and Kay (1981) conducted several controlled experiments in which opossums were fed various diets consisting of high fiber (herbivorous, insectivorous-type), chitin, or grit (pumice) over a 90-day period. The grit-fed animals developed characteristic pits and striations on their molars. Such a microwear pattern was similar to that of the primate grass-eating herbivore *Presbytis johni*. Covert and Kay then examined the teeth of a *Sivapithecus* specimen and found no distinctive microwear pattern, an indication that the animal was excluding grass, other plant parts containing silica, and exogenous grit from its diet shortly before it died. This study has important implications for studying microwear patterns on human dental remains.

Phytoliths and Stone Tool Function

The formation of use-polish on the surfaces of stone tools has long been associated with the working of silica-rich plants. Characteristic silica gloss has been identified on a number of morphologically and functionally dissimilar tool types, including hoes, sickles, and palm pounders (Kamminga, 1979). The finding of silica gloss on the latter tool serves to emphasize that its presence should not be solely associated with the processing of graminaceous plants. A quick survey of Tables 2.2–2.4 provides an indication of the numerous plant species and structures (seeds, fruits, floral bracts, stems) that may produce polish on lithic artifacts. Quantities of silica in bark and timbers as well are probably high enough to produce polish.

Anderson (1980) demonstrated that the morphology of phytoliths is not completely destroyed during polish formation. Using scanning electron microscopy, she identified grass phytoliths adhering to the surfaces of a Mousterian scraper and Neolithic sickle. Further studies of a similar nature hold promise for determining in a precise manner the function of archaeological stone tools.

Other Roles of Phytoliths in Prehistoric Plant Use Reconstructions

The remarkable durability of phytoliths ensures their recovery from a variety of site contexts and artifactual material. Schellenberg (1908) recognized their presence in mud brick of prehistoric sites in Turkestan. More recently, Liebowitz and Folk (1980) identified a 2–3 cm thick phytolith deposit in association with 1300 B.C. iron smelter "factory" areas at Tel Yin'am, Israel. They concluded that phytoliths came from straw, probably wheat, (Twiss, 1983) used as industrial fuel, confirming Pliny's reference to use of straw in smelting because it gave a very intense heat. The second millennium B.C. iron smeltery is the earliest yet known.

An innovative and promising use of phytoliths is found in studies of Bishop *et al.* (1982) and Rands and Bargielski (1986). Phytoliths are a conspicuous component of Mayan ceramics from the Palenque region, where they were probably incorporated into ceramic manufacture as natural inclusions in clays. They are readily observable in analyses of ceramic thin sections. Detection and identification of phytoliths in ceramics may lead to information on areas of clay procurement and pottery manufacture and the nature of plant cover in such areas.

There are a number of other research issues that will benefit from future phytolith studies. Weeds collected from prehistoric fields and other anthropogenic vegetation probably were important components in the diet, as they are in peasant economies today (Bye, 1981). The good picture of secondary-growth herbacious vegetation obtainable from phytolith records will permit study of this issue. Finally, we have seen in Chapters 2 and 3 and the present chapter how the reproductive structures of many species also leave detectable and highly distinctive phytolith residues. The recovery of such phytoliths will aid in the development of a calendar related to seasonal plant use and site occupation.

This chapter has reviewed results of the studies carried out heretofore by modern archaeobotanists and suggested a few areas where future research may prove worthwhile. We will not repeat here the many other potential applications of phytoliths in studies of the domestication and usage of plants by prehistoric peoples, which have been discussed in previous chapters. Suffice it to say, that we may expect a flood of new information when the requisite analyses of modern plants and the recovery and identification of phytolith remains from archaeological deposits in many parts of the world have been achieved.

8

The Role of Phytoliths in Regional Paleoecology

Introduction

In his 1971 paper that did much to infuse life into then moribund archaeological phytolith research, Rovner (1971, pp. 343–344) wrote: "For any fossil system to be useful to the archaeologist at least three criteria must be met. The material must withstand decomposition, exhibit sufficient morphological difference to be of taxonomic significance, and provide sufficient quantities to reflect the nature of the entire assemblage from which it is derived."

It has been shown that phytoliths exhibit all of these attributes, permitting resolution or partial resolution of a number of important issues in prehistoric human behavior. It is further documented from the numerous analyses of assemblages from stratified archaeological horizons and modern soils presented in Chapters 6 and 7 that phytolith profiles with depth representing the phytolith deposition over time are achieved, phytolith spectra closely mirror the existing vegetation, and correlations are possible between other paleobotanical profiles and the phytolith record. That this is true should hardly be surprising, for phytoliths are plant microfossils that, while having their singular traits of production, morphology, and deposition, as do other macro- and microfossils, are a function of (and therefore reflect) vegetation from which they are derived.

The importance of environmental reconstruction in ecologically oriented prehistory dictates a review of phytolith applications in geological contexts—how phytoliths can presently be used to reconstruct paleovegetation and climate, and where future studies may lead.

Geological Phytolith Analysis: Some Consideration of Basic Questions

Perhaps the question most frequently asked of phytolith specialists by interested paleoecologists is whether phytoliths provide reconstructions of vegetational formations that existed in the past. The answer is unequivocally *yes*. We have seen in Chapter 6 how tropical forests, swamps, and agricultural fields and grasslands from Panama contribute their own idiosyncratic soil phytolith assemblages, and we will show in Special Topic 3 how the phytolith record of geological sequences from a nearby region of Panama documented in a very sensitive fashion the history of plant communities and factors that changed them over the past 11,000 years.

Soil scientists and geologists have for quite some time recognized the differences in A horizon phytolith profiles constructed from underneath different vegetational formations. Though phytoliths were isolated from only a part of the total soil fraction, and taxonomy was not developed to the point where many kinds of non-Gramineae opal could be identified, it was clear that soils developed under long periods of grass vegetation commonly contained 5 to 10 times more grass opal than those formed under forested environments (Jones and Beavers, 1964a,b; Witty and Knox, 1964; Verma and Rust, 1969; Wilding and Drees, 1968, 1971; Norgren, 1973). More recently, soils from three plant communities in Kansas, shortgrass prairie, tallgrass prairie, and deciduous forest, have been shown to exhibit statistically different phytolith spectra (Kurmann, 1985). Kurman used the modern spectra to interpret the phytolith assemblage of a Kansas paleosol as deriving from cool-season grassland. Kurman stresses the importance of combined pollen and phytolith studies, because pollen analysis failed to distinguish between tall-grass and short-grass prairie, differences discernable in phytolith spectra.

As has been emphasized in Chapter 6, a considerable amount of modern phytolith sampling is needed from soils underneath different vegetational communities from many parts of the world. The construction of phytolith assemblages from various floristic associations is essential to proper interpretation of assemblages isolated from fossil deposits. Considering what we already know about phytolith morphology in modern plants and phytolith stability in soils, there can be little doubt that corresponding patterns between phytolith spectra and modern vegetation will be found for many floral communities and climatic zones of the world. By reference to phytolith spectra recovered from stratified archaeological deposits, we hold a similar confidence that phytolith assemblages characteristic of a number of vegetation communities and their changes over time will be present in stratified geological sequences. What follows is an analysis of phytoliths from Panamanian geological deposits that are partly interpreted by reference to modern assemblages, and partly to a correlated set of pollen data.

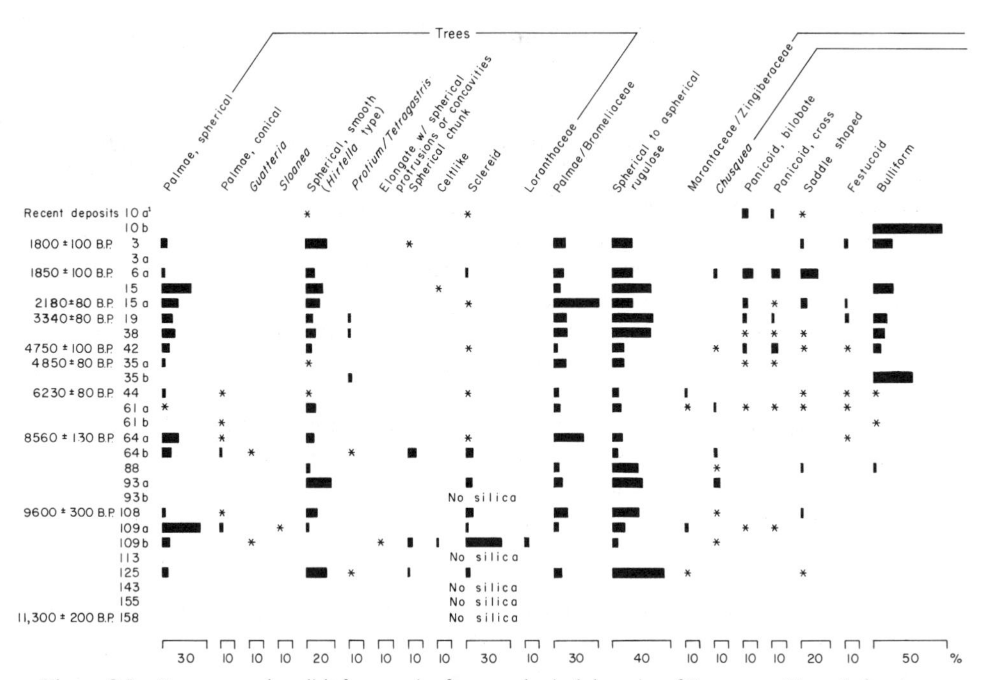

Figure 8.1 Percentage phytolith frequencies from geological deposits of Panama. a, Fine silt fraction; b, coarse silt fraction; no notation, fine and coarse silt; *, observed. Compositae categories: A, abundant in sand fraction; R, rare in sand fraction; NP, not observed in sand fraction.

Special Topic 3: Phytolith Spectra from 11,000–Year–Old Geological Records from the Tropics

One of the few and very exceptional pollen sequences from the humid, lowland tropics of the New World was published by Bartlett and Barghoorn (1973). From terrestrial deposits (now underneath Gatun Lake in the Carribean watershed of eastern Panama, formed by the construction of the Panama Canal), they extracted several very long sediment cores reaching to a depth of 158 feet below sea level which extended back 11,300 years in time. The entire sequence is firmly dated with radiocarbon and shows well-defined vegetational changes from 11,300 B.P. to the present. Early on, tropical forest grew in the study area, later to be replaced by a mangrove and then fresh water swamp. By 3300 B.P. the effect of human influence on the landscape is evident in the pollen record in the form of evidence for maize and slash-and-burn agriculture.

After the pollen analysis was completed, Elso Barghoorn retained unused core samples at Harvard University and kindly permitted me to remove them to Panama

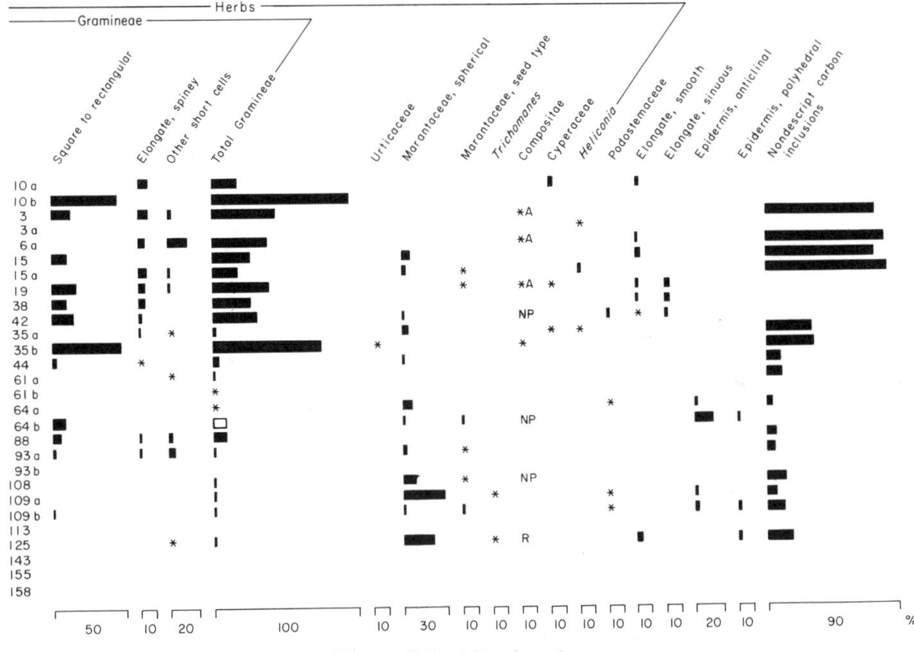

Figure 8.1 (*Continued*)

for phytolith analysis. The coring locations are only about 7 miles from Barro Colorado Island, the site of the modern phytolith rain studies described in Special Topic 2, which, therefore, form suitable modern analogs. The retrieval and analysis of phytoliths from the Gatun sequence (Piperno, 1985c) accomplished a number of important objectives. Phytolith data were compared to and correlated with data from pollen analysis, a technique that, though not without its problems, has been widely accepted and used for decades. Second, the characteristics of past phytolith assemblages representing known, different lowland vegetational associations of the tropics, such as forest, swamp, and disturbed, cut overgrowth, were elucidated and evaluated for the first time. Lastly, the availability of phytolith and pollen data in tandem defined areas where one body of data was more sensitive to landscape changes than the other, indicating that the simultaneous application of both will be a powerful paleoecological tool.

A total of 20 samples from 4 different cores were processed. Fortunately, it was possible to obtain 10 of the radiocarbon-dated samples, which allowed the phytolith sequence to be firmly tied into the pollen record. Figure 8.1 is a phytolith percentage frequency curve revised from that originally presented (Piperno, 1985c).[1]

[1]All levels in Figs. 8.1 and 8.2, except for the uppermost, are in feet below sea level; the uppermost is in feet above sea level.

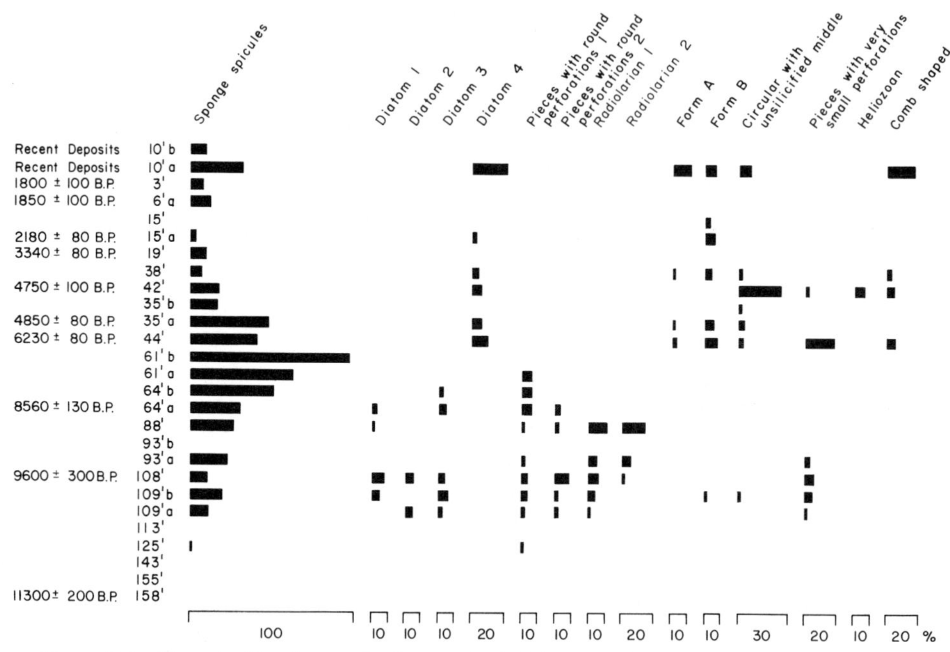

Figure 8.2 Biogenic silica frequencies from geological deposits of Panama. Origin in acquatic plants and animals. a, Fine silt fraction; b, coarse silt fraction.

A number of unknown types have been classified and some general categories have been reorganized and renamed. These will be described later.. Several types of biogenic silica from unicellular aquatic plants and animals (in Fig. 8.2) not classified in the original publication have also been identified. Bartlett and Barghoorn (1973) divided the pollen diagrams into four periods based on inferred vegetational associations present during each time frame. The phytolith diagrams will initially be divided into the same four zones, with some modification occurring.

11,300 B.P. *to 9,000* B.P.

The pollen of this period was mainly that of rainforest genera, such as *Bursera, Virola, Swartzia,* and members of the Bombacaceae, indicating that tropical forest grew near the coring localities. Pollen of *Iriartea,* a stilt palm, *Symplocas,* and the Ericaceae (heaths), thought to grow today only at higher elevations, were used by Bartlett and Barghoorn to argue for a temperature depression of at least 2.5°C. However, since it is now known that all of these taxa grow at lowland elevations, an unequivocal case cannot be presented for substantial temperature depression.

Yet, there are a number of pecularities about the pollen record of this interval that still invite explanation. The pollen of *Iriartea, Alchornea,* and the Urticales are abundant. The latter includes that of *Ficus* and *Cecropia,* genera which colonize open spaces in the canopy. Bartlett and Barghoorn noted that they represented vegetation of disturbed ground. *Iriartea* pollen is found in great frequencies and this palm, though present in low lying habitats today, never achieves any prominence in the modern lowland flora. Hence, it is quite possible that its high fossil frequencies are a result of an altitudinal extension of its range brought about by depressed temperatures. The general pollen picture is one of forested vegetation, yet we are led to wonder about its similarity to present-day lowland forest especially in regard to species prevalence and the degree of canopy closure.

Six samples from this interval were analyzed for phytoliths. Only two, however, from 125 and 109 feet below sea level, produced silica in any appreciable amount. The absence of more phytoliths in the others, which yielded only rare Marantaceae and spherical rugulose forms, plus sponge spicules, is at first glance difficult to explain. Judging from the topography of the coring localities, estimates of where sea level was 11,000 years ago, the nature of the sediments (sands and gravels), and low percentages of *Rhizophora* pollen, Bartlett and Barghoorn felt that deposition took place very close to sea level in front of the coastline rather than behind it in a swamp. Pollen grains and phytoliths, therefore, must have been transported to coring localities by wind, streams, and surface runoff, with the latter two probably moving most of the phytoliths. It may be that transport of phytoliths by erosion and water was less effective than wind in creating a microfossil record early on in the Gatun sequence, although two of the early samples did yield a phytolith record and calculations of pollen influx would be useful in order to estimate how much pollen was actually in the deposits relative to later periods.

The samples with adequate silica, at 129 and 109 feet, are characterized by significant percentages of Palmae, Marantaceae, Palmae/Bromeliaceae, sclereid, and spherical rugulose phytoliths. The latter occur in a number of monocotyledons. As discussed in Special Topic 2, sclereids are produced predominantly by arboreal species and are good indicators of such plant cover. Present in low frequencies are *Trichomanes,* the moist forest fern, possibly *Sloanea terniflora,* a lowland forest tree, *Tetragastris/Protium* and *Guatteria,* trees of moist forests, the parasitic Loranthaceae, the Podostemaceae, which inhabit rapidly flowing streams, and *Chusquea,* the forest bamboo. Other grasses comprising the Panicoid, Festucoid, and Chloridoid subfamilies are rare in occurrence.

By comparison with modern phytolith assemblages from the BCI forest, the phytolith spectra from 11,300 to 9600 years ago can be securely identified as deriving from a tropical forest, one moist enough to have supported such taxa as *Trichomanes,* palms, *Guatteria,* and *Tetragastris* or *Protium.* It is demonstrated that in ancient geological strata phytoliths can be retrieved and classified, and assemblages constructed and identified on the basis of comparisons with modern profiles from known vegetation. Phytolith data independently indicate moist tropical forest and provide information on the presence of a number of genera not evidenced with pollen data—*Chusquea, Trichomanes,* the Marantaceae (possibly *Stromanthe* or *Maranta,* but not *Calathea, Thalia,* or *Ichnosiphon*), *Tetragastris/Protium, Guatteria,* and the Podostemaceae.

It is significant that there is no evidence for climatic aridity in either the pollen or the phytolith record, as pollen sequences from Guatemala (Leyden, 1984) and Venezuela (Bradbury *et al.,* 1981) suggest that Late Pleistocene aridity affected these areas, creating open landscapes, with conditions not ameliorating sufficiently to permit substantial growth of tropical forest flora until around 11,000 B.P. It is possible that the Panama sequence is too recent to document the full Late Glacial climatic and vegetation conditions, which many believe peaked about 13,000 years ago (Livingston and Van der Hammen, 1978; Van der Hammen, 1974; Simpson, 1975). Earlier data from Panama are urgently needed to resolve these questions.

Nevertheless, there may be some important deviations from modern conditions and climate in the pollen record of this period that need to be verified by comparison to extensive modern pollen records. The phytolith record also exhibits a number of significant differences from the BCI modern samples. Spherical rugulose and Marantaceae phytoliths are much more common than in any modern sample and succeeding periods in the cores, while Palmae phytoliths are much less common than in modern samples tested from BCI. Palm phytoliths are especially infrequent in the sample from 125 feet below sea level where phytoliths, generally, and large-sized phytoliths specifically characteristic of trees are far fewer in number than at 109 feet. As Jacobsen and Bradshaw (1981) have quite correctly pointed out, it is probably not appropriate to compare modern and ancient situations having great differences in depositional characteristics, such as a small tract of modern forest where phytolith movement was shown to be inconsequential and a series of deep cores where the size of the microfossil source area is open to question. Yet, it remains to be determined whether differences apparent both in pollen and phytolith records are a result of significantly different climates, cooler if not drier, and forests of different structure and species prevalence (if not general species composition) from those of today.

During this period biogenic silica from unicellular aquatic species such as sponge spicules and diatoms was poorly represented, except in the sample from 109 feet, only 1 foot below a level directly dated to 9600 B.P. Therefore, the aquatic species in the 109-foot level must represent the beginning of the development of the mangrove swamp, the vegetational formation that dominates the next interval.

9600 B.P. *to 7300* B.P.

This period in the pollen record was dominated by the presence of mangrove genera, especially *Rhizophora,* indicating that deposition took place near a well-developed mangrove swamp that formed in response to the slowing of sea level rise. A great variety of tropical forest pollen was still present, indicating that well-developed forest existed at some spot behind the mangrove swamp. Evidence for mangrove species will not be forthcoming in a phytolith record as the mangrove flora *Rhizophora*, *Avicennia*, *Conocarpus*, and *Laguncularia* have no identifiable siliceous remains. However, the biogenic silica record leaves evidence for the tidal inundation of the swamp by the conspicuous presence of a number of marine diatoms (Nos. 1, 2, and 3), Radiolarions, and other salt water organisms. Sponge spicules also occurred in high frequencies during this interval, reaching their highest number in the level at 61 feet, where they accounted for 96% of all silica in the coarse silt fraction.

Tropical forest flora continue to be well represented in the phytolith record. This is a finding of some significance for, though substantial local deposition of biogenic silica obviously took place from within the swamp, phytolith transport to the coring locality was sufficient to reveal the presence of tropical forest that existed at some unknown distance behind the swamp. It appears that we are achieving in geological profiles phytolith records with a greater degree of input from extralocal (within several hundred meters) and regional (beyond several hundred meters) source areas than we had allowed for in our theoretical models of phytolith transport. Surface runoff and stream and river transport have probably all played roles in this regard. The presence of phytoliths from the Podostemaceae, which grow on submerged rocks in rapidly flowing fresh water, is independent evidence of significant phytolith transport to coring localities in rivers and streams.

7300 B.P. *to 4200* B.P.

During this period a transition from mangrove vegetation to freshwater swamp took place as a consequence of the stabilization of sea level rise. *Rhizophora* pollen disappears from the cores and is replaced by abundant pollen of the Cyperaceae, Gramineae, *Typha*, and *Polygonum*. In addition, the presence of pollen from *Myrica* and *Ilex* was thought to signify more pronounced seasonality on the Atlantic watershed because today both grow only in drier formations on the Pacific watershed.

Changes in biosilica frequencies are also apparent during this period. The freshwater nature of the depositional setting is evidenced by the appearance in the 44-foot level of several new aquatic organisms that are also well represented in recent fresh water samples from Gatun Lake. They include Diatom No. 4 and Form A. Also present for the first time are the siliceous skeletons of Heliozoans, fresh water amoeboid protozoans. Sponge spicules continue to be well represented. The abundance and diversity of siliceous skeletons from aquatic species from 6230 B.P. to 4700 B.P. indicates deposition in a well-developed fresh water swamp.

Grass phytoliths increase markedly during this period and include for the first time good representation from all the grass subfamilies, plus the characteristic square to rectangular and bulliform phytoliths (many of the bulliform cell phytoliths were the classic fan-shaped type found only in grasses). This increase may be a result of the swamp formation, or possibly, drier conditions. It is significant, though, that bulliform cell phytoliths are very well represented, perhaps suggesting that grasses were not severely water stressed and that grass phytoliths came in part from swamp grass deposition. Bulliform cells do not normally become silicified in dry habitat grasses (Parry and Smithson, 1958b; Sangster and Parry, 1969).

More importantly, there is evidence that the drastic increase in grass silica at 4850 B.P. is a result of the clearing and firing of the vegetation for agricultural plots, for a number of other associated changes take place in the phytolith record. In core TDS-4 at 42 feet there is a drastic increase in cross-shaped phytoliths, whose three-dimensional structures and sizes indicate they are from maize. They reach their second highest frequency of the entire sequence here, where they account for 4% of the phytolith sum. This evidence for agriculture is supported by changes in the nearly contemporary 35-foot level of core TDS-2, located less than a mile from TDS-4. Phytoliths from grasses reach the highest frequency of the sequence, maize is present, and phytoliths from *Heliconia*, the Cyperaceae, and the Urticaceae, all important components of succession and disturbed ground, appear for the first time. Concurrently, there is a marked increase in the number of nondescript phytoliths with occluded carbon, a product of burnt vegetation, and many identifiable phytoliths show evidence of burning for the first time.

Collectively, these changes indicate that some degree of slash-and-burn agriculture, probably associated with a long fallow system, was initiated in the coring region by 4850 B.P. The extent of the forest clearings is not yet sizeable enough to register in the pollen record, where associated changes, such as the appearance of maize pollen, and increases in Compositae pollen, do not occur until after 4,000 B.P. The phytolith evidence appears to represent smaller, more localized clearings; hence, phytoliths may be more sensitive to the small-scale beginnings of nonintensive, shifting tropical agriculture than are fossil pollen remains.

4200 B.P. to the Present

In the pollen diagrams from this interval the first clear indication of agricultural activity is present. Maize pollen first occurs in a sample from core TDS-4 at 25 feet, which is between levels directly dated to 3340 B.P. (19 feet) and 4750 B.P. (42 feet). At 3340 B.P. Gramineae, Cyperaceae, and Compositae pollen reach high frequencies and finely divided charcoal becomes common somewhat later in time. The pollen evidence from cultivated plants and associated weeds indicates that by

3300 B.P. agriculture was well established in the region. During the later part of this interval nearly all tree pollen disappears, even that of common weed trees, suggesting extensive forest clearance with shorter and shorter fallow periods.

In the phytolith profiles from this interval, grass phytoliths remain high in number. Silicified short cells of all types (Festucoid, Chloridoid, Panicoid) are now well represented and probably reflect the decay of field weeds. Their increasing frequencies document the intensification of deforestation as larger areas of land were cleared for agricultural plots. Other phytoliths of disturbed growth like the Compositae and *Heliconia* increase markedly in frequency as well. All of the Compositae phytoliths were the opaque, perforated platelets described in Chapter 3. There are now massive amounts of phytoliths with occluded carbon, which further reflect the intensive slash-and-burn activity evidenced both in the phytolith and pollen records. Maize phytoliths were isolated from levels dated to 3340 B.P. and 1850 B.P. Some levels yielding maize pollen did not produce maize phytoliths, showing that the phytolith record can also be negatively biased and limited in this regard. During this interval grass phytoliths looked to be worn or eroded, possibly indicating a considerable degree of transport from source areas. Siliceous skeletons from aquatic species decline in frequency, possibly indicating that former swamp land had been reclaimed for use as fields or that the effects of the drastic cultural environmental modification had at any rate destroyed these fresh water formations. A final sample from 10 feet above sea level representing recent lake deposits showed an abundance of silica from aquatic forms and grasses.

Conclusions

This comparison of pollen and phytolith data showed beyond a reasonable doubt that discrete and identifiable phytolith assemblages and other biogenic silica assemblages representing moist tropical forest, marine swamp, fresh water swamp, and anthropogenically disturbed vegetation are present as stratified microfossil associations in geological records. The phytolith record conveyed much of the information that the pollen record conveyed, in terms of the nature of the broad-scale vegetational associations that grew in lowland eastern Panama. In addition, phytoliths were more sensitive to the earliest small-scale introduction of seed agriculture in a tropical forest environment, indicating that the effects of cultural activity on vegetation will leave a well-marked phytolith record. This appears to result partly from the following factors. Grasses and other plants of disturbed growth like *Heliconia* are more prolific phytolith than pollen producers; maize produces abundant identifiable siliceous remains, and phytoliths resulting from the burning of vegetation acquire easily detectable carbon residues. Fossil phytolith assemblages from the humid tropics should, therefore, provide fine-scale resolution of the development and growth of tropical forest agriculture.

More detailed information on the species composition of past vegetation will continue to rely on good pollen records. The lack of phytolith production by many species and problems of taxonomic specificity of silicified forms in many others ensure that a considerable number of rain forest taxa remain silent in fossil assemblages. The phytolith record from forested contexts in the early part of the core sequence reflected very little of the species diversity that is a trademark of the tropical forest system.

Some parts of the phytolith record did not meet expectations with regard to the representation of certain families. For example, Cyperaceae phytoliths, though present, did not attain considerable frequencies in post-4000 B.C. levels when their pollen was deposited in considerable number. There are species in the Cyperaceae (*Fimbristylis annua, Rhynchospora barbaya*) that do not produce the characteristic conical-shaped silica body in high numbers, and we also wonder how many very small (<5 μms) Cyperaceae bodies are lost during the extraction process when clay is removed. These are issues that need further attention.

The representation of phytoliths from other taxa generally met expectations postulated from studies of modern phytolith rain on Barro Colorado Island, Panama, just 7 miles from the core localities. Other than a single solidly silicified nonsegmented hair cell, no other trichomes or hair base phytoliths were recovered. All of the anticlinal and polyhedral epidermal cells retrieved were likewise solidly silicified. The level of silica solubility from various tropical forest taxa is in accord with the number and types of phytoliths recovered from modern A horizons on Barro Colorado Island. We will emphasize again, however, that this situation may be quite different and ameliorated considerably in lake deposits, where phytoliths are fairly quickly washed into protected environments.

Finally, the extent of phytolith movement and transport to the sites appeared to be considerable, comprising significant representation from extralocal and regional source areas. Tropical forest and upland taxa were found in sediments where primary deposition took place in such situations as: (1) in front of a coastline, (2) in mangrove swamp, and (3) in fresh water swamp. Phytolith transport in soils eroded from the hills above the 100-foot contour located about a mile from the sites probably occurred, and the presence of siliceous bodies from the Podostemaceae, which grows in rapidly flowing rivers, provided evidence for phytolith movement via rivers or streams. Bartlett and Barghoorn (1973) pointed out that the nearby Trinidad (and also Chagras) rivers must have meandered over the sedimentary deposits through the centuries, leaving behind their silts and associated microfossils. Any discussion of a "localized" phytolith record is clearly out of place here; indeed, it appears that regional paleoenvironments are fairly well represented in phytolith assemblages.

The Recovery of Phytoliths from Various Depositional Environments

Peat Bogs

Jacobson and Bradshaw (1981) have stated that peat deposits may contain the best record for pollen reconstructions of past regional landscapes and climates. It is doubtful whether this will be the case in phytolith analysis because in bogs the localized character of phytolith deposition might preclude significant input from plants growing outside of the bog, unless stream transport has occurred. Furthermore, if phytolith types produced on peats are often also produced in the surrounding landscape, it may be difficult to judge the size of the phytolith source area. Such situations may be ameliorated by conditions where plants comprising the peat are for the most part nonsilicon accumulators; therefore, phytolith representation would predominantly consist of allocthonous plants deposited from extralocal and regional sources by wind or water. These are issues that must be worked out on a site-by-site and region-by-region basis.

Studies of phytoliths and other biogenic silica from peats have already generated important data bearing on the genesis and depositional history of peat deposits. Cohen (1974) and Andrejko and Cohen (1984) enjoyed a great amount of success in isolating phytoliths and other biogenic silica such as sponge spicules and diatoms from the Okefenokee swamp. This, in itself, is of considerable importance because there has been some question as to whether the environments of highly organic sediments contain enough soluble silicic acid to make possible phytolith formation. However, Upchurch *et al.* (1983) have reviewed the considerable literature indicating that silica is a major sedimentary component derived from the plants and animals associated with some peat-forming environments.

In Okefenokee Swamp cores (Andrejko and Cohen, 1984) phytoliths from grasses and sedges along with biogenic silica of diatoms and sponges were abundant. Grass bulliform cells were used as indicators of relative periods of soil moisture on the basis that silica deposition in them is more common in wetter habitats where the root system has been submerged. Rovner (1983) has also argued persuasively that percent of bulliform cells relative to other phytoliths provide an index of edaphic conditions—wet versus dry—because it has been shown that under conditions of artificially elevated moisture, bulliform cells are silicified in grass taxa where they normally do not occur (Sangster and Parry, 1969). Silicified bulliform cells were recovered in greatest quantity from basal peat layers, which were positively correlated with abundant sponge spicules and negatively correlated with the presence of cypress *(Taxodium)*, suggesting that wetter and very different conditions may have prevailed during the early development of the swamp.

There is also a great deal of promise for documenting the kinds of sedges that dominated the peats through time and, hence, for providing very specific paleoecological information, by the discrimination of sedge genera with scanning electron microscopy of phytoliths. Andrejko and Cohen are here employing criteria based on the number, shapes, and location of satellites, very small projections located on the rim and crown of some conical sedge phytoliths, and the overall shape of the core or crown. These attributes often differ between genera. It appears that SEM is essential in defining fine-scale characteristics of the satellites, although their presence and number can be detected with the biological microscope at a power of 1000×.

Andrejko and Cohen observed a substantial pitting of fossil bulliform cell phytoliths, which appeared to be an end product of boring activities by microorganisms, possibly bacteria or diatoms. In environments where they are plentiful, such microorganisms may play a significant role in recycling biogenic silica and maintaining silica in solution for reabsorption by living systems.

Andrejko's and Cohen's peat studies brought out the importance of applying phytolith studies to these depositional environments. Most paleoecological information involving peat formations has, of course, been derived from pollen analysis. However, pollen is best preserved in acidic, rather than alkaline anaerobic conditions, and fires that commonly occur in peats quickly destroy organic material, including pollen and once recognizable plant fragments. Phytolith studies of peats provide a significant source of independent paleoecological data.

Loess

Fredlund (1986b) has described the phytolith record from a 620,000-year-old section of Nebraskan loess. He points out that such a deep, 35-m section representing continuous deposits over a very long time is the terrestrial cognate for deep-sea sediments. A number of paleosols are exposed in the vertical profile, whose origins have been of some importance to paleoecology in the central United States. The pivotal question is if stable climatic episodes, resulting in soil and paleosol formation, were of moister and cooler or warmer and drier conditions. If the latter proposition was true, then nonglacial regimes were responsible for the palesols, and C4 grasses should have formed the dominant component of the Gramineae cover. If, however, glacial climates were responsible, grasses well adapted to cool conditions, the C3 types, should have prevailed. We have discussed before the possible contribution that phytoliths can make in determining the past distributions of C3 and C4 grasses and climate. A digression is appropriate at this point to summarize in more detail the available evidence bearing on these relationships.

Phytoliths in C3 and C4 Grasses

Grasses, as do other higher plants, possess two photosynthetic pathways in which carbon dioxide is converted into either a three or four carbon atom compound as the first fixed carbon compound. The result is Gramineae taxa having significant physiological and anatomical differences, and they are called C3 and C4 grasses. C3 grasses have large, loosely, and irregularly arranged chlorenchymal cells and a vascular bundle with a double sheath. They are adapted to cool, moist environments, with moisture being a secondary limiting factor. Grasses possessing the C4 pathway have clustered chlorenchyma cells enclosing a single layer of bundle sheath cells. The C4 pathway occurs in plants adapted to warm, dry growing seasons and, hence, probably originated in grasses of tropical origin (Esau, 1977; Downton, 1971).

As it turns out, C3 and C4 grasses tend strongly to have different numbers and shapes of short cell phytoliths. Excellent reviews on the subject have been offered by Twiss (1986b), Brown (1984), and Kaufman *et al.* (1985). In the first articles to state explicitly the potential of these relationships Twiss (1986b) and Brown (1984) noted that grasses of the subfamily Pooideae, which produce Festucoid type phytoliths solely, also possess the C3 photosynthetic pathway, while Chloridoid and Panicoid grasses, producers of saddles, bilobates, and cross shapes, tend strongly to be C4. There are some confuser genera that must be sorted out on a region-by-region basis. The Panicoid subfamily includes both C3 and C4 grasses and even within certain genera like *Panicum*, some species are C3 and some are C4. Many bamboos, all of which are C3 grasses, produce saddle-shaped silica bodies, and *Arundinaria* and *Danthonia*, C3 grasses in the subfamily Arundinoideae, produce saddles and bilobates, respectively. If the presence of confuser genera can be ruled out in fossil assemblages on the basis of phytolith morphology or ecological considerations, the distribution and abundance of grass short cell phytoliths may provide sensitive paleoclimatic indices.

Twiss (1986b) proposes a method of fine resolution using a phytolith index that is a ratio of the Chloridoid types to the total C4 types. Since Chloridoid grasses tolerate heat and drought even better than Panicoid grasses, the index might discriminate between warm and dry versus warm and moister paleoclimates. At the very least, as Brown (1984, p. 22) notes, since the Pooideae (subfamily producing Festucoid-type phytoliths) lack bilobates and crosses, "a paucity of these two phytolith shapes in earth materials, coupled with the abundance of trapezoids, probably indicates a dominance of C3 grasses."

Kaufman *et al.* (1985) demonstrated other differences between C3 and C4 grasses. C4 grasses displayed three times as many silica cells than did C3s and had smaller and thicker long epidermal cells. All of these may contribute to further means of discrimination in phytolith assemblages and serve to further illustrate the disparities between the two groups having different photosynthetic pathways. It is at this time unclear whether phytolith shapes are incidental to the photosynthetic pathway or whether both are related to ecological conditions in which they are found.

Returning again to Fredlund's study of Nebraskan loess, abundant and identifiable phytoliths were found throughout the very ancient deposits. Quantitative phytolith analysis of a column sample showed that phytolith content increased in the upper horizons of most paleosols observed, hence ". . . the translocation of these microfossils is not an acute problem and . . . at least some pedogenic zonation is preserved" (Fredlund, 1986b, p. 16). In most paleosols, especially the best developed, the Sangamon, which is covered by Wisconsinan loess, the relative frequency of Panicoid and Chloridoid phytoliths increased over Festucoid frequencies. Chloridoid types occurred in significantly higher frequencies than in the modern central Nebraskan grass cover. Phytolith evidence suggests that Sangamon climate was warmer and drier than today's, supporting widely held interpretations of paleosol development. Immediately above the Sangamon paleosol the percentage of Panicoid phytoliths increased tremendously, forming a unique phytolith zone that corresponded with a unique vertebrate faunal zone identified from other nearby sections of Nebraskan loess, which is dominated by remains of long-legged grazers adapted for running in open country. The phytolith record suggests that these tall animals lived in tall grass prairie, or perhaps savannah.

Lake Sediments

The enormous potential that phytoliths offer in the paleoecology of lake environments has been discussed at length in Chapter 6. To the author's knowledge, the only published study dealing with biogenic silica in lacustrine sediments is Palmer's (1976), which emphasizes silicified cells as they are found embedded in fossil grass cuticles. Lake core samples from East Africa are rich in grass pollen grains, but their morphologies are one and the same, eliminating the possibility of identification below the family level. It is extremely important in this region to make tribal or genus-level identification of Gramineae fossils because the major shifts in climate and vegetation impacted on the grass cover in ways that could be understood if the composition of the ancient cover were known.

Palmer isolated grass cuticles in large amounts (100/mL of lake mud) from 28,000-year-old East African lake sediments, indicating that disarticulated phytoliths should be recoverable in much higher frequencies yet. The cuticular fragments were mainly charred, implicating fire as the primary depositional agent, although it is unclear whether cuticles were washed into lakes via streams or slopewash or were transported through the air. By association of silica-body types with a number of other anatomical features of the grass epidermis, such as trichomes, papillae, and stomata, and by comparison to an extensive modern collection of native grasses,

Palmer was able to tentatively identify several genera of grasses in the Late Pleistocene lake sediments, grasses that today inhabit savanna environments. Future development of this technique is extremely important in eliciting habitat preferences (woodland, steppe, savanna) of ancient grasses, and hence, ancient landscapes and climate (Palmer and Tucker, 1981, 1983; Palmer *et al.*, 1985).

Disarticulated phytoliths in lake deposits will also offer a means of discrimination of some grass taxa to tribal and genus level. More detailed descriptions of two- and three-dimensional shape attributes along with systematic size measurements need to be made of African grasses. There is every reason to expect the same degree of differentiation of phytoliths from African Bambusoid, Oryzoid, and other grasses that has been found achievable in tropical American representatives, permitting discrimination of forest and open-terrain Gramineae taxa.

Where generic identification is not possible, the dominance of Festucoid (C3) versus Chloridoid and Panicoid (mainly C4) grasses in lake assemblages may offer fine resolution estimates of paleotemperature and altitudinal shifts of vegetation zones. Livingston and Clayton (1980) have documented a very clear altitudinal cline in tropical African grasses. C4 grasses are dominant below 1300 m and start declining in frequency at about 2000 m. C3 grasses, mainly Pooids, increase greatly in frequency above 2000 m and then dominate the flora starting at 3000 m. Above 4000 m only species of the Pooideae subfamily (all C3 grasses) are found. Temperature appears to be the main correlate of the altitudinal cline in grass composition. Chazdon (1978) has documented a similar relationship in Costa Rica. C4 grasses were found primarily in lowland habitats and C3 grasses occurred chiefly in high-altitude environments over 1700 m in elevation. In Panama, the Pooideae genera *Agrostis, Bromus, Festuca,* and *Polypogon* rarely occur at elevations below 2000 m and are not found below 1500 m. All of these grasses contribute large numbers of trapezoids, and *Bromus* and *Polypogon* produce large numbers of the "woven trapezoids" (Brown, 1984, 1986; Twiss *et al.*, 1969, "oblong sinuous" class) that are specific to the Pooideae (Piperno, 1987a). Since the distribution of the Pooideae in tropical environments appears to be primarily correlated with temperature, distribution and abundance of their phytoliths and those of C4 grasses in lake deposits may provide independent estimates of paleotemperature and range extensions of past vegetation.

In an eloquent article written about problems of African lake pollen interpretations Livingston (1982) has enumerated a number of unresolved issues that may benefit from study with phytolith data. The sheer diversity of the flora of tropical Africa with its dominance by insect-pollinated taxa presents formidable barriers to reconstruction of past vegetation. The generation of phytolith data lowers some of these barriers because, as in the American tropics, insect-pollinated plants may be abundant producers of distinctive phytoliths. With phytolith data, many more taxa will be retrievable in significant numbers from the microfossil record, permitting a finer discrimination of floral associations and their climatic correlates and significance. Phytolith analysis should help to resolve, among other important questions, the severeness of aridity in Africa during the Late Pleistocene. In other depositional environments not conducive to good pollen preservation, such as the dry and oxidized upper Pliocene and lower Pleistocene hominid fossil beds, good phytolith records are probably obtainable and may provide valuable data on ancient landscapes.

Deep-Sea Cores

Ehrenberg recognized the significance of phytoliths and fresh-water diatoms he found in North Atlantic dust samples collected by Darwin 300 miles west of the African continent. It meant a considerable degree of aeolian transport from continental land masses. Over time these microindicators of continental plant life will settle and accumulate in deep-sea sediment and when retrieved from deep-sea cores provide valuable paleoecological data. Phytoliths commonly occur in marine sediments in areas close to continents where permanent aeolian transport is possible by strong winds blowing out to sea. Hence, the main areas of phytolith distribution are in the Atlantic Ocean off the coast of Northwest Africa and in the eastern equatorial Pacific where easterly winds prevail. The location and magnitude of biogenic silica particles thus provide an index of past meteorological conditions (Maynard, 1976).

Deep-sea sediments often comprise long chronological records extending back into the Miocene and Pliocene, but only gross temporal correlations between microfossil assemblages and vegetation can be made because millions of years of deposition will often be represented in something like 60–70 m of sediment. Still, a number of intriguing suggestions about past climate and vegetation have been inferred from phytoliths retrieved from deep cores, whose finer details must be worked out with terrestrial paleobotanical data.

Parmenter and Folger (1974) correlated high incidence of phytoliths and fresh-water diatoms in cores from the equatorial Atlantic with cold paleotemperatures as inferred from oxygen isotopes and foraminiferans. They concluded that glacial episodes were coeval with dry continental climates, which reduced North African vegetation and lowered lake levels, permitting phytoliths and diatoms to become a significant part of the aerosol dust load.

Bukry carried out similar analyses of sea core sediments from Northwest Africa (1979) and the tropical Eastern Pacific Ocean (1980). The source area for the latter was considered to be Central America. In the Eastern Pacific core, phytolith occurrence was less frequent in Upper than in Lower Quaternary sediments, possibly suggesting less dry conditions in the Upper Quaternary. Fresh-water diatoms were far scarcer in the Eastern Pacific than in the African assemblages, a factor that argues for a general lack of dried-up lake beds and streams and, therefore, wetter conditions in Central America than in Northwest Africa during the later stages of the Pleistocene. These are intriguing data that urgently need to be correlated with mainland pollen and phytolith profiles, as they bear directly on the nature of the environments that Paleo-Indians encountered when they first settled tropical America.

In sediments from 1000 km off the coast of eastern Australia (Locker and Martini, 1986), grass phytoliths were much more common in middle and upper Miocene zones than in lower Miocene zones, with a sudden increase of grass silica noted in early Pliocene levels. This record independently documented the development of open Australian grassland formations during the middle Miocene, already suggested by pollen studies, and provided a more precise date for this phenomenon than determined from palynological studies.

Other Potential Uses of Phytoliths

At this point in time a few other potential uses of phytoliths in paleoecological reconstruction can be suggested. Just as $^{18}O/^{16}O$ ratios of marine biosilicates have provided paleotemperature estimates, ratios in terrestrial plant phytoliths may do the same. Bombin and Muehlenbachs (1980) found that in controlled experiments with relative humidity held constant ^{18}O levels accurately calculated the temperature under which the plant was grown. Relative humidity in nature is a direct result of both temperature and moisture, hence fluctuations in either may affect isotope ratios.

Lastly, if enough occluded carbon were to be isolated from phytoliths, it might provide a means of direct radiocarbon dating of geological microfossils.

Plates

1. Spherical spinulose phytolith from *Sabal minor*.
2. Spherical spinulose phytoliths from *Bromelia karatas*.
3. Spherical nodular phytoliths from *Maranta arundinaceae*.
4. Spherical nodular phytolith from *Stromanthe lutea*.
5. Spherical rugulose phytoliths from *Canna indica*.
6. Spherical smooth phytoliths from seeds of *Hirtella triandra*.
7. Phytolith with irregularly angled or folded surface from *Calathea violaceae*.
8. Partially armed hair cell phytolith from *Wulffia baccata*.
9. Hair cell phytolith from *Manihot esculenta*.
10. Segmented hair cell phytolith from *Lagenaria siceraria*.
11. Thin, curved hair cell phytoliths from *Odontocarya tamoides*.
12. Mesophyll phytoliths from cotton *(Gossypium barbadense)*.
13. Tracheid phytoliths from avocado *(Persea americana)*.
14. Spherical scalloped phytoliths from *Cucurbita* rinds.
15. Segmented hair phytolith from *Cucurbita pepo*.
16. Segmented hair phytolith from *Cucurbita pepo*.
17. Broken segmented hair phytolith from *Cucurbita palmata*.
18. Segmented hair phytolith from *Luffa aegyptica*.
19. Hair base phytolith from *Luffa aegyptica*.
20. Spined phytoliths from achira *(Canna edulis)*.
21. Conical to hat-shaped phytolith from *Maranta arundinaceae*.
22. Phytoliths with shallow troughs from *Musa*.
23. Variant 1 cross-shaped phytolith from *Zea mays*.
24. Variant 6 and Variant 2 cross-shaped phytoliths from *Tripsacum dactyloides*.
25. Variant 2 cross-shaped phytolith from *Cenchrus echinatus*.
26. Variant 3 cross-shaped phytolith from *Cryptochloa*.
27. Variant 6 cross-shaped phytolith from *Cenchrus echinatus*.
28. Variant 6 cross-shaped phytolith from the husk of *Zea mays*.
29. Variant 6 cross-shaped phytolith from the husk of *Zea mays*.
30. Irregular short cell phytolith from *Paspalum plicatulum*.
31. Irregular short cell phytolith from *Paspalum plicatulum*.
32. Dumbbell from *Oplismenus hirtellus*.

33. Variant 5/6 dumbbell from *Cenchrus echinatus*.
34. Irregular mesophyll (?) cell phytoliths from *Olyra latifolia*.
35. Concave dumbbell from *Streptochaeta*.
36. Concave dumbbell from *Chusquea*.
37. Dumbbell from *Pharus*.
38. Hat-shaped phytolith from *Acrocomia vinifera*.
39. Hat-shaped phytoliths from *Geonoma interuppta*.
40. Hat-shaped phytoliths from *Chaemadorea*.
41. Nodular, conical-shaped phytolith from *Ischnosiphon arouma*.
42. Phytoliths with irregularly angled to folded surfaces from *Dimerocostus uniflorus*.
43. Phytolith with deep, central trough from *Heliconia*.
44. Tongue-shaped phytolith from *Athyrocarpus persecariefolium*.
45. Phytoliths from seed of *Stromanthe lutea*.
46. Nonsegmented hair phytolith from *Laportea aestuans*.
47. Short, deltoid-shaped hair phytolith from *Curatella americana*.
48. Nonsegmented hair phytolith from *Trophis racemosa*.
49. Nonsegmented hair phytolith from *Trophis racemosa*.
50. Hair cell phytoliths from *Piper flagellicuspe*.
51. Long, threadlike nonsegmented hair phytoliths from *Trema micrantha*.
52. Long, threadlike nonsegmented hair phytoliths from *Elephantopus mollis*.
53. Nonsegmented armed hair phytolith from *Castilla elastica*.
54. Segmented armed hair phytolith from *Eclipta alba*.
55. Segmented hair phytolith from *Bidens*.
56. Segmented hair phytolith from *Calea urticifolia*.
57. Segmented hair cell phytoliths from *Melothria guadalupensis*.
58. Segmented hair phytoliths from *Piper pseudoasperi*.
59. Multicelled hair base phytoliths from *Curatella americana*.
60. Hair base phytoliths from *Heliotropium angiospermum*.
61. Hair base phytolith from *Cordia lutea*.
62. Hair base phytolith from *Ficus americana*.
63. Hair base phytolith from *Chlorophora tinctoria*.
64. Hair base phytoliths from *Cecropia peltata*.
65. Spherical verrucose cystoliths from *Boehmeria aspera*.
66. Curved cystolith from *Pilea acuminata*.
67. Elongate, verrucose cystolith from *Laportea aestuans*.
68. Elongate, verrucose cystolith from *Laportea aestuans*.
69. Roughly spherical to irregular cystoliths from *Urera elata*.
70. Cystolith from *Justicia pringlei*.
71. Cystoliths from *Pseuderanthemum davei*.
72. Polyhedral epidermal phytoliths from *Hedyosmum*.
73. Anticlinal epidermal phytoliths from *Hedyosmum*.
74. Silicified epidermis from *Pourouma aspera*.
75. Irregular multifaceted phytolith from *Guatteria dumetorum*.
76. Spherical multifaceted phytolith from *Unonopsis pittieri*.
77. Elliptical multifaceted phytolith from *Protium panamense*.
78. Silicified sclereids from *Hirtella triandra*.
79. Silicified sclereid from *Hirtella triandra*.
80. Double saucer-shape phytolith from *Licania hypoleuca*.
81. Irregular pointed phytolith from *Phthirusa pyrifolia*.
82. Irregular pointed phytolith from *Odontocarya tamoides*.

83. Fruit phytolith from *Tetragastris panamensis*.
84. Fruit phytolith from *Tetragastris panamensis*.
85. Fruit phytolith from *Protium panamense*.
86. Seed phytolith from *Bursera simaruba*.
87. Seed phytolith from *Trattinickia aspera*.
88. Seed phytolith from *Mendoncia retusa*.
89. Opaque, perforated platelet from floral bracts of *Zinnia elegans*.
90. Elongate phytoliths from *Adiantum fructosum*.
91. Bowl-shaped phytolith from *Trichomanes osmundoides*.
92. Phytoliths from *Selaginella arthritica*.
93. Elongate phytoliths from modern soils underneath Panamanian tropical forest.
94. *Curatella americana* multicelled hair base phytolith from the Aguadulce shelter.
95. Spherical spinulose Palmae phytolith from the Aguadulce shelter.
96. Elongate multifaceted phytolith from *Guatteria amplifolium*.

1. SEM photograph of spherical spinulose phytolith from *Sabal minor* (size range: 3–27 μm, diam).

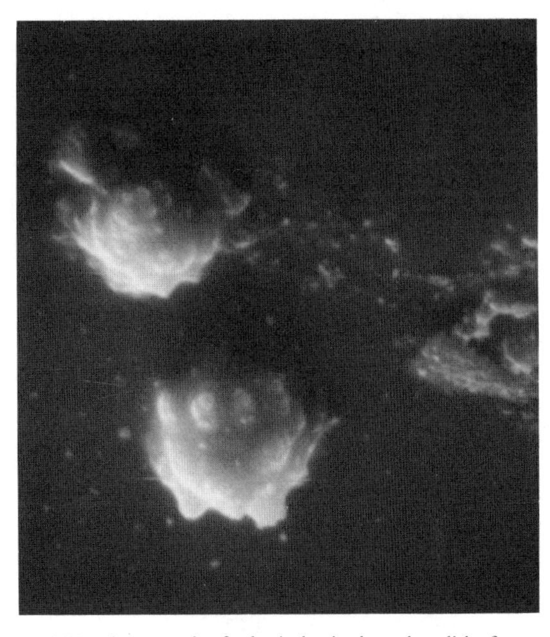

2. SEM photograph of spherical spinulose phytoliths from *Bromelia karatas* (size range: 2–8 μm, diam).

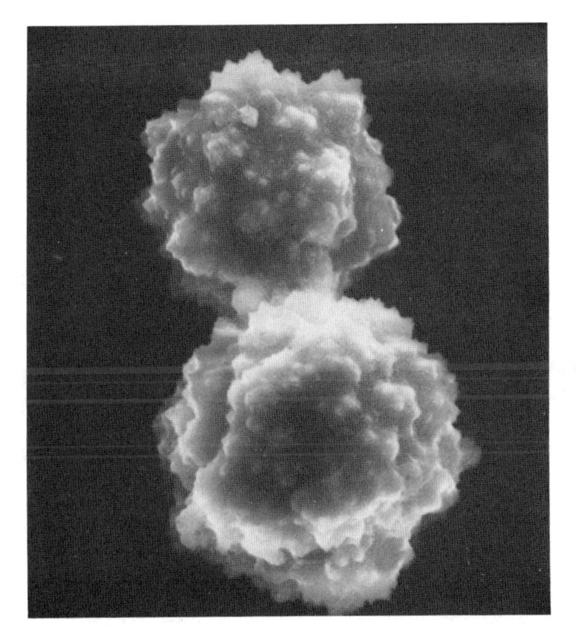

3. SEM photograph of spherical nodular phytoliths from *Maranta arundinaceae* (size range: 9–18 μm, diam).

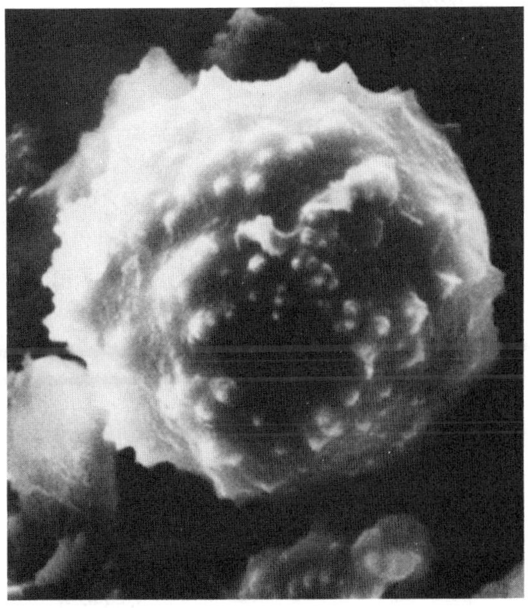

4. SEM photograph of spherical nodular phytolith from *Stromanthe lutea* (size range: 12–27 μm, diam).

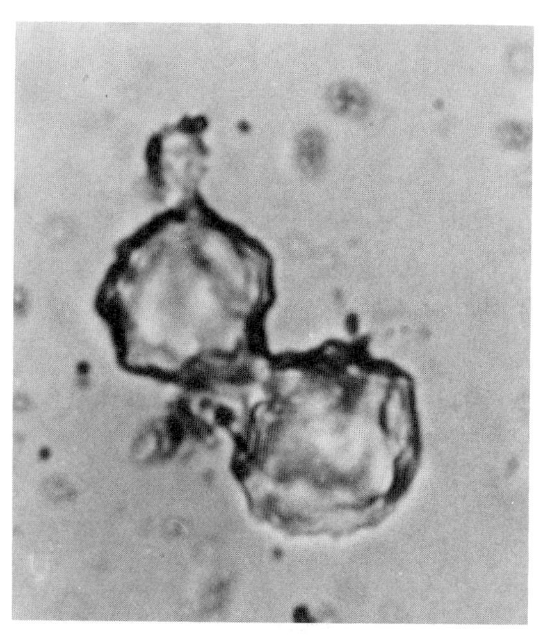

5. Spherical rugulose phytoliths from *Canna indica* (size range: 9–30 μm, diam).

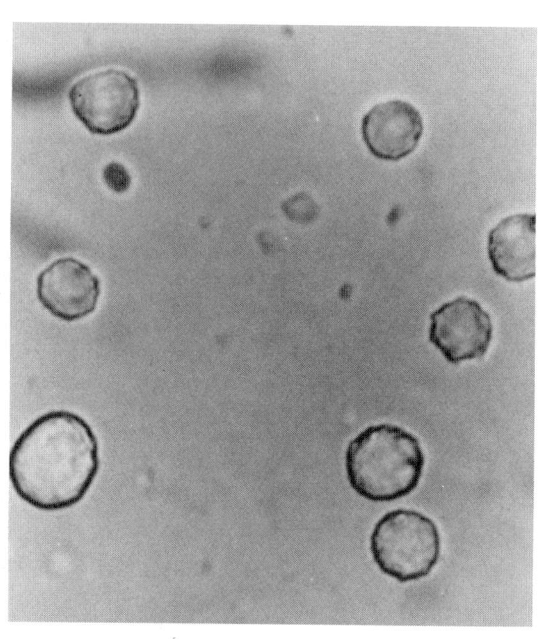

6. Spherical smooth phytoliths from the seeds of *Hirtella triandra* (size range: 6–13 μm, diam). A few spherical rugulose forms are also visible.

7. SEM photograph of phytolith with irregularly angled or folded surface from *Calathea violaceae* (size range: 6–18 μm, diam).

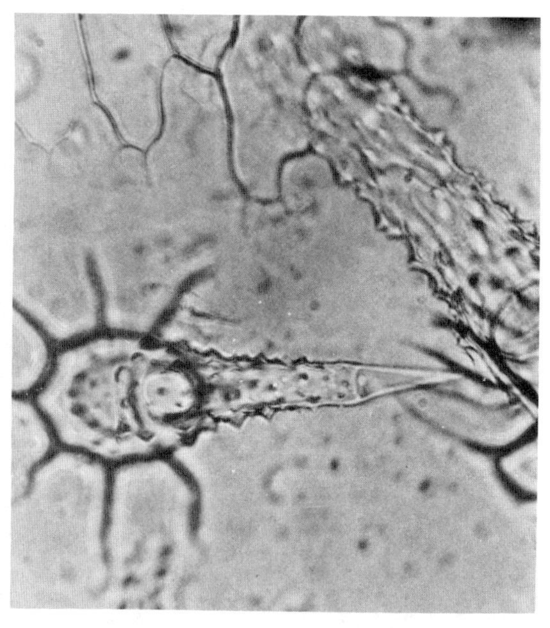

8. Partially armed hair cell phytolith from *Wulffia baccata* (size range: 78–455 μm long).

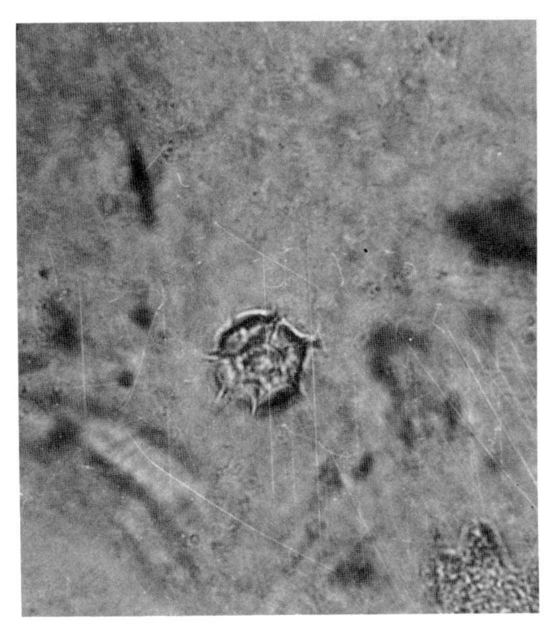

9. Hair cell phytolith from *Manihot esculenta* (size range: 16–26 μm, diam).

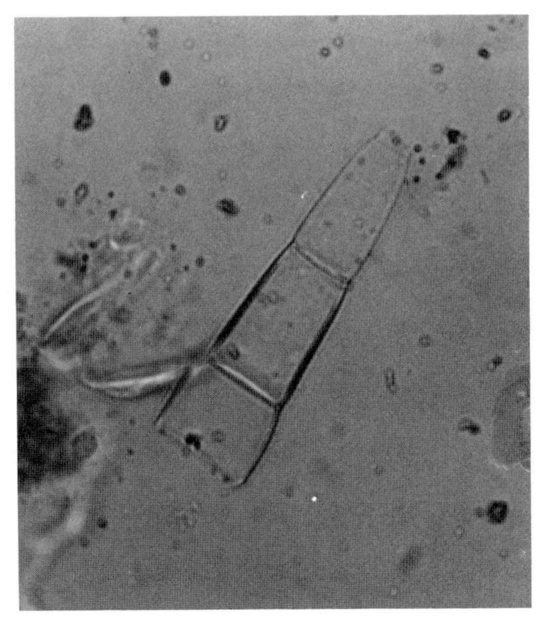

10. Segmented hair cell phytolith from *Lagenaria siceraria* (size range: 66–225 × 9–36 μm).

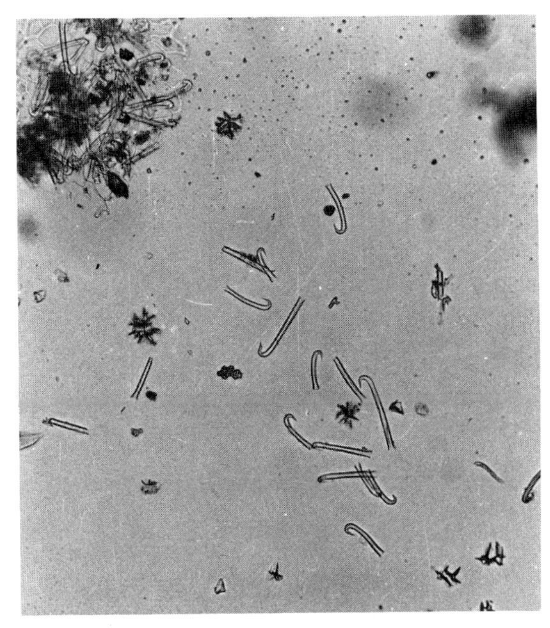

11. Thin, curved hair cell phytoliths from *Odontocarya tamoides* (size range: 48–108 × 6–12 μm). Also present are irregular, pointed phytoliths.

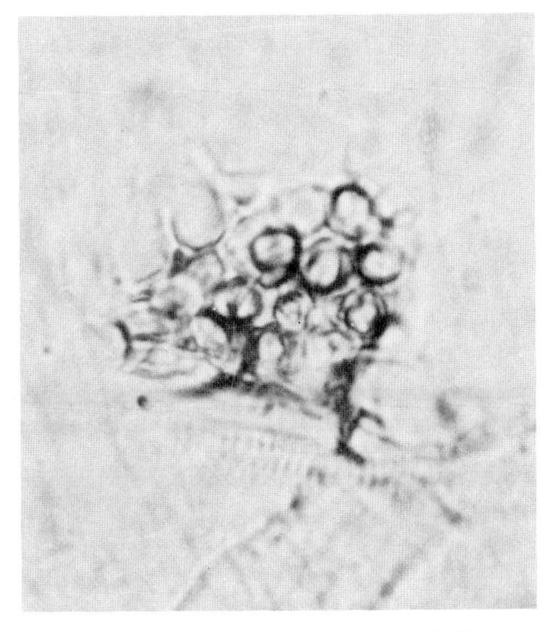

12. Mesophyll phytoliths from cotton *(Gossypium barbadense)*.

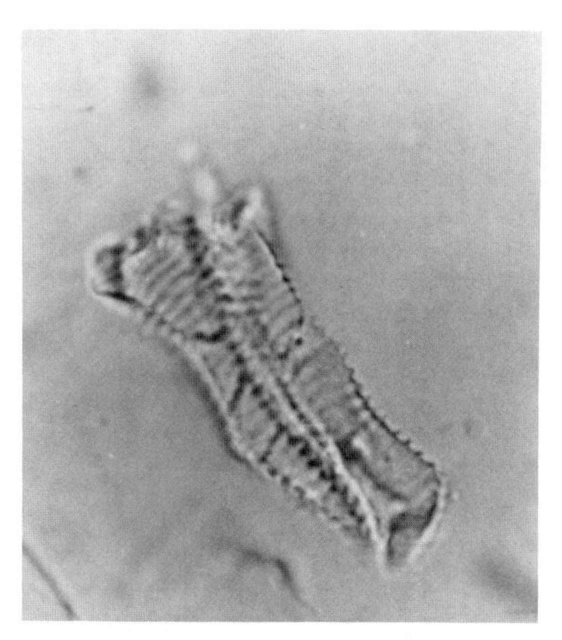

13. Tracheid phytoliths from avocado *(Persea americana)*.

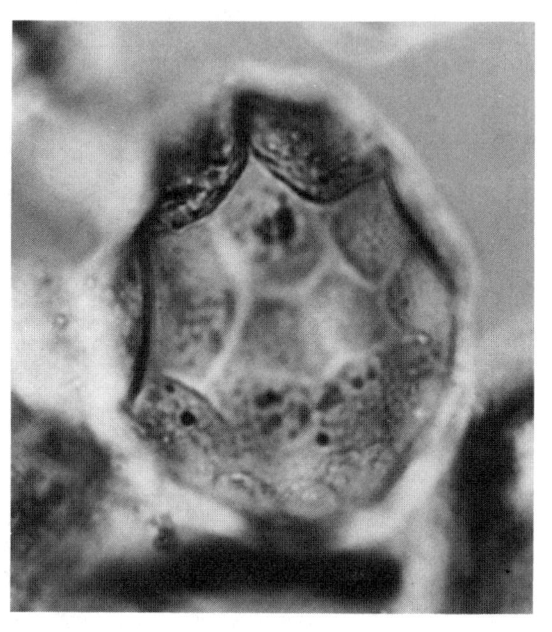

14. Spherical scalloped phytoliths from *Cucurbita* rinds (size range: 48–78 μm, diam).

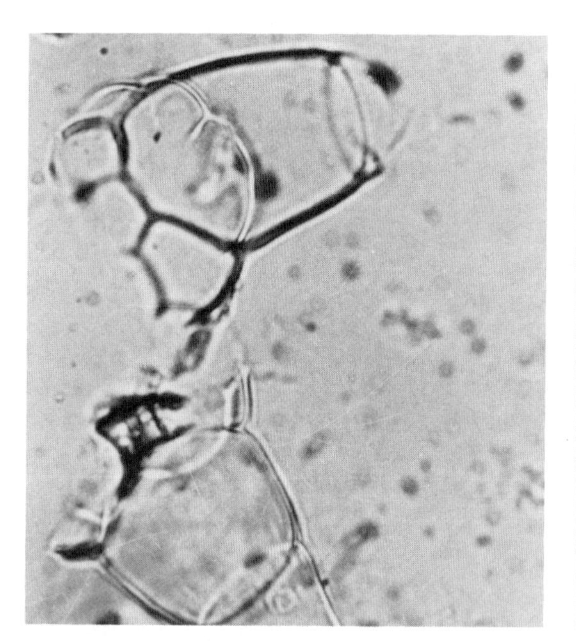

15. Segmented hair phytolith from *Cucurbita pepo* with two segments and a small, round apical segment.

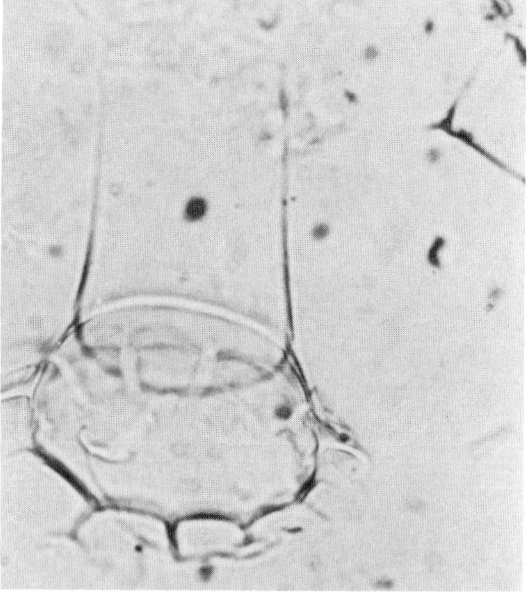

16. Segmented hair phytolith from *Cucurbita pepo* with somewhat tapered basal and intermediate segment and small, rounded apical segment (size range: 99–135 × 59–69 μm).

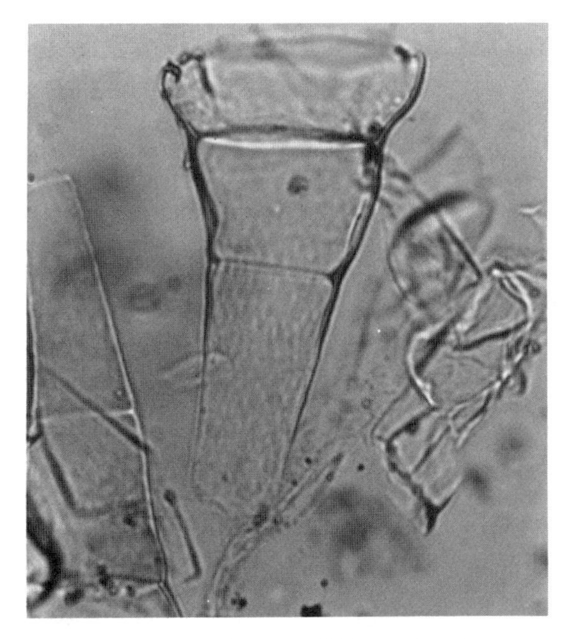

17. Broken segmented hair phytolith from *Cucurbita palmata* with the surface pattern of short, fine striations.

18. SEM photograph of segmented hair phytolith from *Luffa aegyptica* that has small spines; with light microscopy the spines appear as short, fine striations (size range: 63–120 × 21–45 μm).

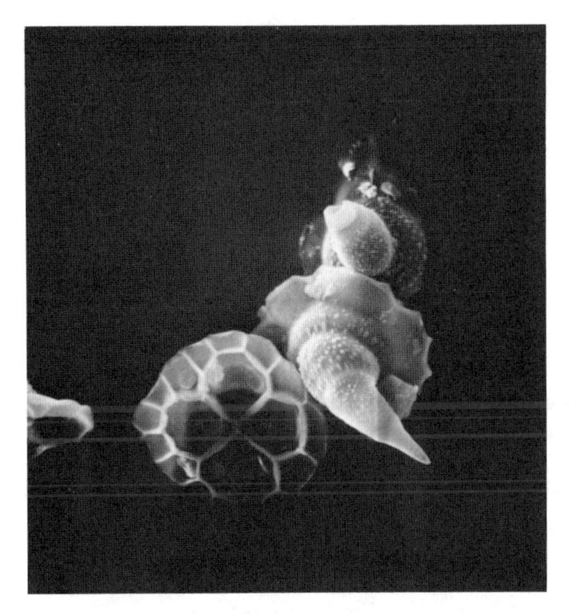

19. SEM photograph of hair base phytolith from *Luffa aegyptica* consisting of four conjoined spheres. Also shown are hair cell phytoliths still attached to the hair base and hair base phytoliths bearing the remnants of armed hairs.

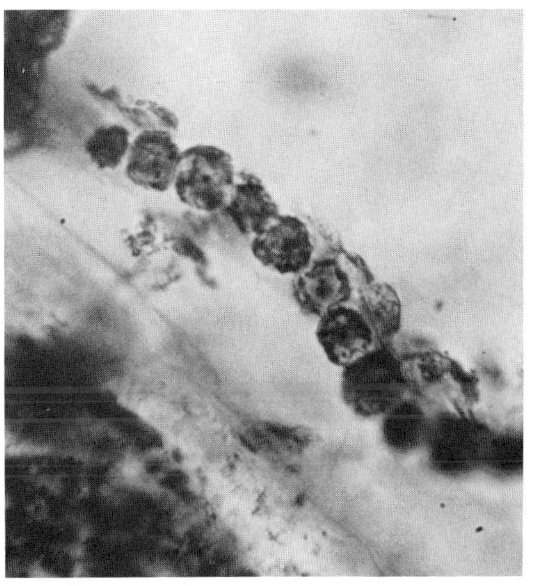

20. String of "spined" phytoliths from achira *(Canna edulis)*.

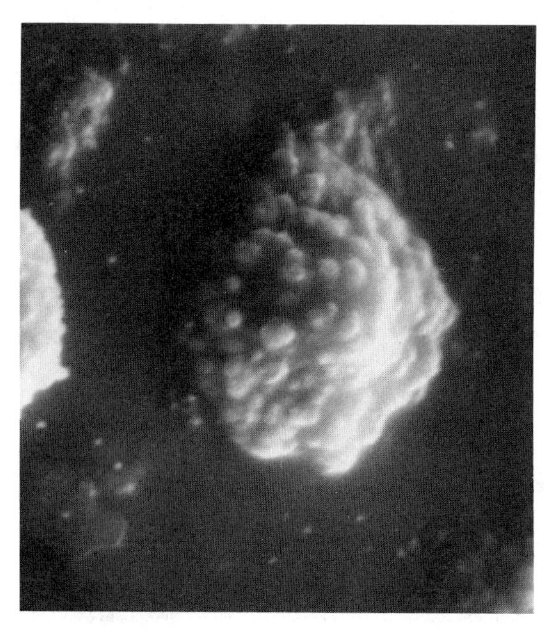

21. SEM photograph of conical to hat-shaped phytolith from *Maranta arundinaceae* (size range: 6–15 μm, base diam).

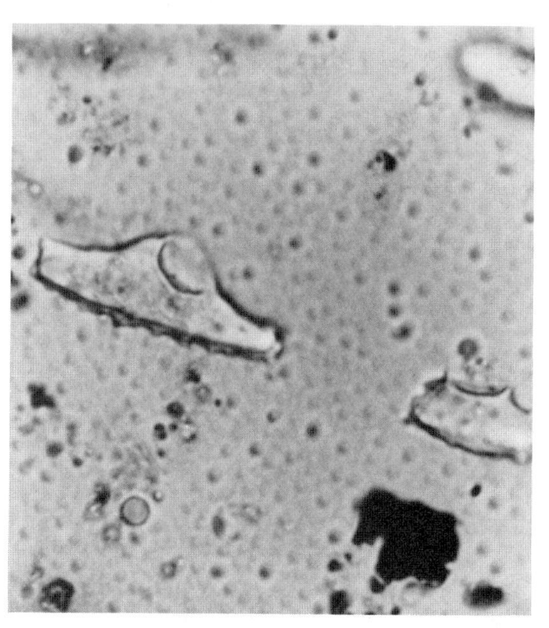

22. Phytoliths with shallow troughs from *Musa*.

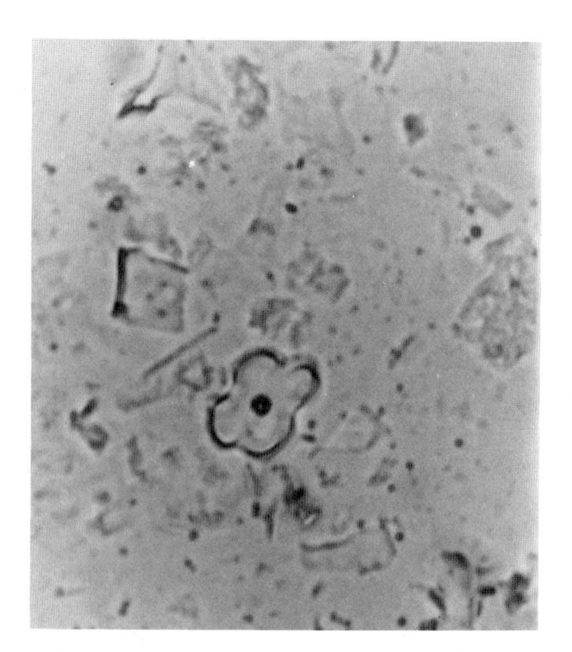

23. Variant 1 cross-shaped phytolith from *Zea mays*.

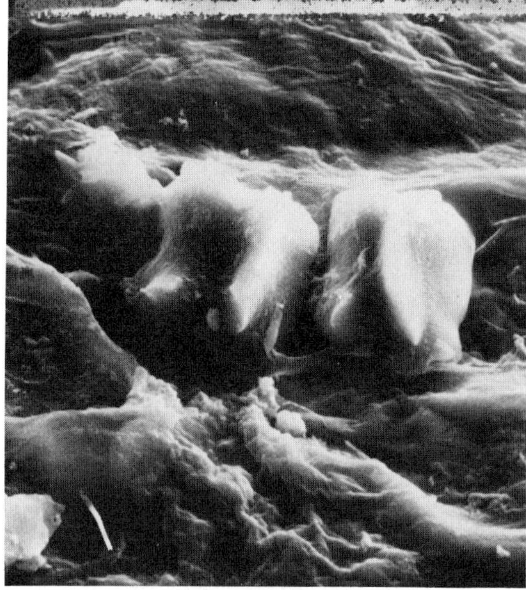

24. SEM photograph of Variant 6 (left) and Variant 2 (right) cross-shaped phytoliths from *Tripsacum dactyloides*.

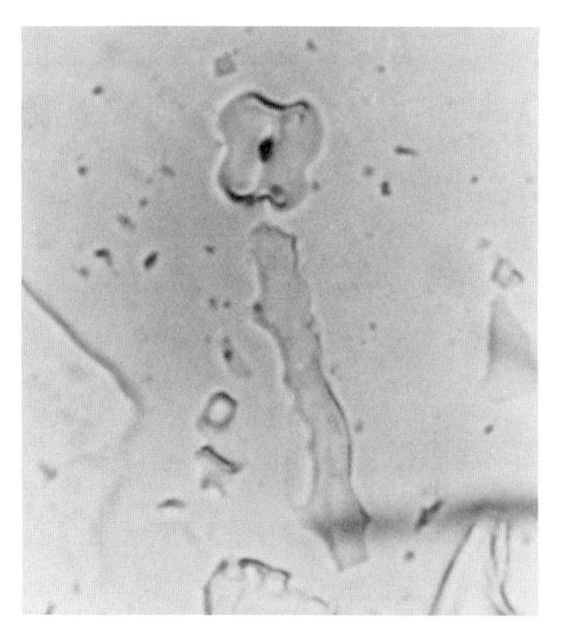

25. Variant 2 cross-shaped phytolith from *Cenchrus echinatus*.

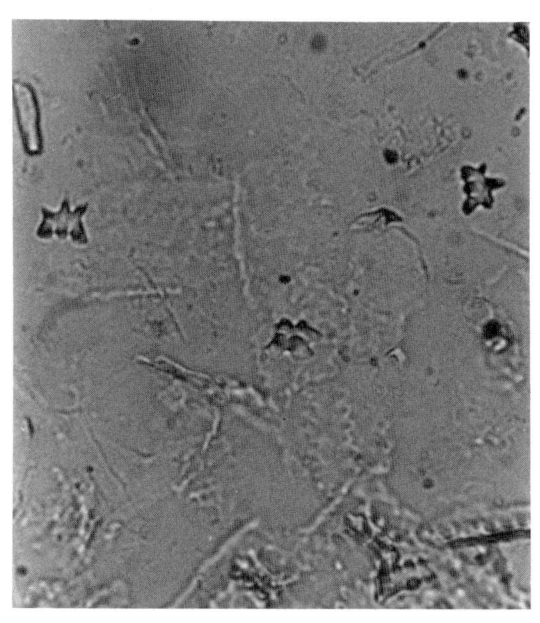

26. Variant 3 cross-shaped phytolith (center) from *Crypto-chloa*. Also shown are irregular mesophyll (?) cell phytoliths.

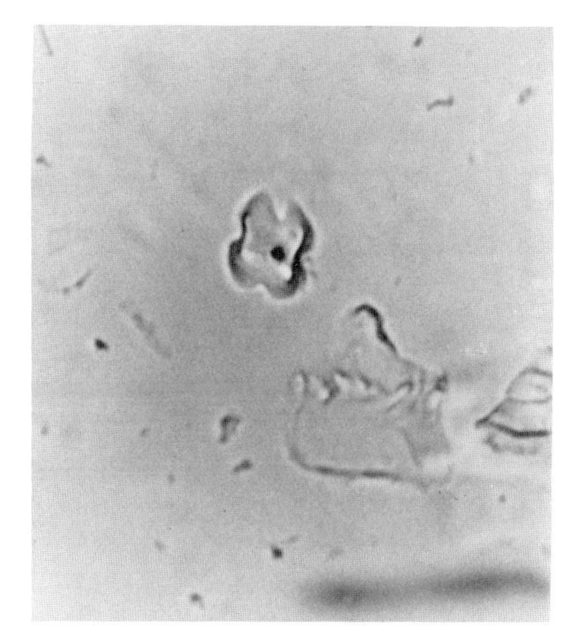

27. Variant 6 cross-shaped phytolith from *Cenchrus echinatus*.

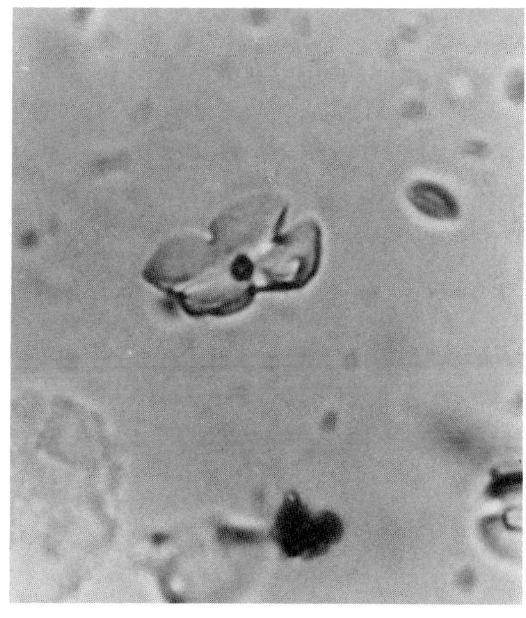

28. Variant 6 cross-shaped phytolith from the husk of *Zea mays*.

29. SEM photograph of a Variant 6 cross-shaped phytolith from the husk of *Zea mays*. The phytolith is on its side, demonstrating its extreme thickness.

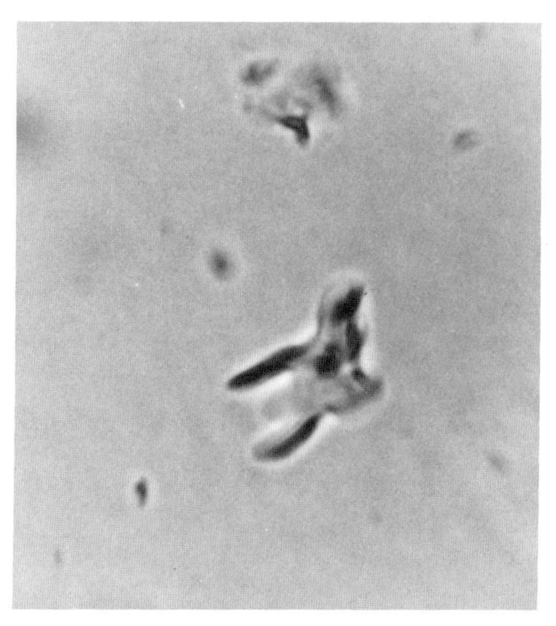

30. Irregular short-cell phytolith from *Paspalum plicatulum*.

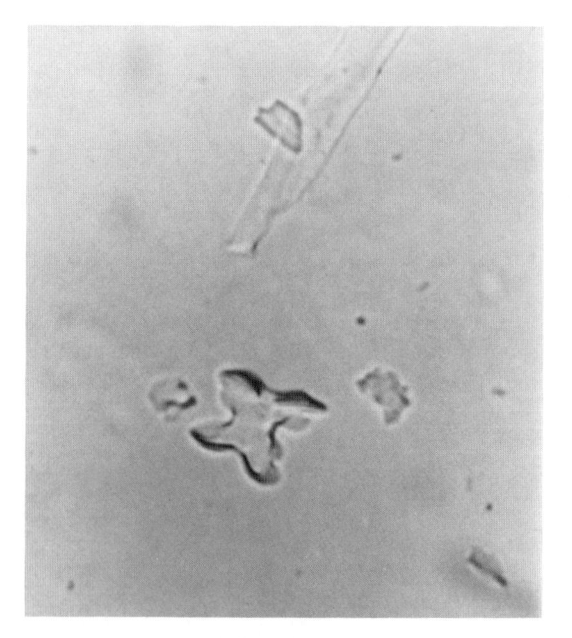

31. Irregular short-cell phytolith from *Paspalum plicatulum*.

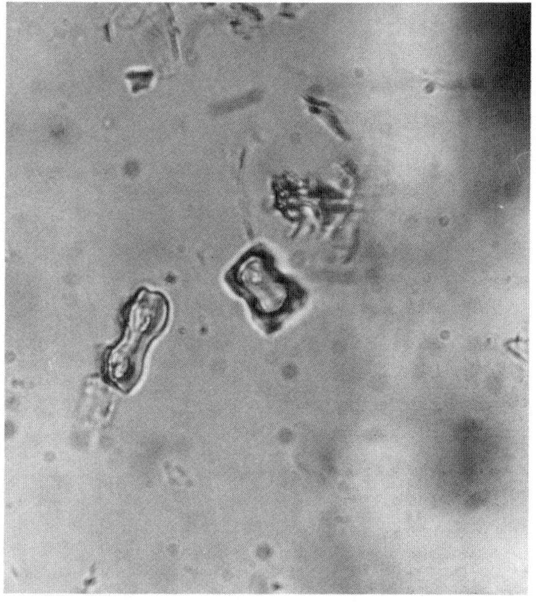

32. Dumbbell (center) from *Oplismenus hirtellus* with thick, rectangluoid structure on its opposite face.

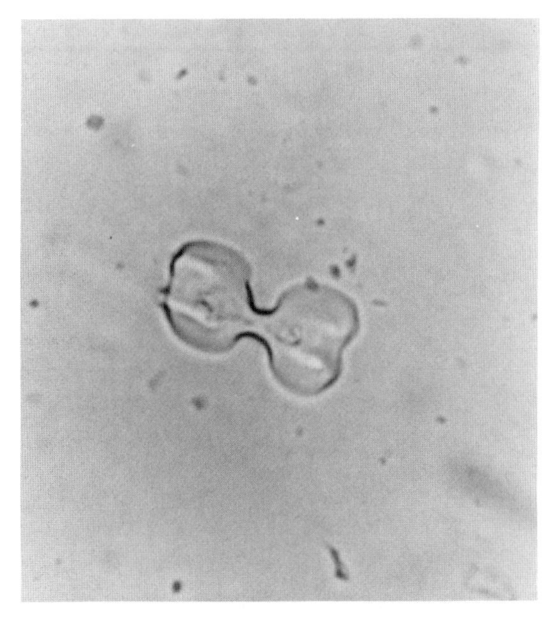

33. Variant 5/6 dumbbell from *Cenchrus echinatus*.

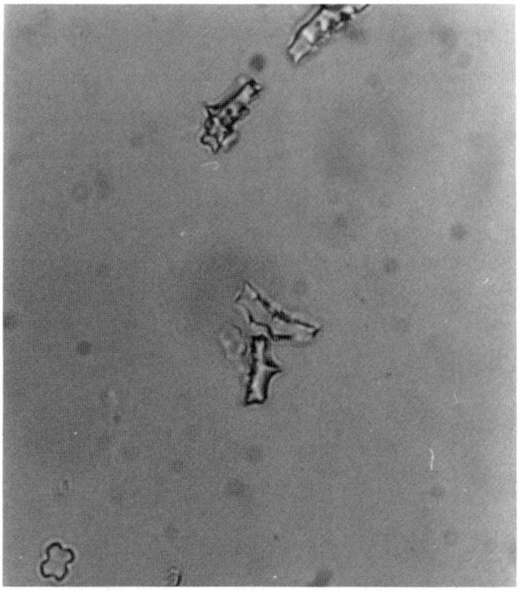

34. Irregular mesophyll (?) cell phytoliths from *Olyra latifolia*.

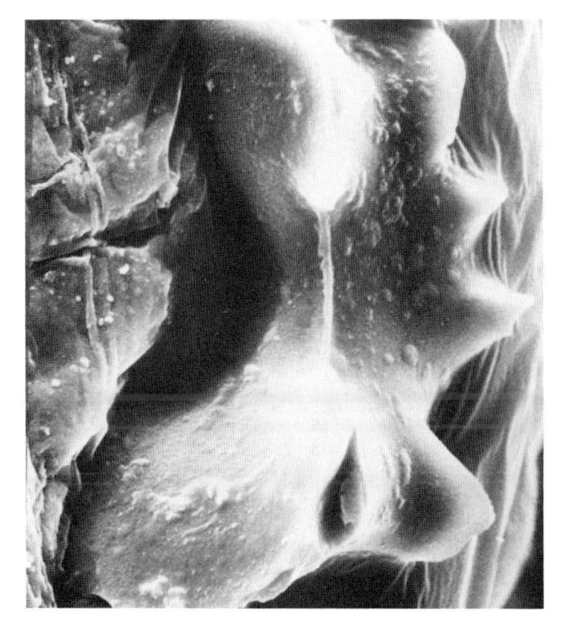

35. SEM photograph of concave dumbbell with several saddle-shaped structures on its opposite face from *Streptochaeta*.

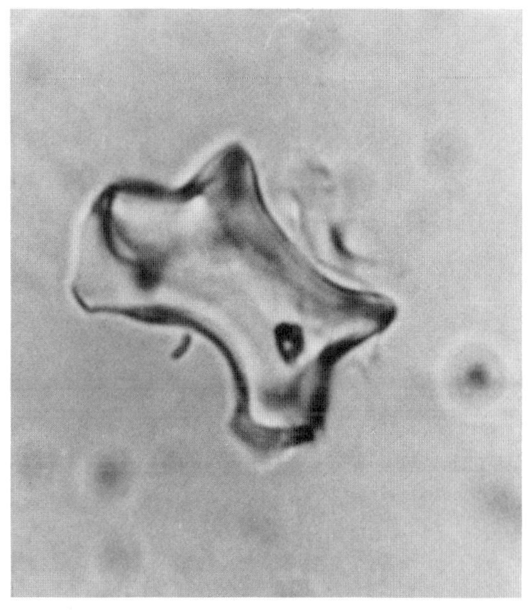

36. Concave dumbbell from *Chusquea* with single saddle-shaped structure on opposite face.

37. Side view of dumbbell from *Pharus* showing the thin and high rectanguloid structure on the opposite face (left side of the phytolith).

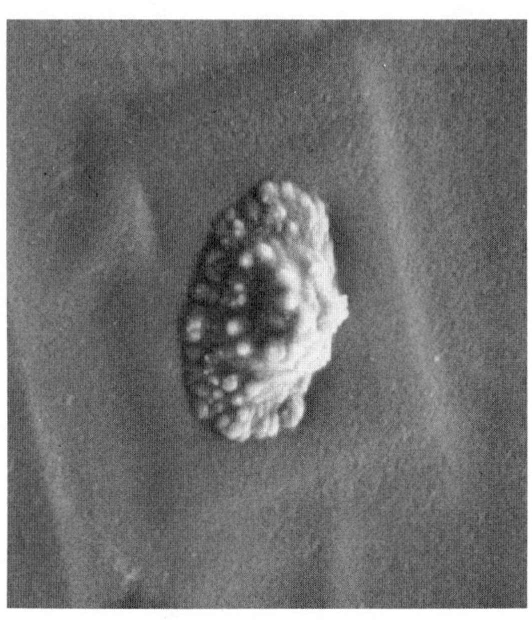

38. SEM photograph of hat-shaped phytolith from *Acrocomia vinifera* (size range: 3–15 μm, base diam).

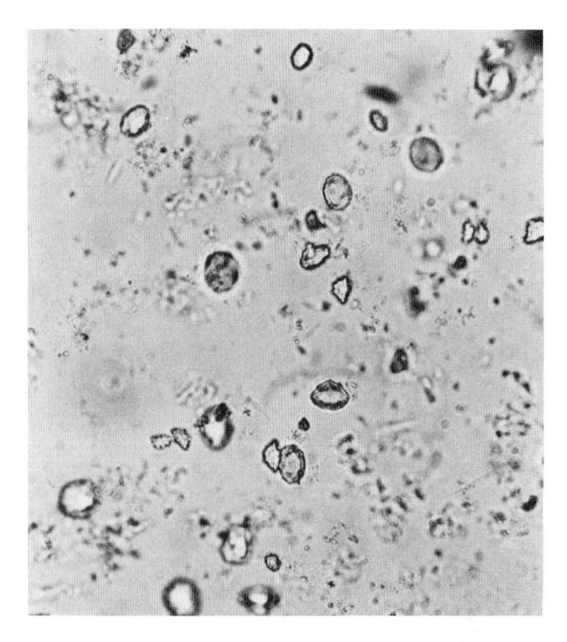

39. Hat-shaped phytoliths from *Geonoma interuppta* showing bulky, rugulose appearance (size range: 6–20 μm, base diam).

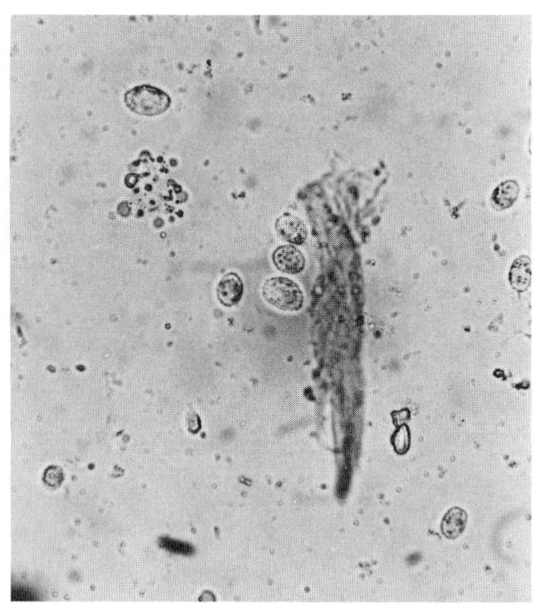

40. Hat-shaped phytoliths from *Chaemadorea*.

41. SEM photograph of a nodular, conical-shaped phytolith from *Ischnosiphon arouma* (size range: 9–18 μm, base diam).

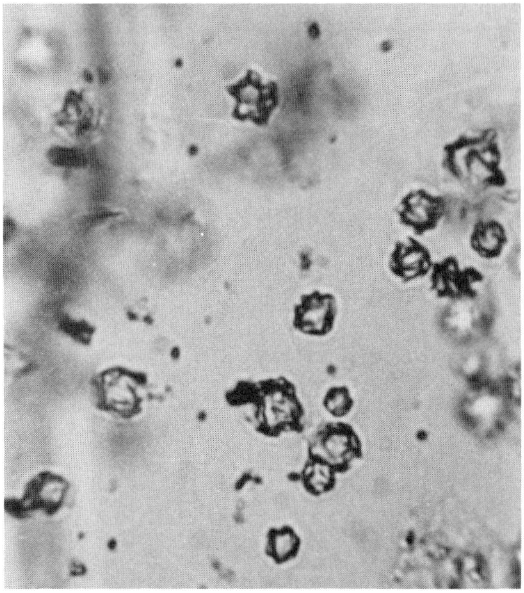

42. Phytoliths with irregularly angled to folded surfaces from *Dimerocostus uniflorus* (size range: 9–24 μm, diam).

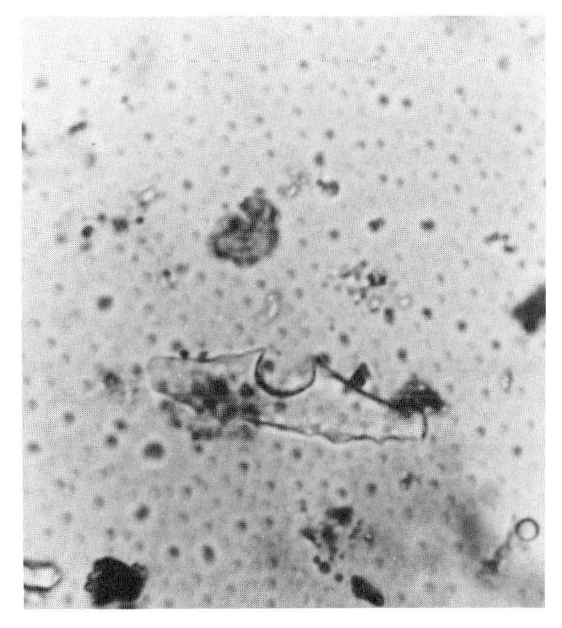

43. Phytolith with deep, central trough from *Heliconia* (size range: 13–25 × 4–10 μm).

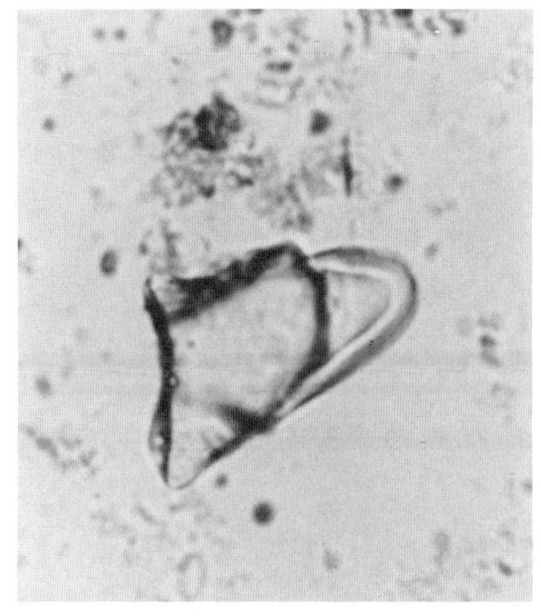

44. Tongue-shaped phytolith from *Athyrocarpus persecariefolium* (size range: 63–90 × 29–57 μm).

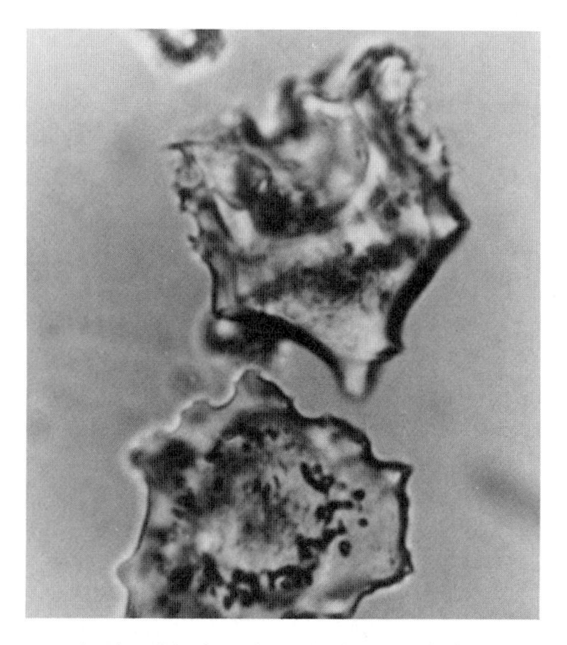

45. Phytoliths from the seed of *Stromanthe lutea*.

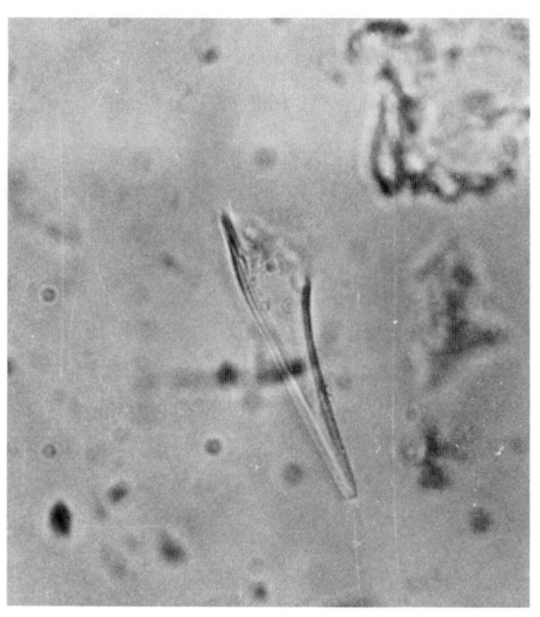

46. Nonsegmented hair phytolith from *Laportea aestuans*.

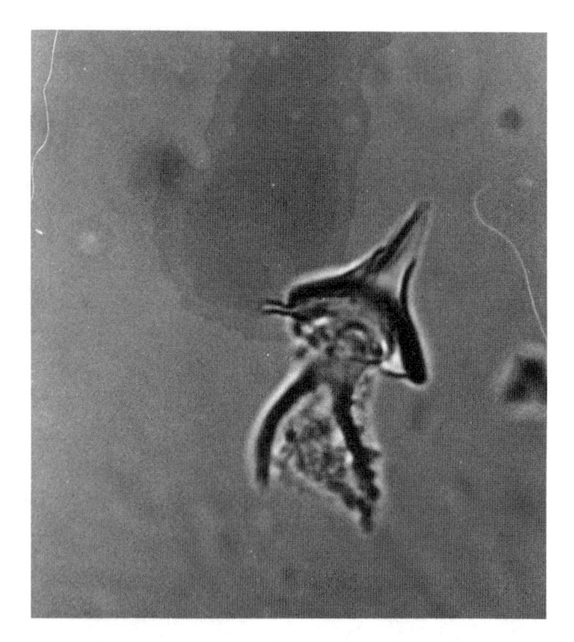

47. Short, deltoid-shaped hair phytolith from *Curatella americana* (size range: 17–21 × 15–18 μm).

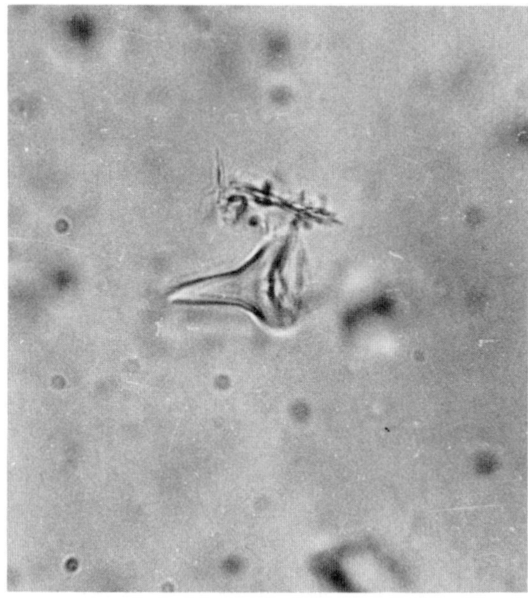

48. Nonsegmented hair phytolith from *Trophis racemosa*.

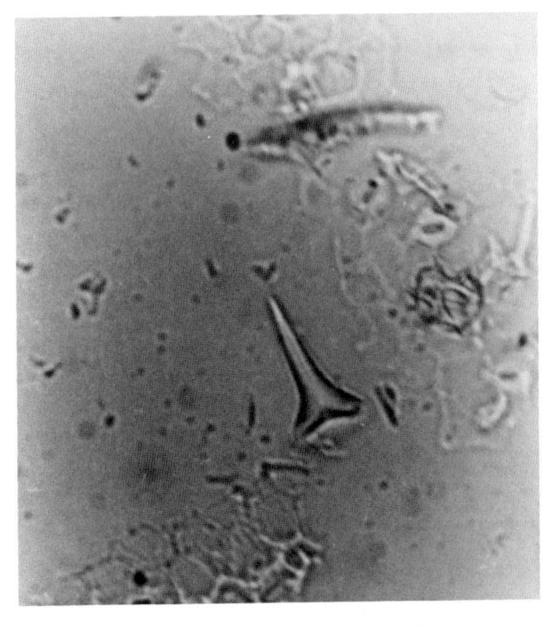

49. Nonsegmented hair phytolith from *Trophis racemosa*.

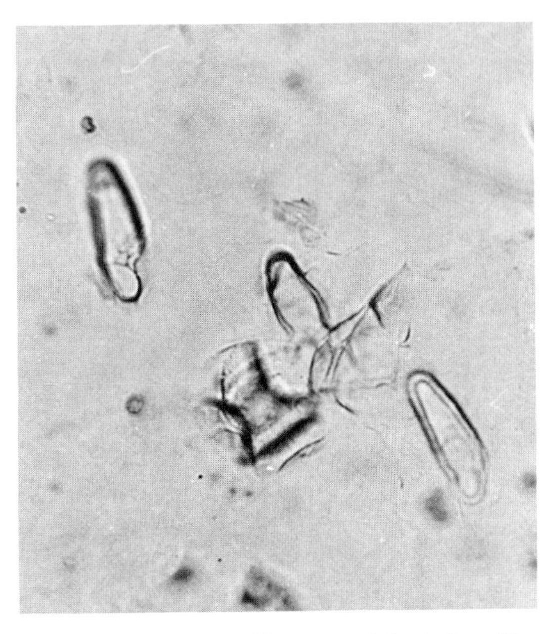

50. Hair cell phytoliths with two rounded ends from *Piper flagellicuspe* (size range: 27–46 × 9–12 μm).

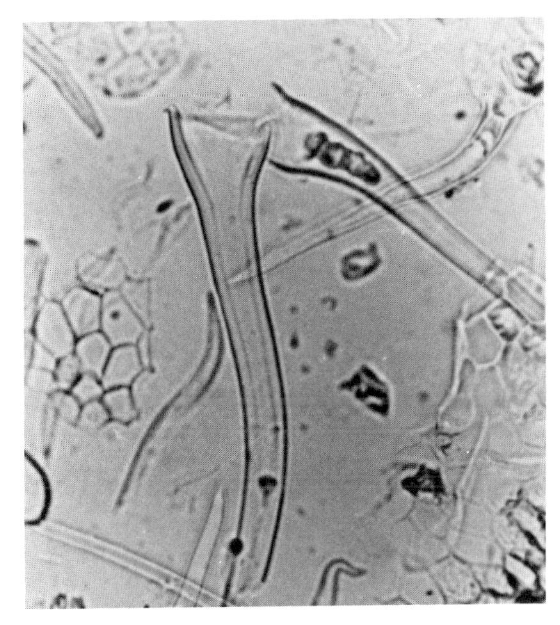

51. Long, threadlike nonsegmented hair phytoliths from *Trema micrantha*. Also present are polyhedral epidermal phytoliths.

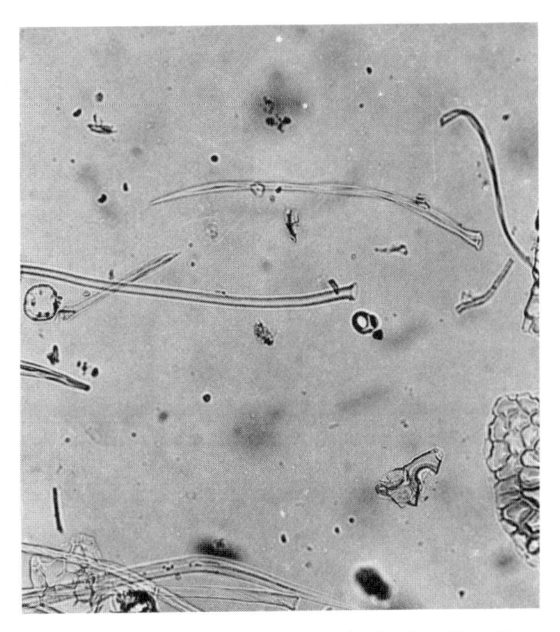

52. Long, threadlike nonsegmented hair phytoliths from *Elephantopus mollis* (size range: 330–670 μm).

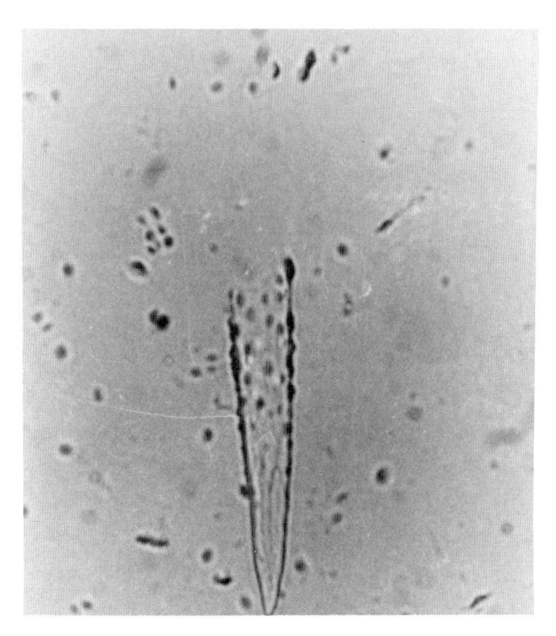

53. Nonsegmented armed hair phytolith with unarmed apex from *Castilla elastica*.

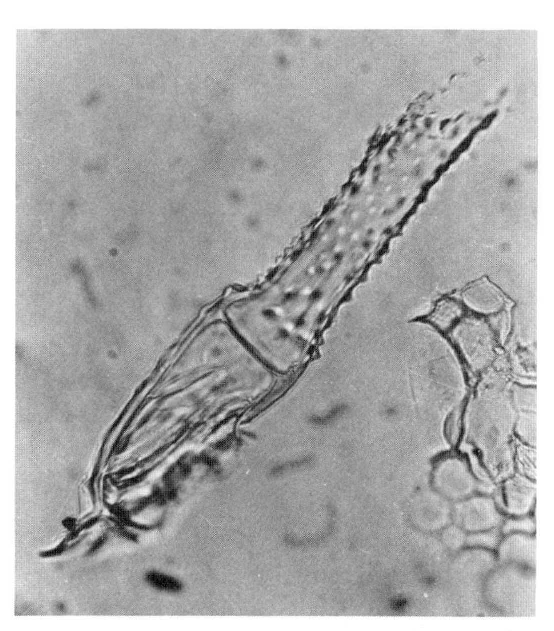

54. Segmented armed hair phytolith with unarmed neck and base from *Eclipta alba* (size range: 90–363 μm).

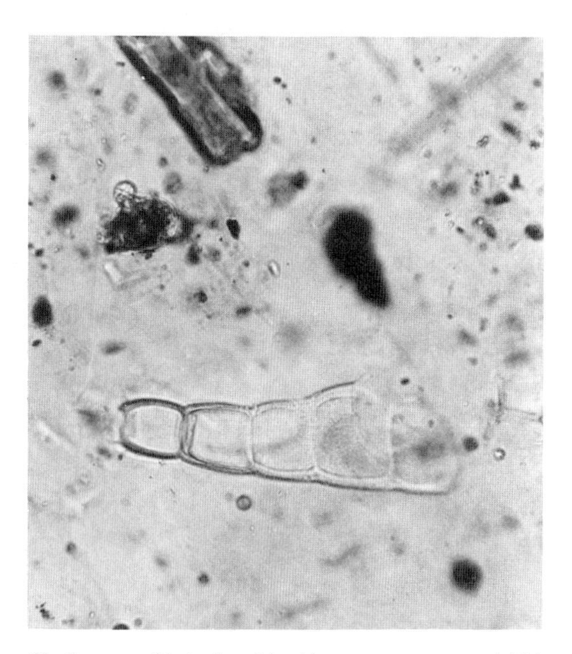

55. Segmented hair phytolith with square segments and thick cell walls from *Bidens* (size range: 53–198 μm).

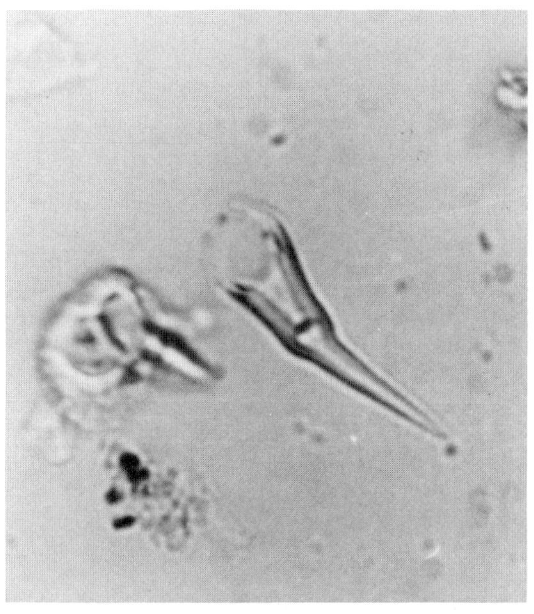

56. Segmented hair phytolith from *Calea urticifolia* with spherical base, tapered shaft and apex, and thick cell walls.

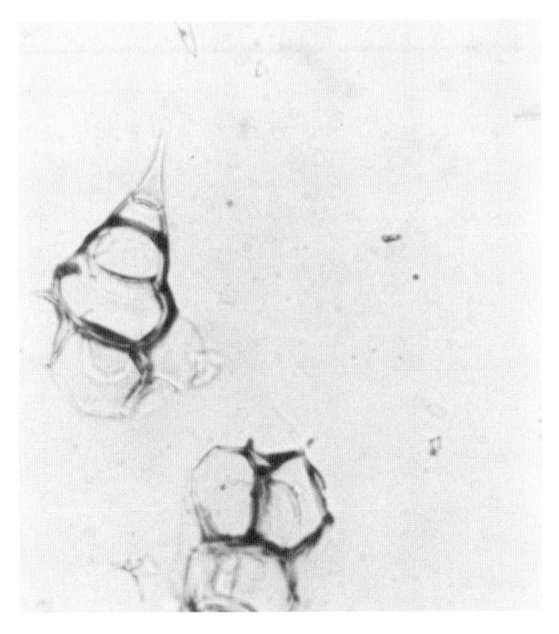

57. Segmented hair cell phytoliths with circular segments from *Melothria guadalupensis*.

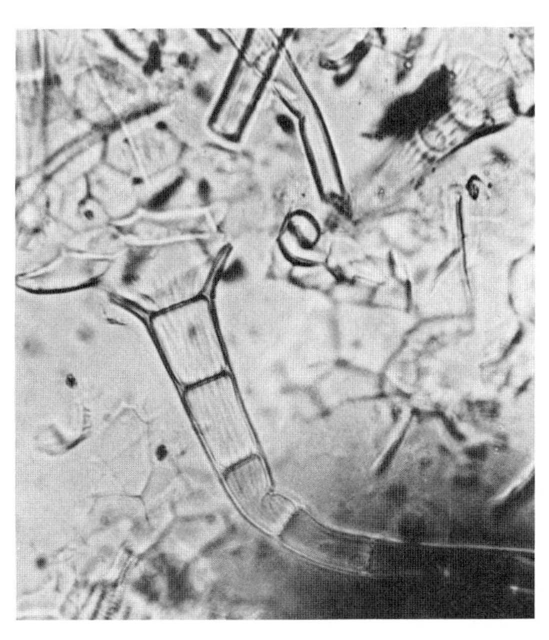

58. Segmented hair phytoliths with long, distinct surficial striations from *Piper pseudoasperi* (size range: 68–360 × 12–37 μm).

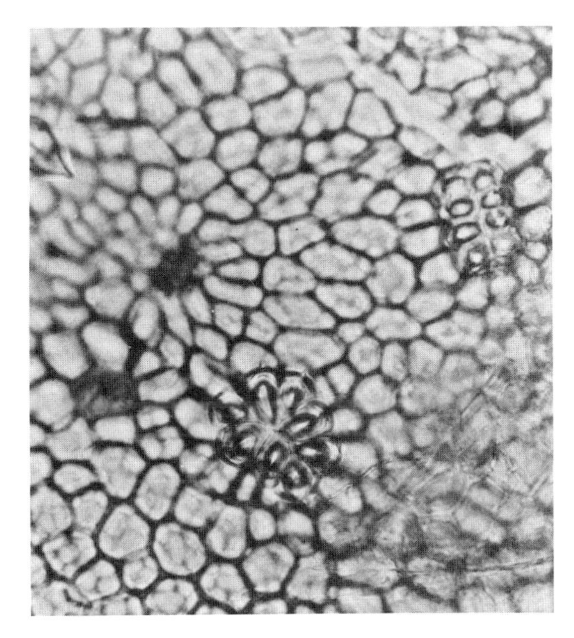

59. Multicelled hair base phytoliths from *Curatella americana* (size range: 46–56 μm).

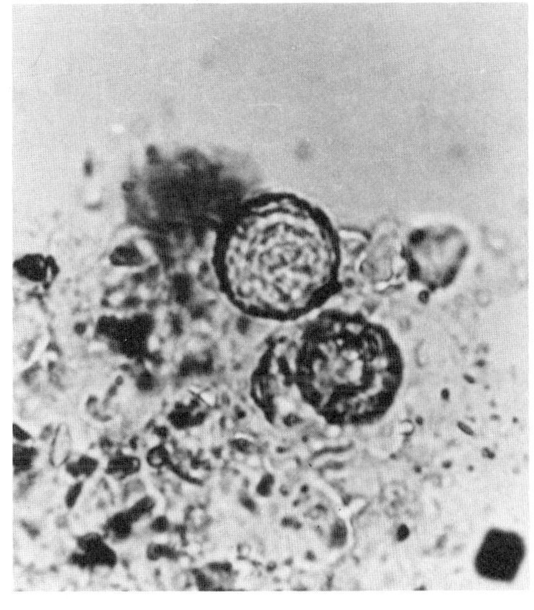

60. Hair base phytoliths from *Heliotropium angiospermum* with rugulose surfaces and concentric ring patterns (size range: 25–53 μm, diam).

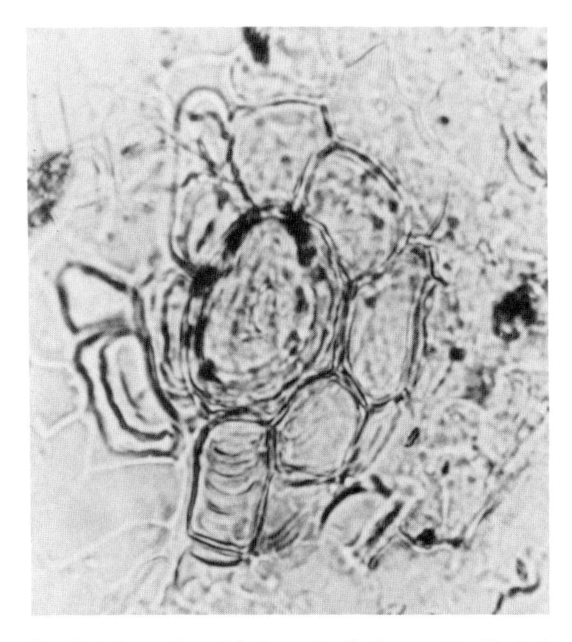

61. Hair base phytolith from *Cordia lutea* with elliptical central pore and surrounded by epidermis with spherical inclusions (size range: 57–60 μm, diam).

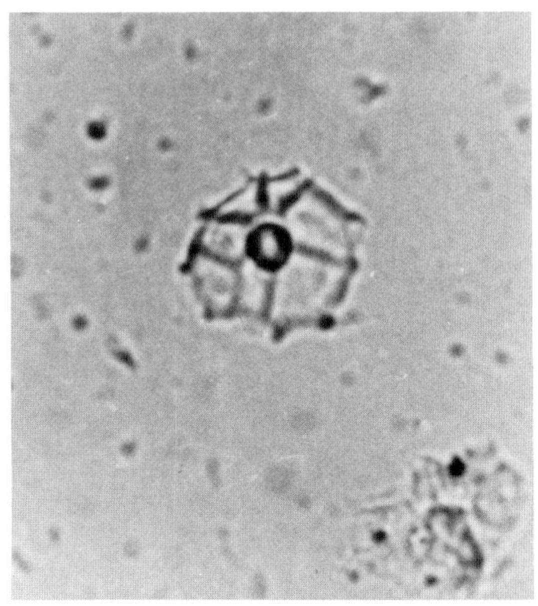

62. Hair base phytolith with irregular striations radiating from center of cell from *Ficus americana*.

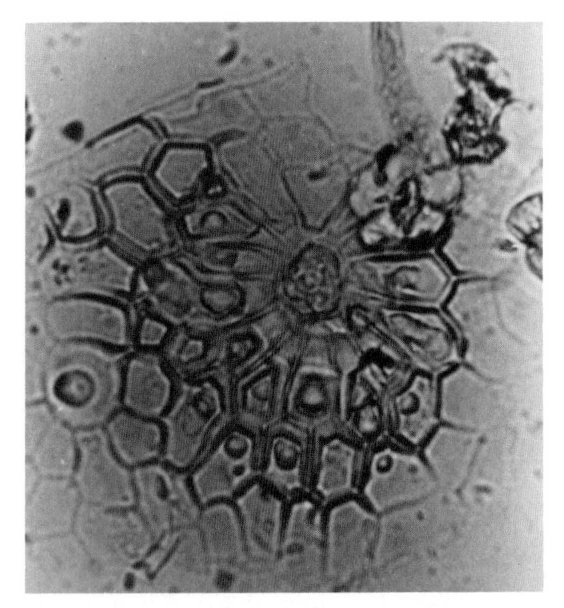

63. Hair base phytolith from *Chlorophora tinctoria* surrounded by silicified epidermis with spherical inclusions. Striations emanate from inside, but not from center of base, to cell periphery. The hair base center is occupied by what would technically be called a "hair cystolith" (size range: 33–80 μm, diam).

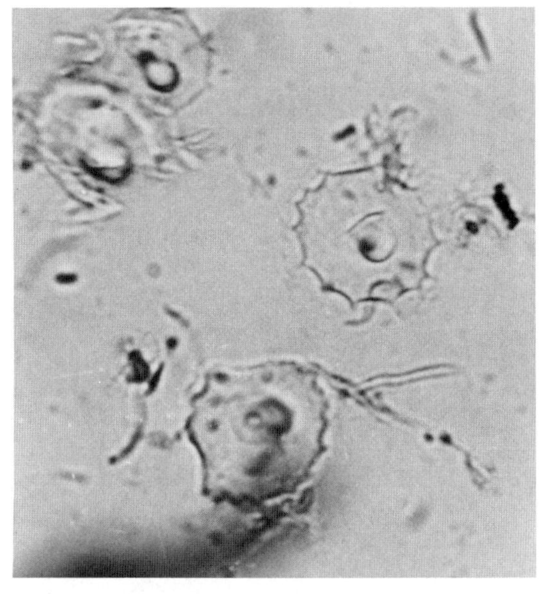

64. Hair base phytoliths with short lines projecting from the perimeter from *Cecropia peltata* (size range: 27–81 μm, diam).

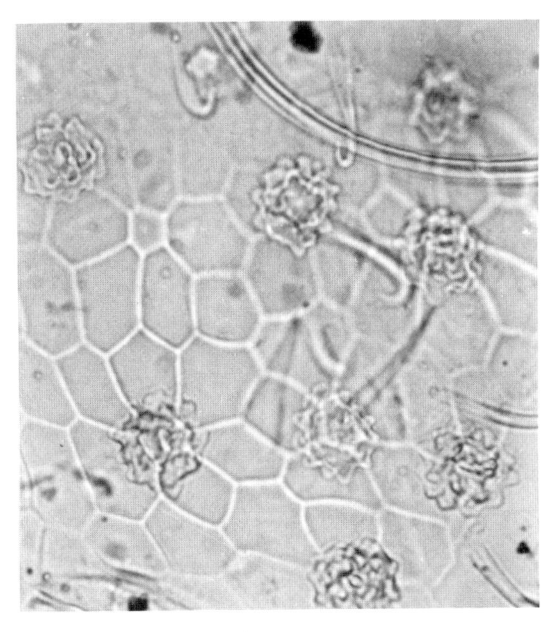

65. Spherical verrucose cystoliths from *Boehmeria aspera* attached to polyhedral epidermal phytoliths (size range: 24–56 μm, diam).

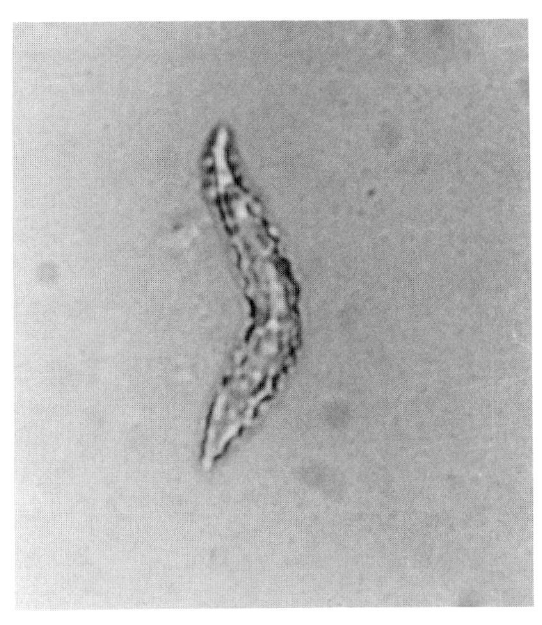

66. Curved cystolith with two tapered ends from *Pilea acuminata* (size range: 59–185 × 10–30 μm).

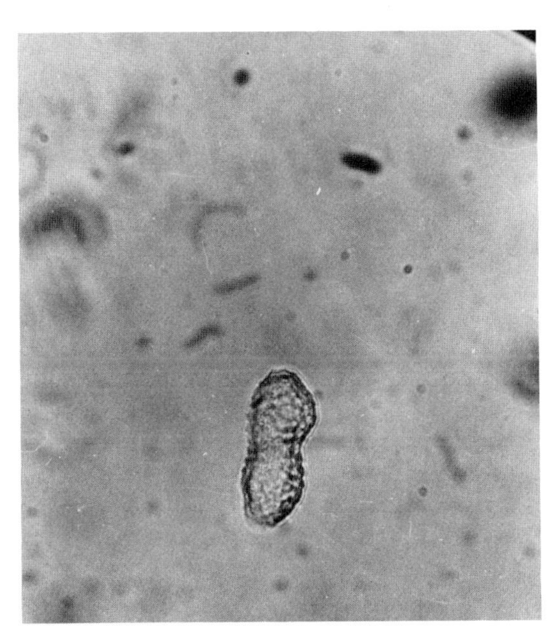

67. Elongate, verrucose cystolith from *Laportea aestuans*.

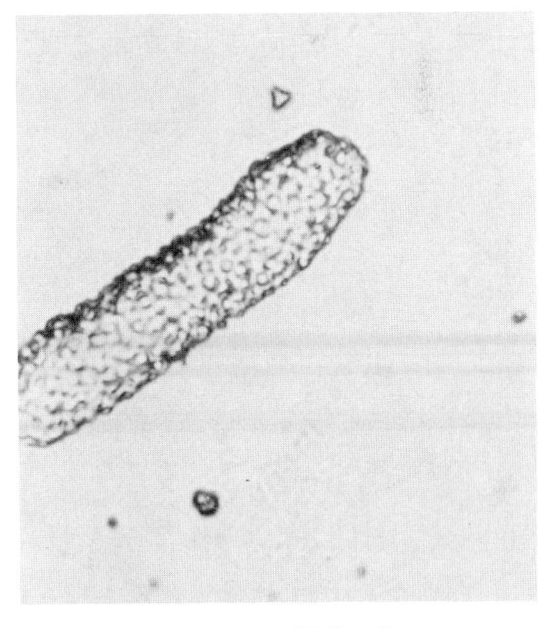

68. Elongate, verrucose cystolith from *Laportea aestuans*.

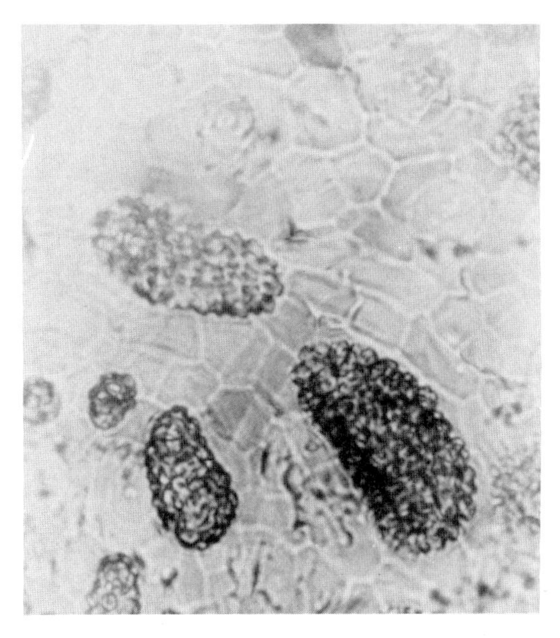

69. Roughly spherical to irregular cystoliths from *Urera elata* (size range: 45–255 × 27–60 μm).

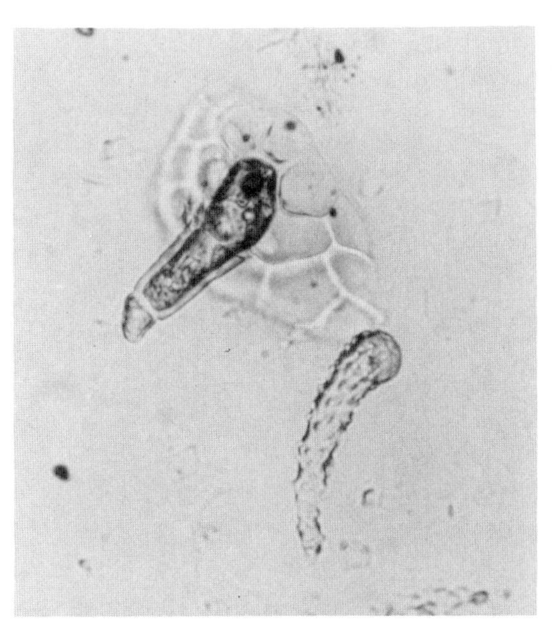

70. Cystolith (right) from *Justicia pringlei* showing the tuberculate surface decoration.

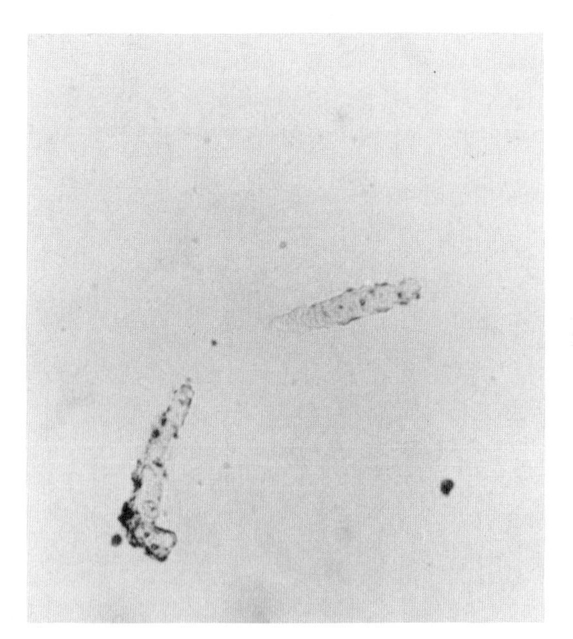

71. Cystoliths from *Pseuderanthemum davei;* the right cystolith has seashell-like surface decoration (size range: 51–108 × 12–18 μm).

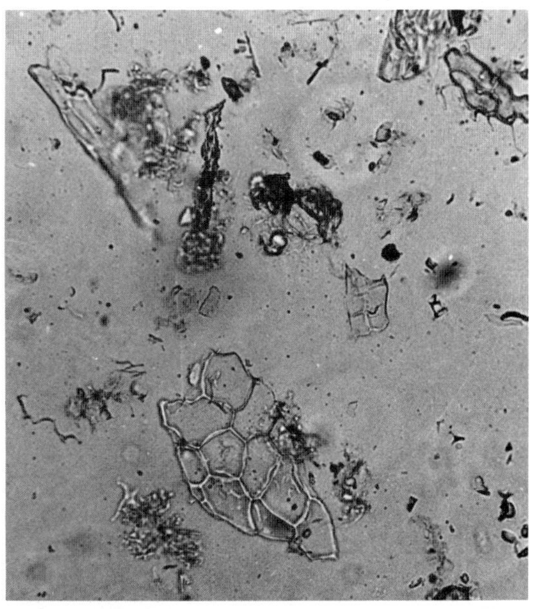

72. Polyhedral epidermal phytoliths from *Hedyosmum*.

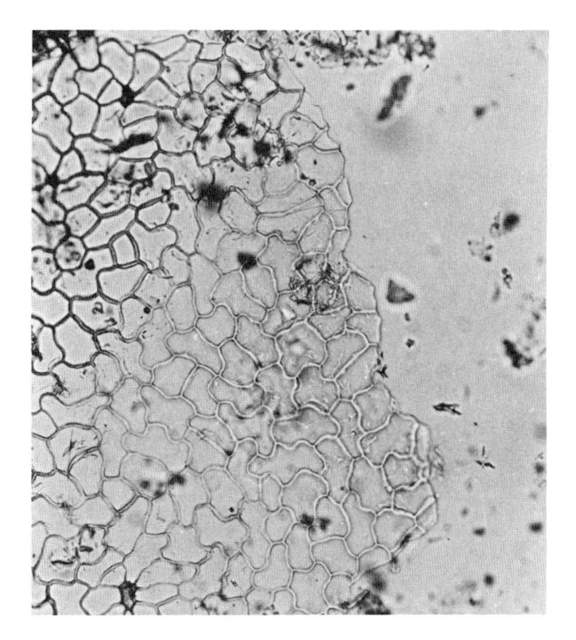

73. Anticlinal epidermal phytoliths from *Hedyosmum*.

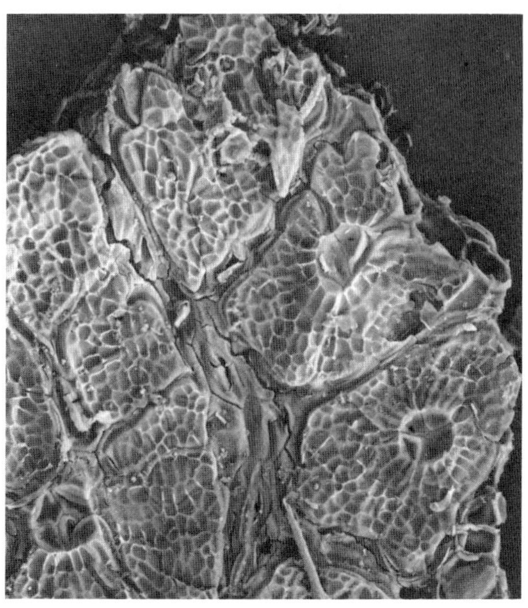

74. SEM photograph of silicified epidermis from *Pourouma aspera* showing polyhedral epidermis, hair bases, and nonsegmented hair cell.

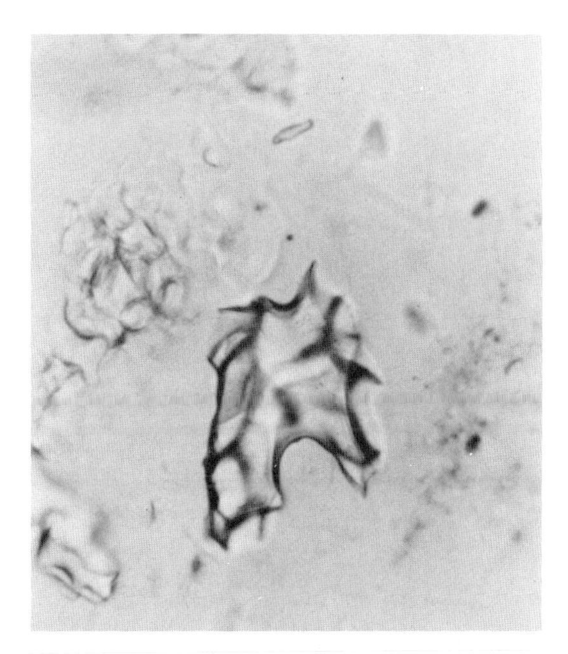

75. Irregular multifaceted phytolith from *Guatteria dumetorum* (size range: 50–65 × 30–48 μm).

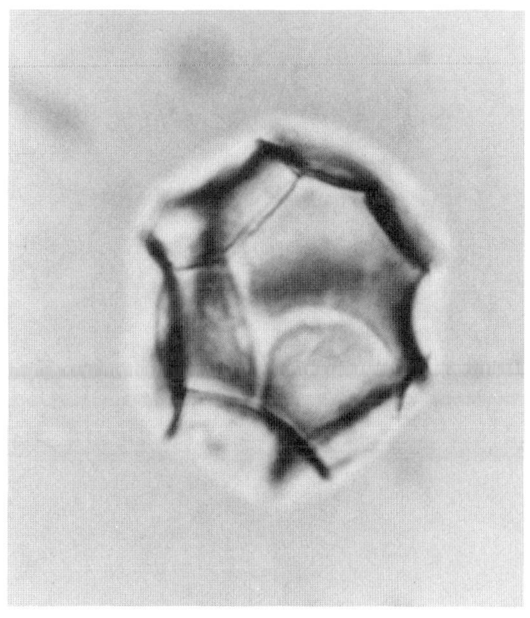

76. Spherical multifaceted phytolith from *Unonopsis pittieri* (size range: 48–65 μm, diam).

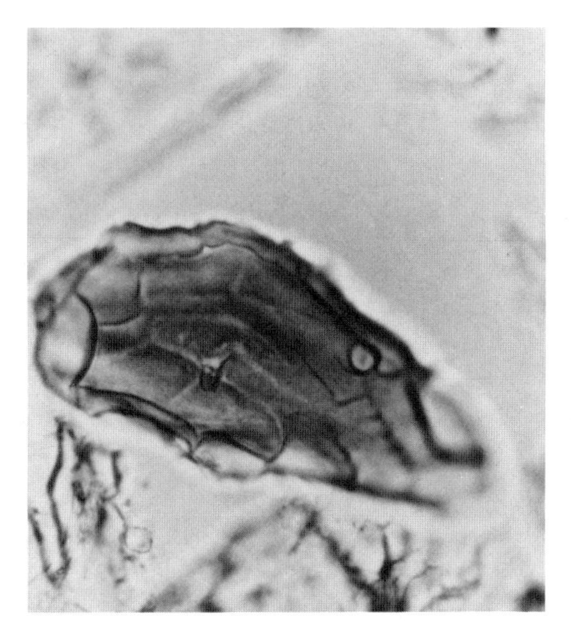

77. Elliptical multifaceted phytolith from *Protium panamense* (size range: 90–153 × 45–69 μm).

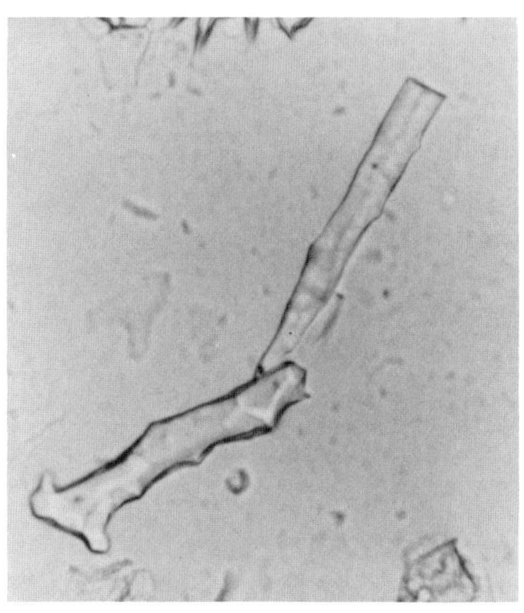

78. Silicified sclereids from *Hirtella triandra*.

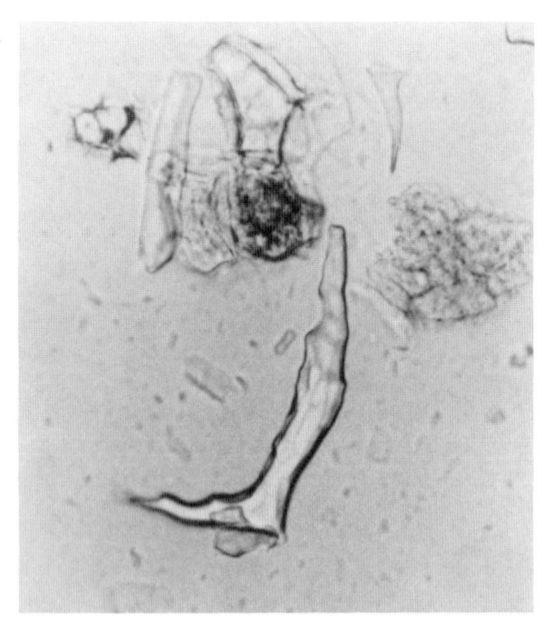

79. Silicified sclereid from *Hirtella triandra*.

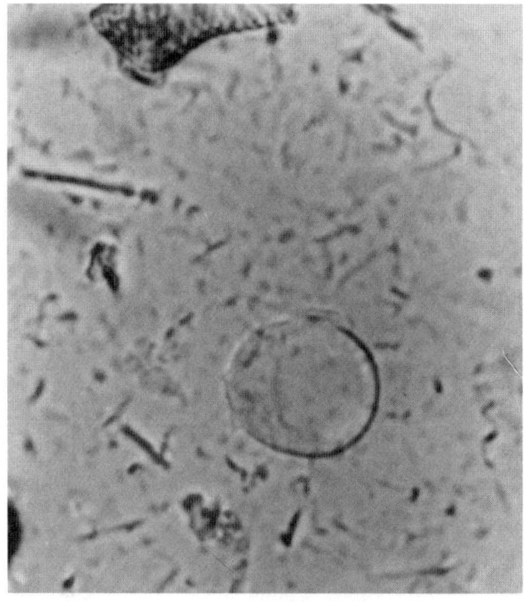

80. Double saucer-shaped phytolith from *Licania hypoleuca* lying flat so three-dimensional structure is not clearly apparent (size range: 21–30 μm, diam).

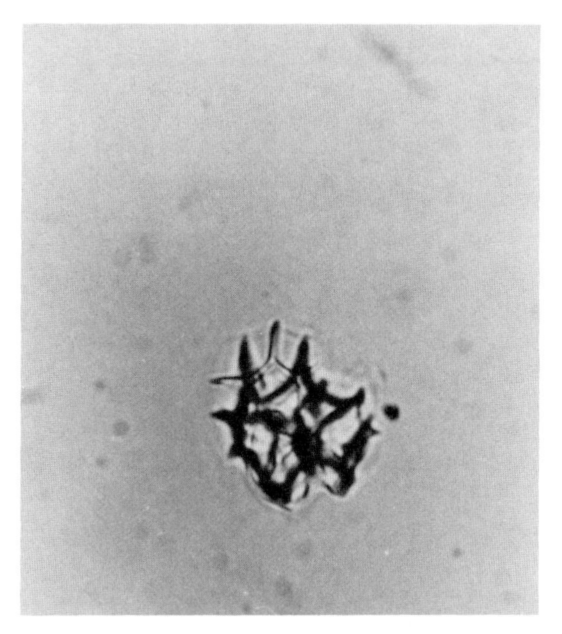

81. Irregular pointed phytolith from *Phthirusa pyrifolia* (size range: 66–99 × 33–79 μm).

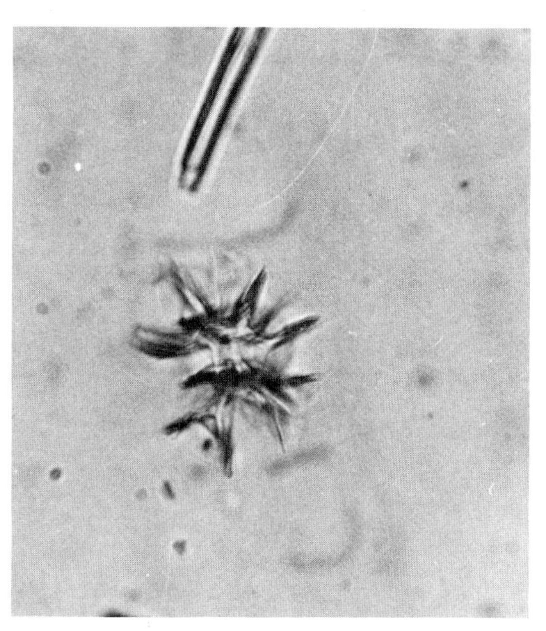

82. Irregular pointed phytolith from *Odontocarya tamoides* (size range: 36–51 × 36–46 μm).

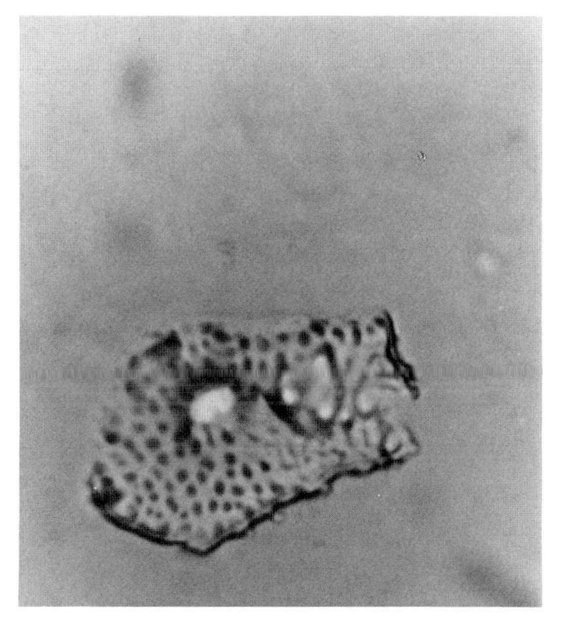

83. Fruit phytolith from *Tetragastris panamensis* in surface view (size range: 30–48 × 15–30 μm).

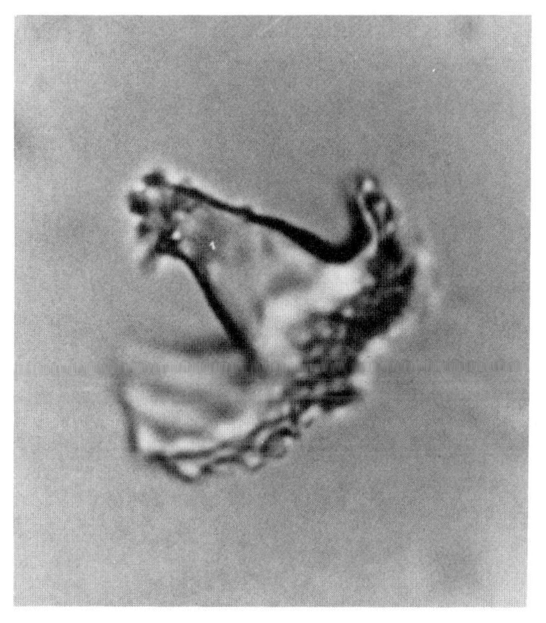

84. Fruit phytolith from *Tetragastris panamensis* in side view.

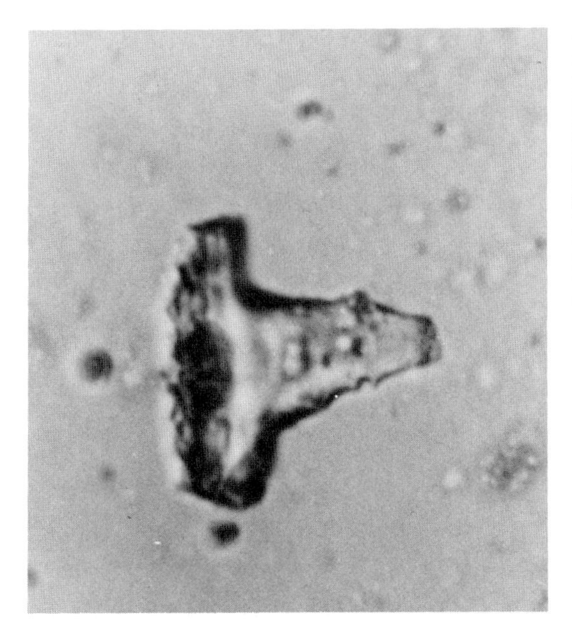

85. Fruit phytolith from *Protium panamense* in side view [size range (when flat): 36–57 × 27–57 μm].

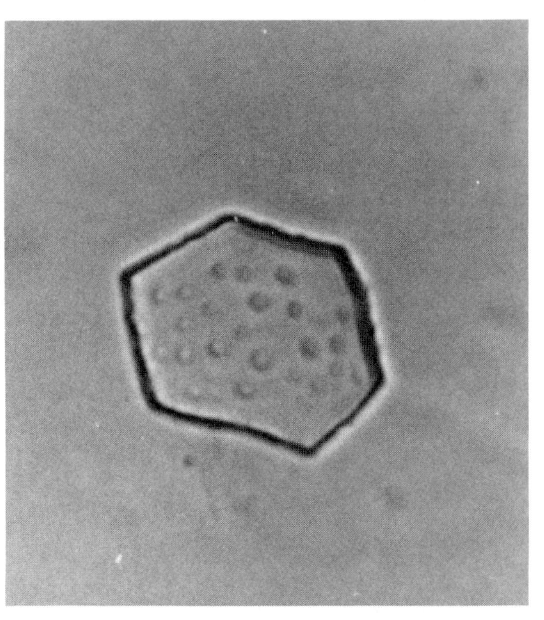

86. Seed phytolith from *Bursera simaruba* in surface view.

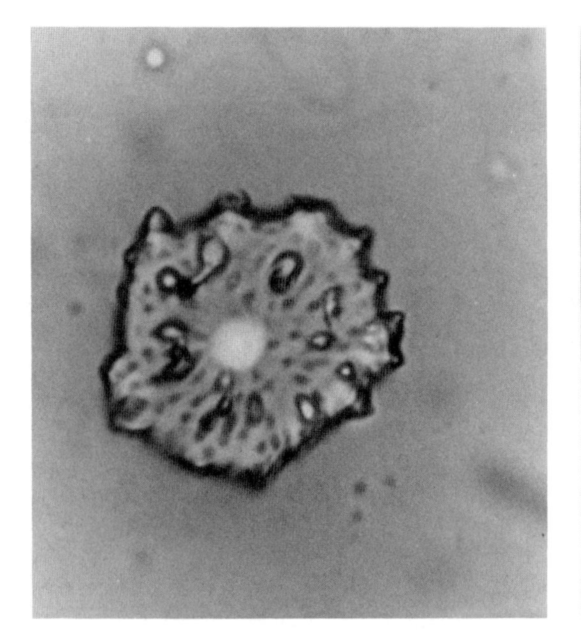

87. Seed phytolith from *Trattinickia aspera* in surface view.

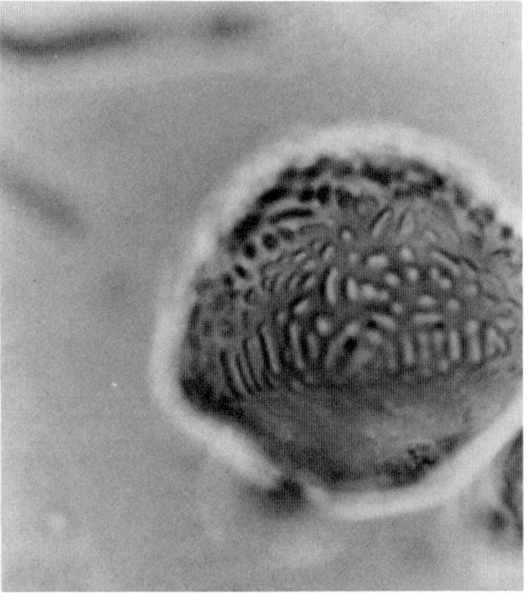

88. Seed phytolith from *Mendoncia retusa* with the surface pattern of short striations on one hemisphere only.

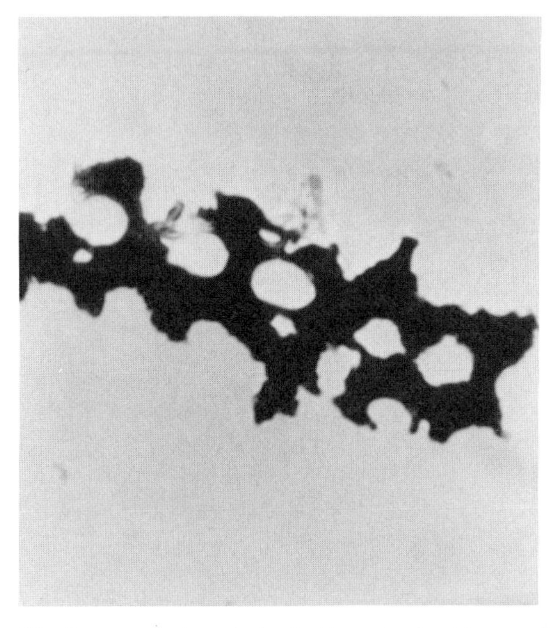

89. Opaque, perforated platelet from the floral bracts of *Zinnia elegans*.

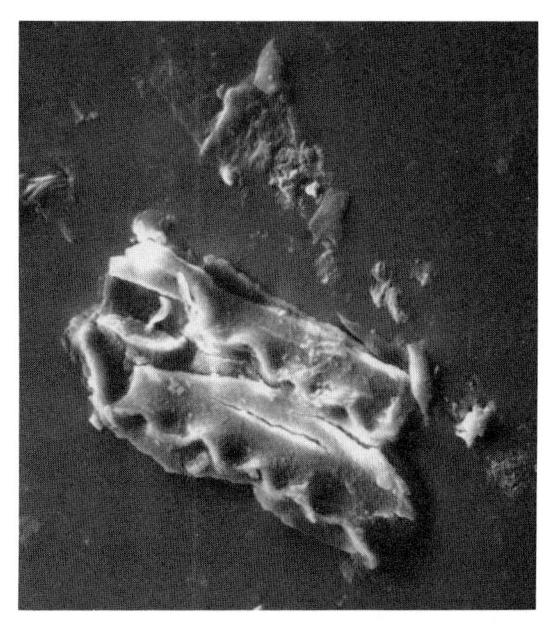

90. SEM photograph of elongate phytoliths with undulating ridges from *Adiantum fructosum* (size range: 65–728 × 13–26 μm).

91. SEM photograph of bowl-shaped phytolith from *Trichomanes osmundoides* (size range: 15–39 μm, diam).

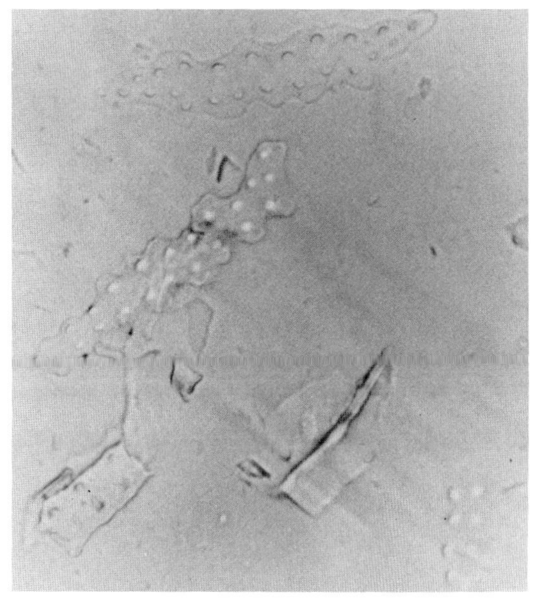

92. Phytoliths from *Selaginella arthritica*.

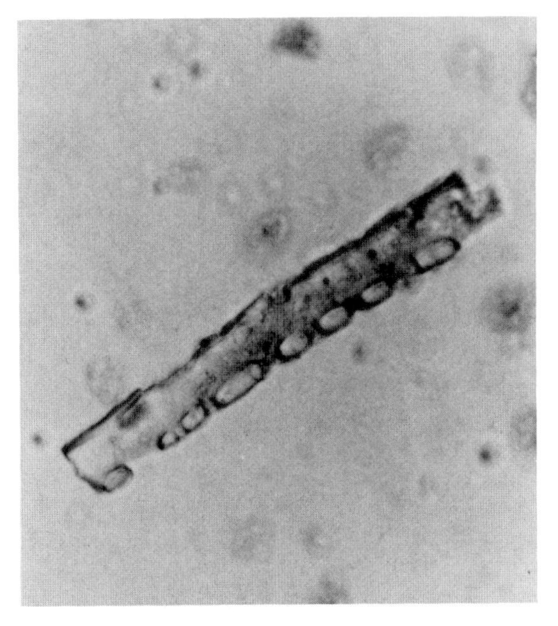

93. Elongate phytoliths with concavities or protrusions from modern soils underneath Panamanian tropical forest.

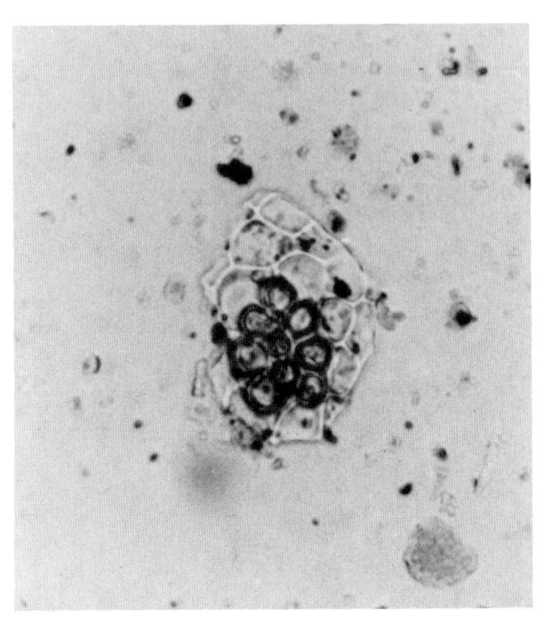

94. *Curatella americana* multicelled hair base phytolith from the Aguadulce shelter surrounded by polyhedral epidermal phytoliths.

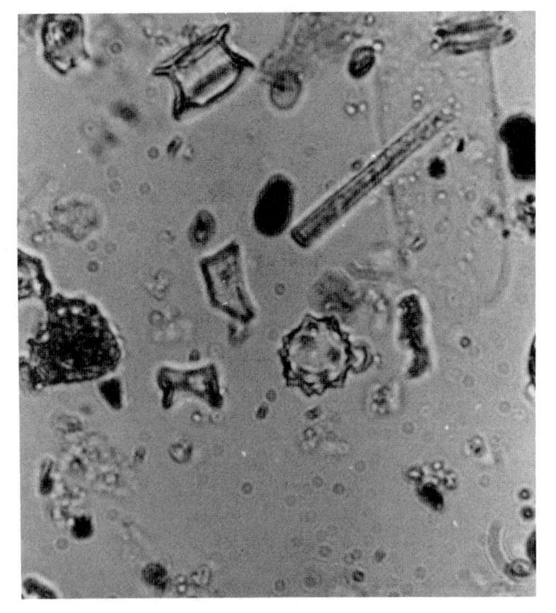

95. Spherical spinulose Palmae phytolith from the Aguadulce shelter.

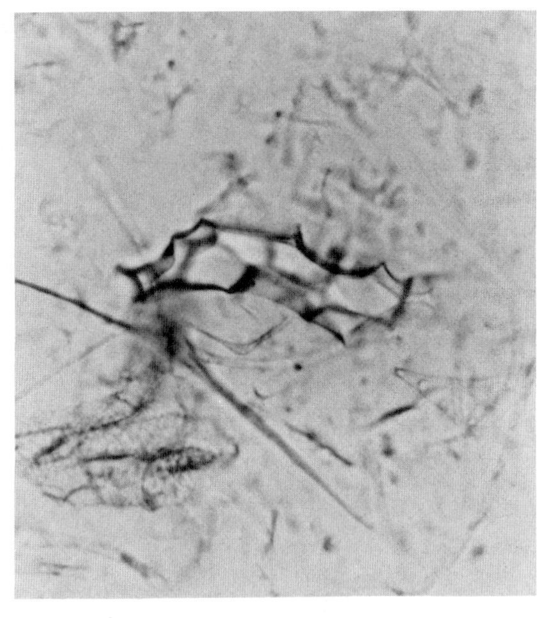

96. Elongate multifaceted phytolith from *Guatteria amplifolium* (size range: 35–90 × 21–30 μm).

Appendix

Phytolith Keys

This appendix presents two different kinds of phytolith keys. One type, comprising a large number of plants and plant families from the New World tropics, is based on a generic classification whereby phytoliths are organized wherever possible by origin in living tissue. It was developed from the author's study of 1000 species of plants comprising over 70 families from the Republic of Panama. The other type of key comprises two specialized morphological grass keys developed by Brown (1984) and the author for Gramineae of the Central North American Plains and Lower Central America, respectively. As was discussed in Chapter 3, a classification system that combines generic (by specific cellular origin) and morphological (size and shape) attributes appears to be a sound approach to ordering large numbers of silicified taxa. Categories are consistent and repeatedly determinable from plant to plant, fulfilling two prime requisites of classification systems.

Two generic keys for nongraminaceous tropical plants are presented. One lists all known families with distinct (to at least the level of family) phytoliths, plus phytoliths shared only by a few families or closely related families; for example, belonging to the same order of plants. In these cases ecological considerations may permit the exclusion of many alternatives in the fossil phytolith record. Smaller units below the family level are entered if it appears that categories are restricted to subfamily or occur in only one to several genera of the family. When a phytolith type has been found to cross-cut subfamilies or occur in more than three genera, subcategorization of phytoliths is not listed. In addition to this key specifying very distinct forms of phytoliths, one is presented for phytoliths shared by several or more unrelated families. This will provide examples of highly redundant morphologs and their distribution in tropical plants.

Keys are not exhaustive. In some cases it is possible to identify smaller units, and future work with certain categories, especially cystoliths, hair cells, and hair bases, will develop further taxonomic precision. Keys should be not considered fundamental as future work will probably isolate more areas of shared morphology. Especially in tropical plants, the diversity of modern species and numbers whose phytoliths have not yet been classified ensure that some areas will need modification.

Phytolith keys are complicated entities requiring large numbers of categories. Because many plants produce more than one kind of silicified cell, a single species will often be entered more than once. Those just beginning in phytolith science should not expect to be able to use keys presented here at once. In phytolith identification there is no substitute for one's own modern comparative collection and a great deal of practice. Except where noted, all phytoliths are derived from leaf specimens.

General Phytolith Key for Plants of the New World Tropics

I. Segmented hairs
   A. Nonarmed
      1. With squarish segments and thick cell walls
         Compositae and Cucurbitaceae (Plate 55)
      2. With circular segments
         Cucurbitaceae *(Melothria)* (Plate 57)
      3. With long fine striations on the surface
         Cucurbitaceae
      4. With faint spines that appear as short fine striations on the surface
         Cucurbitaceae (Plates 17 and 18)
      5. With long distinct striations on the surface
         Piperaceae *(Piper pseudoasperi, Piper hispidum)* (Plate 58)
      6. With shortly divided segments close to the base of the hair
         Boraginaceae *(Hackelia mexicana)*
      7. With cushion cells attached at the base of the hair
         Compositae
      8. With very long (>90 μm) segments
         Cucurbitaceae *(Ğurania)*
      9. With longitudinally divided segments near the base of the hair
         Cucurbitaceae *(Cyclanthera, Rystidostylis)*
      10. With blackish silica inclusions in the segments
         Compositae *(Bidens)*, Cucurbitaceae *(Pittieria, Sicyos)*
   B. Armed
      1. With shortly divided segments near the base
         Boraginaceae *(Hackelia mexicana)*

2. With an unarmed neck, apex, or base
   Compositae
   a. nonarmed base, neck, and apex *(Eclipta alba)* (Plate 54)
   b. nonarmed apex, no neck present *(Wulffia baccata)* (Plate 8)
   c. nonarmed apex, base usually armed, hair differently shaped than *Eclipta* and *Wulffia (Melanthera hastata)*
3. With an unarmed apex, apex attached to the base at an angle
   Acanthaceae *(Thunbergia alata)*

II. Nonsegmented hairs
   A. Nonarmed
      1. Short (not greater than 25 μm if prong is not measured), deltoid-shaped, with and without prongs
         Dilleniaceae and Burseraceae *(Trattinickia aspera)*
         a. With a more flared base and greater number of prongs *(Trattinickia aspera)*
         b. With a less flared base and fewer prongs (Dilleniaceae) (Plate 47)
      2. Short (not greater than 30 μm) and thin (not greater than 7 μm), with and without prongs; hair is twice as long as it is wide
         Aristolochiaceae*(Aristolochia chapmaniana)*
      3. Not short and deltoid-shaped, with prongs and lateral projections
         Dilleniaceae*(Davilla)*
      4. Solidly silicified, with elongated flat or indented bases
         Moraceae(Plate 49)
      5. Solidly silicified, with rounded indented and nonindented bases
         Moraceae(Plate 48)
      6. Trident shaped
         Euphorbiaceae*(Dalechampia gruingiana, Tragia volubilis)*
      7. Long and thread-like, with weave pattern and narrow base
         Compositae*(Elephantopus mollus, Vernonia argyropoppa)* (Plate 52)
      8. Long and threadlike, with weave pattern and a wider base
         Malvaceae*(Malvaviscus)*

III. Hair Bases
   A. Multicelled
      Dilleniaceae
      1. 4–10 cells, less than 5 cells infrequent, surrounded by polyhedral epidermis *(Curatella americana)* (Plate 59)
      2. 2–8 cells, more than 4 cells rare, surrounded by anticlinal epidermis *(Tetracera volubilis)*
   B. Heavily silicified, with rugulose surface, slightly wavy edges, a concentric ring pattern, and a pore in the center.
      Boraginaceae*(Heliotropium)* (Plate 60)

C. Not heavily silicified, with a nonrugulose surface, a concentric ring pattern, and a pore in the center.
 Boraginaceae *(Heliotropium indicum, Tournefortia volubilis)*

D. With a large, central elliptical pore; surrounded by epidermis with spherical inclusions
 Boraginaceae*(Cordia lutea)* (Plate 61)

E. With an irregular pattern of striations inside of the base
 Moraceae(Plate 62)

F. With regular, whiteish striations radiating from inside, but not the center of the base, to the periphery of the base; sometimes attached to epidermis with spherical inclusions
 Moraceae*(Chlorophora, Ficus)* (Plate 63)

G. Four half- or entire spheres joined together
 Cucurbitaceae(Plate 19)

H. Two half-spheres joined together
 Compositae, Cucurbitaceae

IV. Cystoliths

A. Spherical, with a verrucose surface
 Urticaceae*(Boehmeria, Pouzolzia, Urera)* (Plate 65)

B. Curved, with a verrucose surface
 Urticaceae*(Pilea)* (Plate 66)

C. Elongate, with a verrucose surface, verrucae densely distributed
 Urticaceae*(Myriocarpa, Urera, Laportea)* (Plates 67 and 68)

D. Spherical to irregular with a verrucose surface, verrucae densely distributed
 Urticaceae*(Urera)* (Plate 69)

E. Elongate, two blunt ends, with surficial tubercles that are not densely distributed
 Acanthaceae

F. Elongate, tapered at one end, with surficial tubercles that are not densely distributed
 Acanthaceae(Plate 70)

G. Elongate, two blunt ends, with a surface pattern that resembles lines on a seashell
 Acanthaceae(Plate 71)

H. Elongate, tapered at one end, with a solid piece of silica on the nontapered end
 Acanthaceae*(Justicia pringlei, Odontonema bracteolata)* (Plate 70)

I. Torpedo shaped
 Moraceae*(Ficus)*

J. Spherical to elliptical to irregularly shaped with a rough surface, often attached to stellate-shaped hair bases
 Moraceae *(Ficus, Poulsenia, Perebea)*

V. Epidermal or subepidermal (usually mesophyll) tissue. Phytoliths arising as small (5–25-μm) solid plugs from interior of cells, which are often "stegmata"
   A. Conical to hat-shaped
      1. Nonspinulose surface, cone pointed, single to multicrowned body
         Cyperaceae
      2. Nonspinulose surface, cone rounded
         Orchidaceae
      3. Surface spinulose
         a. With distinct spinules on the top of the hat
            Palmae (genera in the subfamilies Bactroid, Chamaedoroid, Iriartoid, and Nypoid)
            1. Hat shapes more rugulose and bulky in general appearance
               Geonoma(Plate 39)
            2. With 7–17 spinules arranged in no order on the top of the hat
               Chaemadorea(Plate 40)
         b. Spinules less defined, not distinct on the top of the hat
            Palmae*(Acrocromia, Desmoncus, Cocus)* Marantaceae *(Maranta)* (Plates 21 and 38)
      4. Surface nodular
         Marantaceae*(Ichnosiphon, Stromanthe, Calathea)* (Plate 41)
   B. Spherical to aspherical
      1. Surface spinulose
         a. Spinules well defined, many in number
            Palmae(genera in the subfamilies Arecoid, Borassoid, Cocoid, Lepidocaryoid, Phytelephantoid, and Sabaloid) (Plate 1)
         b. Spinules not well defined, smaller, often fewer in number
            Bromeliaceae, Palmae (Plate 2)
      2. Surface nodular
         Marantaceae*(Maranta, Stromanthe)* (Plates 3 and 4)
      3. Surface rugulose
         Marantaceae, Cannaceae, Heliconiaceae, Chrysobalanaceae *(Hirtella*-seed, *Licania morii)* (Plate 6)
      4. Surface smooth
         a. 12 μm and larger in diameter
            Cannaceae
         b. Less than 12 μm in diameter
            Chrysobalanaceae*(Hirtella*-seed, *Licania morii)* (Plate 6)
   C. With irregularly angled to folded surfaces
      Cannaceae, Marantaceae, Zingiberaceae (Plate 7)
   D. With troughs
      1. With deep troughs that are centrally located on the phytolith
         Heliconiaceae*(Heliconia)* (Plate 43)

2. With shallow troughs, often not centrally located on the phytolith; trough often on an elevated part of the phytolith
   Musaceae*(Musa)* (Plate 22)

VI. Epidermal, phytoliths arising as large incrustations of cell walls or chunks of silica filling the entire cell.
   A. Multifaceted bodies
      1. Spherical to aspherical
         Annonaceae*(Unonopsis pittieri)* (Plate 76)
      2. Irregularly shaped
         Annonaceae *(Guatteria dumetorum)* (Plate 75)
      3. Elongated, thin
         Annonaceae*(Guatteria)* (Plate 96)
      4. Elliptical
         Burseraceae *(Protium panamense)* (Plate 77)
   B. Elongate, with a flat base and a surface having two, parallel undulating ridges
      Polypodiaceae(Plate 90)
   C. Elongate, undulating margins, with small, conical-shaped surficial projections
      Selaginellaceae(Plate 92)
   D. With sloping margins and a stippled surface
      Burseraceae(*Protium costaricense*-fruit)
   E. With nonsloping margins, a lightly stippled surface, and a protuberance (stalk) emanating from the center of the cell into the ground tissue; may be pentagonal or hexagonal in surface view
      Burseraceae(*Protium panamense*-fruit) (Plate 85)
   F. With a central protuberance emanating from a smooth, dome-shaped structure
      Burseraceae(*Protium tanuuifolium*-fruit)
   G. With greatly sloping margins, a central protuberance, and a stippled surface; may be pentagonal or hexagonal in surface view
      Burseraceae(*Tetragastris panamensis*-fruit) (Plate 84)
   H. With jagged edges and small nodules on the surface
      Burseraceae(*Trattinickia aspera*-seed) (Plate 87)
   I. With a faintly stippled surface and a small protuberance emanating from the center of the cell; may be pentagonal or hexagonal in surface view
      Burseraceae(*Bursera simaruba*-seed) (Plate 86)
   J. Spherical with a surface pattern of short striations on one hemisphere only
      Acanthaceae(*Mendoncia retusa*-seed) (Plate 88)
   K. With numerous small projections on the surface of the cell
      Moraceae(*Celtis schippi*-seed)

L. Long and narrow, with undulating margins
Heliconiaceae *(Heliconia)*, Marantaceae *(Calathea)*

VII. Origin in Plant Tissue Unknown

A. Roughly bowl-shaped, with a shallow depression
Hymenophyllaceae *(Trichomanes)* (Plate 91)

B. Tongue-shaped
Commelinaceae *(Athyrocarpus persicariaefolium)* (Plate 44)

C. Irregular pointed bodies
Loranthaceae, Menispermaceae (Plates 81 and 82)

D. Rectangular Chunks with heavily granulated surfaces
Podostemaceae

E. Double saucer-shaped
Chrysobalanaceae *(Licania hypoleuca)* (Plate 80)

F. Rectangular to aspherical, with cavities of various shapes
Elaeocarpaceae *(Sloanea terniflora)*

G. Aspherical, with crenated margins and fingerlike projections
Podostemaceae

H. Opaque, perforated plates
Compositae-infloresence(Plate 89)

Phytolith Key for Plants of the New World Tropics Encompassing Redundant Phytolith Shapes

As in the preceding key, listing of genera after a family entry indicates phytoliths, at this point in time, are limited in occurrence to such taxa. It should be noted that further differentiation of certain kinds of phytoliths will probably be possible, e.g., in the case of segmented hairs with rounded apices. In addition, segregation of plant taxa by growth habit (arboreal, herbacious), as appears possible with silicified sclereids, may result from future comparative work.

I. Nonsegmented hairs

A. Nonarmed

1. Thin and curved
Urticaceae, Moraceae, Boraginaceae, Aristolochiaceae, Leguminoseae *(Desmodium, Clitoria)*, Menispermaceae *(Odontocarya)* (Plate 11)

2. With rounded apices
Moraceae *(Brosimum)*, Piperaceae, Labiatae *(Hyptis suaveolens)*, Urticaceae *(Pouzolzia)* (Plate 50)

3. Long and threadlike
Compositae, Malvaceae *(Malvaviscus, Kosteletzkya)*, Moraceae *(Cecropia, Dorstenia)*, Urticaceae, Ulmaceae (Plate 51)

4. Undifferentiated (in part)
   Compositae, Cucurbitaceae, Piperaceae *(Piper)*, Acanthaceae, Sterculiaceae, Verbenaceae, Labiatae, Moraceae, Euphorbiaceae, Leguminoseae, Urticaceae, Malvaceae, Ulmaceae

B. Armed
1. With an unarmed apex
   Boraginaceae *(Ehretia)*, Urticaceae *(Boehmeria, Pouzolzia)*, Moraceae, Ulmaceae (Plate 53)
2. With entire shaft armed
   Boraginaceae, Urticaceae, Ulmaceae

II. Segmented Hairs, undifferentiated (in part)
   Compositae, Cucurbitaceae, Piperaceae *(Piper)*, Acanthaceae, Sterculiaceae *(Guazuma)*, Labiatae

III. Hair Bases
A. With remnants of armed hairs
   Compositae *(Melanthera, Wulffia)*, Moraceae *(Olmedia, Poulsenia)*, Ulmaceae, Boraginaceae *(Cordia)*, Cucurbitaceae (Plate 19)
B. Surrounded by epidermal cells with spherical inclusions
   Compositae *(Eclipta, Tridax)*, Cucurbitaceae, Boraginaceae, Dilleniaceae *(Tetracera)*, Moraceae, Sterculiaceae *(Waltheria)*, Verbenaceae *(Lantana)* (Plates 61 and 63)
C. With a nonirregular pattern of striations inside the base
   Moraceae, Dilleniaceae *(Davilla)*
D. With regular whiteish striations emanating from inside of the base to the periphery of the cell
   Moraceae *(Ficus, Trophis)*, Sterculiaceae, Dilleniaceae *(Davilla)*, Piperaceae *(Pothomorphe)*

IV. Sclereids (in part)
   Leguminoseae, Piperaceae, Moraceae, Combretaceae, Sterculiaceae, Annonaceae *(Unonopsis)*, Dilleniaceae, Euphorbiaceae, Connaraceae *(Rourea)*, Flacourtiaceae *(Hasseltia, Tetrathylacium)*, Capparidaceae, Acanthaceae, Chrysobalanaceae (Plates 78 and 79).

V. Epidermis, arising as large incrustations of cell walls or chunks of silica filling the entire cell
A. Polyhedral and Anticlinal, with spherical inclusions
   Compositae, Euphorbiaceae, Boraginaceae, Acanthaceae, Asclepidaceae, Leguminoseae, Verbenaceae, Moraceae *(Chlorophora)*, Cucurbitaceae, Sterculiaceae
B. Anticlinal (in part)
   Polypodiaceae, Schizeaceae, Cyatheaceae, Cucurbitaceae, Burseraceae, Annonaceae, Menispermaceae, Verbenaceae, Leguminoseae, Boraginaceae, Euphorbiaceae, Compositae, Flacourtiaceae, Dilleniaceae (Plate 73)

C. Polyhedral (in part)

Piperaceae, Moraceae, Cucurbitaceae, Boraginaceae, Compositae, Verbenaceae, Dilleniaceae, Chrysobalanaceae, Bombacaceae, Annonaceae, Burseraceae, Leguminoseae, Elaeocarpaceae, Bignoneaceae, Anacardiaceae, Euphorbiaceae, Commelinaceae (Plates 72 and 74)

Special Keys for the Gramineae

In this section two specialized grass keys are presented, one constructed by Brown (1984, 1986) for over 100 species comprising the native cover of the central United States, and the other constructed by the author for the Panicoid and Bambusoid grasses of Panama. As will be seen, very different approaches can be taken in making phytolith keys. Brown and I emphasize quite different aspects of phytolith morphology and varying numbers of phytolith categories.

Central North American Grasses

Brown's key comprises all subfamilies of grasses and divides phytoliths into eight major shape categories, which in turn are subdivided by minor shape categories. The eight major shape categories and their distributions in grasses are presented in Table A. 1. Table A. 2 presents more specific shape details on trapezoid short cell phytoliths, found mainly in Festucoid grasses. Parentheses indicate the range of occurrence seen among species examined.

Lower Central American Grasses

This key emphasizes a smaller number of attributes of phytoliths from the Panicoid and Bambusoid grasses of Panama. Greater weight is placed on more distinctive characteristics of grass genera. Three-dimensional shape attributes, sizes, and proportions of phytoliths play major roles in phytolith description.

I. Short Cell Phytoliths

A. Cross shaped

1. Variant 3 in three-dimensional structure (having one side with large nodules projecting from each lobe)

   Cryptochloa, Lithachne (Plate 26)

2. Predominantly (greater than 70% of all cross shapes) Variant 8 in three-dimensional structure

   Olyra

Table A.1

General Phytolith Shape Characteristics for Selected Tribes and Subfamilies of Grasses[a]

	Number of species	Elongates plates with protusions	Trichomes	Thick rectangle and round plates	Double outline irregular oval	Double outlines-regular	Trapezoids	Bilobate	Saddles
Poeae	22	68	95	45	0	36	82	0	0
Aveneae	4	100	100	25	0	25	100	0	0
Phalarideae	1	0	100	100	0	0	100	0	0
Agrostoideae	8	63	100	13	0	0	88	0	0
Triticeae	13	92	92	0	0	8	100	0	0
Meliceae	1	100	100	100	0	0	100	0	0
Stipeae	7	57	100	86	57	29	100	0	0
Pooideae total	56	73	96	38	7	21	91	0	0
Arundineae	1	100	0	0	0	0	100	0	100
Danthonieae	2	100	100	0	0	50	100	50	0
Aristideae	7	100	86	0	0	29	100	100	14
Arundinoideae total	10	100	80	0	0	30	100	80	20
Aeluropodeae	1	0	100	100	0	0	100	0	0
Eragrosteae	11	73	91	27	0	73	64	45	64
Sporoboleae	7	43	57	43	0	29	57	0	43
Chlorideae	13	62	92	23	0	16	38	8	85
Eragrostoideae total	32	59	84	31	0	37	53	19	66
Paniceae	8	62	75	13	0	13	13	88	0
Andropogoneae	6	33	67	17	0	0	0	100	0
Panicoideae total	14	50	71	14	0	7	7	93	0

[a]Numbers are percentage of species within the taxonomic group having various types of phytoliths. From Brown (1984).

Table A.2

Specific Shape Details on Trapezoid Short Cell Phytoliths[a]

Trapezoids	Reference species
A. Sinuous	
1. Long (length >3 width)	
a. Square ends (or flattened)	
1. plateau top	*Trisetum spicatum*
2. wavy top (*Agrostis scabra*)	*Calamagrostis rubescens*
3. ridge or round top	*Agrostis scabra*
b. Round ends	
1. plateau top (*Poa secunda*)	*Vulpia octoflora*
2. wavy top (*Bromus catharticus, B. commutatus, Poa secunda*)	*Trisetum spicatum*
c. Intermediate ends	
1. plateau top	*Panicum obtusum*
2. wavy top	*Poa compressa*
d. Mixed or irregular ends	
1. plateau top	*Poa interior*
2. wavy top (*Bromus catharticus, B. commutatus, B. diandrus, Dactylis glomerata*)	*Poa secunda*
3. ridge or round top	*Bromus pumpellianus*
2. Short	
a. Square end	
1. plateau top (*Agrostis exerata, A. scabra, Phleum alpinum, Calamagrostis montanensis*)	*Oryzopsis hymenoides*
2. ridge (reference only)	*Phalaris arundinacea*
b. Round end	
1. plateau top (*Phalaris arundinacea, Calamagrostis scopulorum, Agropyron spicatum, Danthonia spicata, Eragrostis oxylepis, Stipa viridula*)	*Poa compressa*
2. wavy top (reference only)	*Oryzopsis hymenoides*
3. ridge on round top (*Stipa lettermanii, Calamovilfa longifolia, Eragrostis ciliansis, E. secundiflora, E. peotinacea, Calamagrostis montanensis, Vulpia octoflora, Phalaris arundinaca, Buchloe dactyloides*)	*Poa compressa*
4. sloping plain top (reference only)	*Buchloe dactyloides*
c. Intermediate ends	
1. plateau top (*Melica bulbosa*)	*Agropyron spicatum*
2. wavy top (reference only)	*Agrostis exerata*
3. ridge on round top (*Aristida fendlerana*)	*Calamagrostis montanensis*
d. Mixed or irregular ends	
1. plateau top (*Bromus pumpellianus, B. diandrus, Stipa comata, S. lettermanii, Monroa squarrosa, Dactylis glomerata, Poa secunda, Agrostis scabra*)	*Bromus rubens*
2. wavy top (*Agropyron smithii, Agrostis scabra, Poa secunda, Bromus pumpellianus*)	*Vulpia octoflora*
3. ridge or round top (*Bromus inermis, Stipa comata, S. spartea, Avena barbata, Poa compressa*)	*Oryzopsis hymenoides*
4. sloping plain top (*C. montanensis*)	*Calamagrostis scopulorum*

(*continues*)

Table A.2 (*Continued*)

Trapezoids	Reference species
B. Nonsinuous	
1. Long	
a. Square ends	
1. wavy top (reference only)	*Poa glauca rupicola*
2. ridge or round (*Bromus mollis, B. tectorum, Hordeum leporinum*)	*Calamagrostis rubescens*
b. Round ends, plateau top (reference only)	*Calamagrostis rubescens*
c. Intermediate ends, ridge (reference only)	*Phleum pratense*
d. Mixed ends flat top (*Poa pratensis*)	*Agropyron repens*
2. Short	
a. Square ends	
1. flat top (*Aristida fendlerana, A. longiseta robusta, Bouteloua hirsuta, Calamogrostis rubescens, Vahlodea autopurpurea, Bromus mollis, Phleum pratense, Eragrostis curvula, E. spectabilis*)	*Agropyron spicatum*
2. ridge or round top (*Aristida tuberculosa, Bouteloua hirsuta, B. simplex*)	*Agropyron subsecundum*
b. Round ends	
1. flat top (28 species of 11 genera)	*Elymus canadensis*
2. wavy top (*Hordeum jubatum*)	*Bromus inermis*
3. ridge or round top (34 sp. of 19 genera)	*Agropyron spicatum*
4. sloping plain (26 sp. of 15 genera)	*Agropyron spicatum*
c. Intermediate ends	
1. flat top (8 species of 5 genera)	*Stipa comata*
2. ridge or round top (*Agropyron pseudorepens, Aristida purpurea, Bromus inermis*)	*Melica bulbosa*
3. sloping plain top (*Aristida tuberculosa*)	*Eragrostis trichodes*
d. Mixed or irregular ends	
1. plateau top (16 species of 10 genera)	*Agropyron spicatum*
2. ridge or round top (*Bouteloua hirsuta, Festuca occidentalis, F. rubra*)	*Aristida purpurea*
3. sloping plain top (*Calamagrostis montanensis, Phleum alpinum*)	*Koeleria cristata*

[a]From Brown (1986).

3. Predominantly (greater than 70% of all cross shapes) Variant 6 and/or Variant 2 in three-dimensional structure; Variant 6 is more than twice as frequent than Variant 2
   Cenchrus, Oplismenus, Tripsacum (Plates 24, 25, 27, and 28)
   a. Mean size of Variant 1 cross shapes 13 μm or greater
      Oplismenus hirtellus, Cenchrus echinatus
4. Predominantly (greater than 70% of all cross shapes) Variant 1 in three-dimensional structure

a. Mean size of cross shapes less than 12 μm
Hymenache amplexicaulis, Axonopus compressus, Paspalum virgatum, Andropogan scoparius
b. Mean size of cross shapes often greater than 13 μm
Zea mays (Plate 23)

B. Bilobates
1. One side a concave bilobate, the other with 2 to 3 adjoining saddle shaped forms
Streptochaeta (Plate 35)
2. One side a concave bilobate, the other a single saddle shape
Chusquea (Plate 36)
3. One side a bilobate, the other a very wide tent-shaped or rectanguloid structure
Pharus (Plate 37)
4. Bilobates having a thick, rectanguloid structure on the opposite face
Oplismenus (Plate 32)
5. Predominantly Variant 1 in three-dimensional structure
Andropogon, Panicum, Digitaria, Isachne, Trachypogan
6. High frequency of Variant 2 and/or 6 three-dimensional structures
Cenchrus, Pennisetum, Oplismenus (Plate 33)

C. Irregular short cells with three or four lobes often having a sharp apex
Paspalum plicatulum (Plates 30 and 31)

II. Mesophyll (?) Phytoliths
A. One side with two parallel sets of lobes, the other with a central, pointed prong
Olyra, Bambusa (Plate 34)
B. One side with two parallel sets of lobes, the other with a central, extremely pointed prong
Cryptochloa, Lithachne (Plate 26)
C. Elliptical, entire margin consisting of undulations
Olyra, Cryptochloa

Bibliography

Allen, S. E., W. M. Gimshaw, J. A. Parkinson, C. Quamby, and J. D. Roberts, 1976. Chemical analysis. In *Quantitative Plant Ecology* (S. B. Chapman, ed.), pp. 411–463. New York: Wiley.

Amos, G. L., 1952. *Silica in Timbers.* Bulletin No. 267, Commonwealth Scientific and Industrial Research Organization, Melbourne.

Anderson, S. T., 1970. The relative pollen productivity and pollen representation of North European trees, and correction factors for tree pollen spectra. *Danmarks Geologiske Undersaegelse, Raekke,* 96:1–99.

Anderson, P., 1980. A testimony of prehistoric tasks: Diagnostic residues on stone tool working edges. *World Archaeology* 12:181–194.

Andrejko, M., and A. Cohen, 1984. Scanning electron microscopy of silicophytoliths from the Okefenokie swamp–marsh complex. In *The Okefenokie Swamp: Its Natural History, Geology and Geochemistry* (A. D. Cohen, D. J. Casagrande, M. J. Andrejko, and G. R. Best, eds.), pp. 468–491. Los Alamos, New Mexico: Wetland Surveys.

Armitage, P. L., 1975. The extraction and identification of opal phytoliths from the teeth of ungulates. *Journal of Archaeological Science* **2**:187–197.

Ayensu, E. S., 1972. *Anatomy of the Monocotyledons. VI. Dioscoreales.* London and New York: Oxford Univ. Press (Clarendon).

Baker, G., 1959a. Opal phytoliths in some Victorian soils and "Red Rain" residues. *Australian Journal of Botany* 7:64–87.

Baker, G., 1959b. Fossil opal-phytoliths and phytolith nomenclature. *Australian Journal of Science* 21:305–306.

Baker, G., 1960a. Fossil opal-phytoliths. *Micropaleontology* 6:79–85.

Baker, G., 1960b. Phytoliths in some Australian dusts. *Proceedings of the Royal Society of Victoria* 72:21–40.

Baker, G., L. W. P. Jones, and I. D. Wardrop, 1959. Cause of wear in sheep's teeth. *Nature* 184:1583–1584.

Bamber, R. K., and J. W. Lanyon, 1960. Silica deposition in several woods of New South Wales. *Tropical Woods* 113:48–54.

Barber, D. A., and M. G. T. Shone, 1966. The absorption of silica from aqueous solutions by plants. *Journal of Experimental Botany* 17:569–578.

Bartlett, A. S., and E. S. Barghoorn, 1973. Phytogeographic history of the Isthmus of Panama during the past 12,000 years (a history of vegetation, climate and sea-level change). In *Vegetation and Vegetational History of Northern Latin America* (A. Graham, ed.), pp. 203–209. New York: Elseview.

Bartoli, F., and L. P. Wilding, 1980. Dissolution of biogenic opal as a function of its physical and chemical properties. *Soil Science Society of America Proceedings* 44:873–878.
Beadle, G., 1972. The mystery of maize. *Field Museum of Natural History Bulletin* 43:2–11.
Beadle, G., 1980. The ancestry of corn. *Scientific American* 242:112–119.
Beard, J. S., 1944. Climax vegetation in tropical America. *Ecology* 25:127–158.
Beavers, A. H., and I. Stephen, 1958. Some features of the distribution of plant opal in Illinois soils. *Soil Science* 86:1–5.
Beavers, A. W., and R. L. Jones, 1962. A fractionator for silt. *Soil Science Society of America Proceedings* 26:303.
Bennett, D. M., and A. G. Sangster, 1981. The distribution of silicon in the adventitious roots of the bamboo *Sasa Palmata*. *Canadian Journal of Botany* 59:1680–1684.
Bennett, D. M., and A. G. Sangster, 1982. Electron-probe microanalysis of silicon in the adventitious roots and terminal internode of the culm of *Zea mays*. *Canadian Journal of Botany* 60:2024–2031.
Benson, L., 1957. *Plant Classification*. Indianapolis, Indiana: Heath.
Bertoldi de Pomar, H. 1972. Opalo organogeno en sedimentos superficiales de la llanura santafesina. *Ameghiana* 9:265–279.
Bigalke, H., 1933. Die blattspodogramme der urticaceae und ihre verwendbarkeit für die systematik. *Beiträge zur Biologie der Pflanzen* 21:1–56.
Birks, W. J. B., 1981. Late Wisconsin vegetational and climatic history at Kylen Lake, northeastern Minnesota. *Quaternary Research* 16:322–355.
Birks, H. J. B., and A. D. Gordon, 1985. *Numerical Methods in Quaternary Pollen Analysis*. London: Academic Press.
Bird, J. B., and R. G. Cooke, 1978. La cueva de los ladrones: Datos preliminares sobre la ocupacion formativa. *Actas del V Simposium National de Antropologia, Arqueologia y Etnohistoria de Panama*, pp. 283–304. Universidad de Panama: Instituto Nacional de Cultura.
Bishop, R. L., R. L. Rands, and G. R. Holley, 1982. Ceramic composition analysis in archaeological perspective. In *Advances in Archaeological Methods and Theory* (M. B. Schiffer, ed.), Vol. 5, pp. 275–330. New York: Academic Press.
Blackman, E., 1968. The pattern and sequence of opaline silica deposition in rye (*Secale cereale* L.). *Annals of Botany* 32:207–218.
Blackman, E., 1969. Observations on the development of the silica cells of the leaf sheath of wheat *(Triticum aestivum)*. *Canadian Journal of Botany* 47:827–838.
Blackman, E., 1971. Opaline silica in the range grasses of southern Alberta. *Canadian Journal of Botany* 49:769–781.
Blackman, E., and D. W. Parry, 1968. Opaline silica deposition in rye (*Secale Cereale* L.). *Annals of Botany* 32:199–206.
Blackwelder, R., 1967. *Taxonomy: A Text and Reference Book*. New York: Wiley.
Bombin, M., and K. Muehlenbachs, 1980. *Potential of* ^{18}O ^{16}O *ratios in opaline plant silica as a continential paleoclimatic tool*. Abstracts, Sixth Biennial Meeting, American Quaternary Association, Toronto.
Bonnett, O. T., 1972. Silicified cells of grasses: A major source of plant opal in Illinois soils. *Agricultural Experiment Bulletin* 742. Urbana: University of Illinois.
Bonny, A. P., 1978. The effect of pollen recruitment processes on pollen distribution over the sediment surface of a small lake in Cumbria. *Journal of Ecology* 66:385–416.
Bozarth, S., 1985. *Distinctive phytoliths from various dicot species*. Paper presented at the 2nd Phytolith Research Workshop, Duluth, Minnesota.
Bozarth, S., 1986. Morphologically distinctive *Phaseolus, Cucurbita,* and *Helianthus Annuus* Phytoliths. In *Plant Opal Phytolith Analysis in Archaeology and Paleoecology* (I. Rovner, ed.), pp. 56–66. Occasional Papers No. 1 of the Phytolitharien, Raleigh: North Carolina State University.
Bradbury, J. P., B. Leyden, M. Salgado-Labouriau, W. Lewis, Jr., C. Schubert, M. Binford, D. Frey, D. Whitehead, and F. Weibezohn, 1981. Late Quaternary environmental history of Lake Valencia, Venezuela. *Science* 214:1299–1305.

Bradshaw, R. H. W., 1981. Modern pollen representation factors for woods in south–east England. *Journal of Ecology* 69:45–70.

Bray, W., L. Herrera, and M. Cardale-Schrimpff, 1985. Report on the 1982 field season in Calima. *Pro Calima* 4:2–26. Basil, Switzerland: Vereinigung Pro Calima.

Brown, D., 1984. Prospects and limits of a phytolith key for grasses in the central United States. *Journal of Archaeological Science* 11:221–243.

Brown, D., 1986. Taxonomy of a midcontinent grasslands phytolith key. In *Opal Phytolith Analysis in Archaeology and Paleoecology* (I. Rovner, ed.), pp. 67–85. Occasional Papers No. 1 of the Phytolitharien. Raleigh: North Carolina State University.

Bryant, V. M., 1974. The role of coprolite analysis in archaeology. *Texas Archaeological Society Bulletin* 45:1–28.

Bryant, V. M., and G. Williams-Dean, 1976. The coprolites of man. In *Avenues to Antiquity* (B. Fagan, ed.), pp. 257–266. San Francisco: Freeman.

Brydon, J. E., W. G. Dove, and J. S. Clark, 1963. Silicified plant asterosclereids preserved in soil. *Proceedings of the Soil Science Society of America* 27:476–477.

Bukry, D., 1979. Comments on opal phytoliths and stratigraphy of neogene silicoflagellates and coccoliths at deep sea drilling project site 397 off northwest africa. In *Initial Reports of the Deep Sea Drilling Project* (B. P. Luyendyk and J. R. Canns, eds.), Vol. XLIX, pp. 977–1009. Washington, D.C.: U.S. Printing Office.

Bukry, D., 1980. Opal phytoliths from the tropical eastern Pacific Ocean, deep sea drilling project leg 54. In *Initial Reports of the Deep Sea Drilling Project.* (B. R. Rosendahl and R. Hekinian, eds.), Vol. LIV, pp. 575–589. Washington, D.C.: U.S. Printing Office.

Bye, R. A., 1981. Quelites—ethnoecology of edible greens—past, present, and future. *Journal of Ethnobiology* 1:109–123.

Carbone, V., 1977. Phytoliths as paleoecological indicators. *Annals of the New York Academy of Science* 288:194–205.

Caseldine, C. J., 1981. Surface pollen studies across bankhead moss, Fife, Scotland. *Journal of Biogeography* 8:7–25.

Caseldine, C. J. and A. D. Gordon, 1978. Numerical analysis of surface pollen spectra from Bankhead Moss, Fife. *New Phytologist* 80:435–453.

Chen, C. H., and J. C. Lewin, 1969. Silicon as a nutrient of *Equisetum arvense. Canadian Journal of Botany* 47:125–131.

Chazdon, R. L., 1978. Ecological aspects of the distribution of C4 grasses in selected habitats of Costa Rica. *Biotropica* 10:265–269.

Cohen, A. D., 1974. Petrography and paleoecology of Holocene peat from the Okefenokee Swamp–Marsh complex of Georgia. *Journal of Sedimentary Petrology* 44:716–726.

Colinvaux, P., and F. West, 1984. The Beringian ecosystem. *Quarterly Review of Archaeology* 5: 10–16.

Commoner, B., and M. L. Zucker, 1953. Cellular differentiation: An experimental approach. In *Growth and Differentiation in Plants,* (W. E. Loomes, ed.), pp. 339–92. Iowa: Ames.

Cooke, R. G., 1984. Some current research problems in central and eastern Panama: A review of some problems. In *The Archaeology of Lower Central America* (F. Lange and D. Stone, eds.), pp. 263–302. School of American Research, Albuquerque: University of New Mexico Press.

Cooke, R., and A. J. Ranere, 1984. The "Proyecto Santa Maria." A multidisciplinary analysis of prehistoric adaptations to a tropical watershed in Panama. In *Recent Advances in Isthmian Archaeology* (F. Lange, ed.), pp. 3–30. Proceedings of the 44th International Congress of the Americanists, British Archaeological Reports, International Series 212, Oxford.

Covert, W. W., and R. F. Kay, 1981. Dental microwear and diet: Implications for determining the feeding behaviors of extinct primates, with a comment on the dietary pattern of *Sivapithecus. American Journal of Physical Anthropology* 55:331–336.

Croat, T. 1978. *The Flora of Barro Colorado Island.* Stanford: Stanford University Press.

Cutler, H. C., and T. W. Whitaker, 1968. A new species of cucurbita from Ecuador. *Annals of the Missouri Botanical Garden* 55:392–396.

Dahlgren, R. M. T., and W. T. Clifford, 1982. *The Monocotyledons: A Comparative Study*. London: Academic Press.

Darwin, Charles, 1909. *The Voyage of the Beagle*. New York: Collier. First published in 1845.

Davis, M. B., 1963. On the theory of pollen analysis. *American Journal of Science* 261:897–912.

Davis, M. B., 1969. Climatic changes in southern Connecticut recorded by pollen deposition at Rogers Lake. *Ecology* 50:409–422.

Davis, M. B., L. B. Brubaker, and T. Webb III, 1973. Calibration of absolute pollen influx. In *Quaternary Plant Ecology* (H. J. B. Birks and R. G. West, eds.), pp. 9–25. Oxford: Blackwell.

Dayanandan, P., and P. B. Kaufman, 1976. Trichomes of *Cannabis Sativa* L. (Cannabaceae). *American Journal of Botany* 63:578–591.

Dayanandan, P., P. B. Kaufman, and C. I. Franklin, 1983. Detection of silica in plants. *American Journal of Botany* 70:1079–1084.

DeBary, A., 1884. *Comparative Anatomy of the Vegetative Organs of the Phanerograms and Ferns*. (F. O. Bower and D. H. Scott, trans.) London and New York: Oxford Univ. Press (Clarendon).

Deflandre, G., 1963. Les phytolithaires (Ehrenberg) *Protoplasma* 57:234–259.

Denevan, N. 1982. Hydraulic agriculture in the American tropics: forms, measures, and recent research. In *Maya Subsistence: Studies in Memory of Dennis E. Puleston,* (K. Flannery, ed.), pp. 181–203. New York: Academic Press.

Dimbleby, G., 1967. *Plants and Archaeology*. London: John Baker.

Dimbleby, G., 1978. *Plants and Archaeology,* 2nd Ed. Atlantic Highlands, New Jersey: Humanities Press.

Dimbleby, G., 1985. *The Palynology of Archaeological Sites*. London: Academic Press.

Dinsdale, D., A. H. Gordon, and S. George, 1979. Silica in the mesophyll cell walls of Italian rye grass (*Lolium multiflorum* Lam. cv. RvP). *Annals of Botany* 44:73–77.

Doebley, J. F., 1983. The maize and teosinte male inflorescence: A numerical taxonomic study. *Annals of the Missouri Botanical Garden* 70:32–70.

Doebley, J. F., and H. H. Iltis, 1980. Taxonomy of *Zea* (Gramineae) I. A subspecific classification with key to taxa. *American Journal of Botany* 67:982–993.

Doebley, J. F., M. M. Goodman, and C. W. Stuber, 1984. Isoenzymatic variation in *Zea* (Gramineae). *Systematic Botany* 9:203–218.

Dormaar, J. F. and L. E. Lutwick, 1969. Infrared spectra of humic acids and opal phytoliths as indicators of paleosols. *Canadian Journal of Soil Science* 49:29–37.

Downton, W. J. S., 1971. Adaptive and evolutionary aspects of C4 photosynthesis. In *Photosynthesis and Photorespiration* (M. D. Hatch, C. B. Osmond, and R. O. Slatyer, eds.), pp. 3–17. New York: Wiley (Interscience).

Duke, J. A., 1968. *Darien Ethnobotanical Dictionary*. Columbus, Ohio: Battelle Memorial Institute, Columbus Laboratories. Unpublished manuscript.

Duke, J. A., 1975. Ethnobotanical observations on the Cuna Indians. *Economic Botany* 29:278–293.

Dunn, M. Eubanks, 1983. Phytolith analysis in archaeology. *Midcontinental Journal of Archaeology* 8:287–297.

Dunne, T., 1978. Rates of chemical denudation of silicate rocks in tropical catchments. *Nature* 274:244–246.

Edman, G., and E. Söderberg, 1929. Auffindung von Reis in einer Tonscherbe aus einer etwas fünftausendjährigen Chinesischen Siedlung. *Bulletin of the Geological Society of China* 8:363–365.

Ehrenberg, C. G., 1841. Über Verbreitung und Einfluss des mikroskopischen Lebens in Süd- und Nordamerika. *Monatsberichte der Königlich Preussischen Akademie der Wissenschaften zu Berlin,* pp. 139–44.

Ehrenberg, C. G., 1846. Über die Vulkanischen Phytolitharien der Insel Ascension. *Monatsberichte der Königlich Preussischen Akademie der Wissenschaften zu Berlin,* pp. 191–202.

Ehrenberg, C. G., 1854. *Mikrogeologie*. Leipzig: Leopold Voss.

Esau, K. 1977. *Anatomy of Seed Plants* (2nd Ed.) New York: Wiley.

Faegri, K., and J. Iversen, 1975. *Textbook of Pollen Analysis* (3rd Ed.). New York: Hafner.

Folger, D. W., L. H. Burckle, and B. C. Heezen, 1967. Opal phytoliths in a North Atlantic dust fall. *Science* 155:1243–1244.

Foster, R., and N. Brokaw, 1982. Structure and history of the vegetation of Barro Colorado Island. In *The Ecology of a Tropical Forest: Seasonal Rhythms and Long-term Changes* (E. Leigh, Jr., A. S. Rand, and D. Windsor, eds.), pp. 67–81. Washington, D.C.: Smithsonian Institution Press.

Formanek, J., 1899. Über die Erkennung der in den Nahrungs - und Futtermitteln vorkommenden Spelzen. *Zeitschrift für Untersuchung der Nahrungs - und Genussmittel* 11:833–843.

Fredlund, G., 1986a. Problems in the simultaneous extraction of pollen and phytoliths from clastic sediments. In *Plant Opal Phytolith Analysis in Archaeology and Paleoecology.* (I. Rovner, ed.), pp. 102–111. Occasional Papers No. 1 of the Phytolitharien. Raleigh: North Carolina State University.

Fredlund, G., 1986b. A 620,000 year opal phytolith record from central Nebraskan loess. In *Plant Opal Phytolith Analysis in Archaeology and Paleoecology* (I. Rovner, ed.), pp. 12–23. Occasional Papers No. 1 of the Phytolitharien. Raleigh: North Carolina State University.

Frey-Wyssling, A., 1930a. Über die Ausscheidung der Kieselsäure in der Pflanze. *Berichte der Deutschen Botanischen Gesellschaft* 48:179–83.

Frey-Wyssling, A., 1930b. Vergleich zivischen der Ausscheidung von Kieselsäure und Kalziumsalzen in der Pflanze. *Berichte der Deutschen Botanischen Gesellschaft* 48:184–91.

Frohnmeyer, M., 1914. Die Entstehung und Ausbildung der Kieselzellen bei Gramineen. *Bibliotheca Botanica* 21:1–41.

Fujiwara, H., R. Jones, and S. Brockwell, 1985. Plant opals (Phytoliths) in Kakadu archaeological sites. In *Archaeological Research in Kakadu National Park* (R. Jones, ed.), Special Publication No. 13, pp. 155–164, Australian National Parks and Wildlife Service, Canberra.

Galinat, W., 1971. The origin of maize. *Annual Review of Genetics* 5:447–448.

Geis, J. W., 1972. *Biogenic silica in selected plant materials.* Unpublished Ph.D. thesis, State University of New York College of Forestry at Syracuse University. Syracuse, New York.

Geis, J. W., 1973. Biogenic silica in selected species of deciduous angiosperms. *Soil Science* 116:113–130.

Geis, J. W., 1978. Biogenic opal in three species of gramineae. *Annals of Botany* 42:1119–1129.

Geis, J., 1983. *Classification of phytoliths from angiosperm and coniferous trees.* Paper presented at the annual meeting of the American Association for the Advancement of Science, Detroit.

Gill, E. D., 1967. Stability of biogenetic opal. *Science* 158:810.

Golson, J., 1977. No room at the top: Agricultural intensification in the New Guinea highlands. In *Sunda and Sahul: Prehistoric Studies in Southeast Asia, Melanesia and Australia* (J. Allen, J. Golson, and R. Jones, eds.), pp. 601–638. London: Academic Press.

Gregory, W., 1855. On the presence of Diatomaceae, Phytolitharia, and sponge spicules in soils which support vegetation. *Proceedings of the Botanical Society, Edinburgh*: 69–72.

Gordon, B., 1982. *A Panama Forest and Shore.* Pacific Grove, California: Boxwood Press.

Grob, A., 1896. Beiträge zur anatomie der epidermis der gramineenblätter. *Bibliotheca Botanica,* 36.

Haberlandt, G., 1914. *Physiological Plant Anatomy.* London: Macmillan.

Haeckel, E., 1899–1904. *Kunstformen der Natur.*

Hammond, N., 1978. The myth of the Milpa: Agricultural expansion in the Maya lowlands. In *Pre-Hispanic Maya Agriculture,* (P. D. Harrison and B. L. Turner, eds.), pp. 23–34. Albuquerque: University of New Mexico Press.

Handreck, K. A., and L. H. P. Jones, 1968. Studies of silica in the oat plant. IV. Silica content of plant parts in relation to stage of growth, supply of silica and transpiration. *Plant and Soil* 24:449–459.

Hayward, D. M., and D. W. Parry, 1973. Electron-probe microanalysis studies of silica distribution in barley (*Hordeum sativum* L.). *Annals of Botany* 37:579–91.

Hayward, D. M., and D. W. Parry, 1975. Scanning electron microscopy of silica deposition in the leaves of barley (*Hordeum sativum* L.). *Annals of Botany* 39:1003–1009.

Hayward, D. M., and D. W. Parry, 1980. Scanning electron microscopy of silica deposits in the culms, floral bracts and awns of barley (*Hordeum sativum* L.). *Annals of Botany* 46:541–548.

Helbaek, H., 1961. Studying the diet of ancient man. *Archaeology* 14:95–101.

Helbaek, H., 1969. Palaeo-ethnobotany. In *Science in Archaeology,* (D. Brothwell and E. Higgs, eds.), pp. 177–185. London: Thames and Hudson.

Herrera, C. M., 1985. Grass/grazer radiations: An interpretation of silica body diversity. *Oikos* 45:446–447.

Hodder, I., 1982. *The Present Past.* London: Batsford.

Hodson, M. J., 1986. Silicon deposition in the roots, culm and leaf of *Phalaris canariensis* L. *Annals of Botany* 58:167–177.

Hodson, M. J., A. G. Sangster, and D. W. Parry, 1982. Silicon distribution in the inflorescence bristles and macrohairs of *Setaria italica* (L.) Beauv. *Annals of Botany* 50:843–850.

Hodson, M. J., A. G. Sangster, and D. W. Parry, 1984. An ultrastructural study on the development of silicified tissues in the lemma of *Phalaris canariensis* L. *Proceedings of the Royal Society of London B*222:413–425.

Hodson, M. J., A. G. Sangster, and D. W. Parry, 1985. An ultrastructural study on the developmental phases and silicification of the glumes of *Phalaris canariensis* L. *Annals of Botany* 55:649–665.

Hoffman, F. M., and C. Hillson, 1979. Effects of silicon on the life cycle of *Equisetum Hyemale* L. *Botanical Gazette* 140:127–132.

Hubbell, S., 1979. Tree dispersion, abundance, and diversity in a tropical dry forest. *Science* 203:1299–1309.

Hubbell, S., and R. Foster, 1983. Diversity of canopy trees in a neotropical forest and implications for conservation. In *Tropical Rain Forest: Ecology and Management,* (S. Sutton, T. Whitmore, and A. Chadwick, eds.), pp. 25–41. Oxford: Blackwell.

Hubbell, S., and R. Foster, 1987. Commonness and rarity in a neotropical forest: Implications for tropical tree conservation. In *Conservation Biology: The Science of Scarcity and Diversity* (M. Soule, ed.). Sunderland, Massachusetts: Sinauer Associates.

Iler, R., 1979. *The Chemistry of Silica: Solubility, Polymerization, Colloid and Surface Properties, and Biochemistry.* New York: Wiley.

Iltis, H. H., 1983. From teosinte to maize: The catastrophic sexual transmutation. *Science* 222:886–894.

Iltis, H. H., and J. F. Doebley, 1980. Taxonomy of *Zea* (Gramineae) II. Subspecific categories in the *Zea mays* complex and a generic synopsis. *American Journal of Botany* 67:994–1004.

Iltis, H. H., and J. F. Doebley, 1984. *Zea*—A biosystematical odyssey. In *Plant Biosystematics* (W. F. Grant, ed.), pp. 587–616. Toronto: Academic Press.

Jacobson, G. L., and R. H. W. Bradshaw, 1981. The selection of sites for paleovegetation studies. *Quaternary Research* 16:80–96.

Jackson, M. L., 1956. *Soil Chemical Analysis-Advanced Course.* Mimeograph published by the Department of Soil Science, University of Wisconsin, Madison.

Jones, L. H. P., and K. A. Handreck, 1963. Effects of iron and aluminum oxides on silica in soils. *Nature* 198:852–853.

Jones, L. H. P., and K. A. Handreck, 1965. Studies of silica in the oat plant. III. Uptake of silica from soils by the plant. *Plant and Soil* 23:79–96.

Jones, L. H. P., and K. A. Handreck, 1967. Silica in soils, plants and animals. *Advances in Agronomy* 19:107–149.

Jones, L. H. P., and K. A. Handreck, 1969. Uptake of silica by *Trifolium incarnatum* in relation to the concentration in the external solution and to transpiration. *Plant and Soil* 30:71–80.

Jones, L. H. P., and A. A. Milne, 1963. Studies of silica in the oat plant. I. Chemical and physical properties of the silica. *Plant and Soil* 18:207–220.

Jones, L. H. P., A. A. Milne, and S. M. Wadham, 1963. Studies of silica in the oat plant. II. Distribution of silica in the plant. *Plant and Soil* 18:358–371.

Jones, R. L., 1964. Note on occurrence of opal phytoliths in some Cenozoic sedimentary rocks. *Journal of Paleontology* 38:773–775.

Jones, R. L., and A. H. Beavers, 1963. Some mineralogical and chemical properties of plant opal. *Soil Science* 96:375–379.

Jones, R. L., and A. H. Beavers, 1964a. Aspects of catenary and depth distribution of opal phytoliths in Illinois soils. *Soil Science Society of America Proceedings* 28:413–416.

Jones, R. L., and A. H. Beavers, 1964b. Variation of opal phytolith content among some great soil groups in Illinois. *Soil Science Society of America Proceedings* 28:711–712.

Jones, R. L., W. W. Hay, and A. H. Beavers, 1963. Microfossils in Wisconsin loess and till from western Illinois and eastern Iowa. *Science* 140:1222–1224.

Kalicz, P. J. and E. L. Stone, 1984. The Longleaf Pine Islands of the Ocala National Forest, Florida: A soil study. *Ecology* 65:1743–1754.

Kamminga, J. 1979. The nature of use-polish and abrasive smoothing on stone tools. In *Lithic Use-Wear Analysis* (B. Hayden, ed.), pp. 143–157. New York: Academic Press.

Kanno, I., and S. Arimura, 1958. Plant opal in Japanese soils. *Soil and Plant Food* 4:62–67.

Kaplan, L., and M. B. Smith, 1980. *Procedures for phytolith reference materials.* Unpublished manuscript distributed at the 45th annual meeting of the Society for American Archaeology, Philadelphia.

Karlstrom, P., 1978. Epidermal leaf structures in species of Strobilantheae and Petalidieae (Acanthaceae). *Botaniska Notiser* 131:423–433.

Karlstrom, P., 1980. Epidermal leaf structures in species of Asystasieae, Pseuderanthemeae, Graptophylleae and Odontonemeae (Acanthaceae). *Botaniska Notiser* 133:1–16.

Kaufman, P. B., P. Takeoka, W. C. Bigelow, R. Sehmid, and N. S. Ghosheh, 1971. Electron microprobe analysis of silica in epidermal cells of *Equisetum*. *American Journal of Botany* 58:309–316.

Kaufman, P. B., P. Takeoka, W. C. Bigelow, J. D. Jones, and R. Iler, 1981. Silica in shoots of higher plants. In *Silicon and Siliceous Structures in Biological Systems* (T. L. Simpson and B. E. Volcani, eds.), pp. 409–449. New York: Springer-Verlag.

Kaufman, P. B., P. Dayanandan, and C. I. Franklin, 1985. Structure and function of silica bodies in the epidermal system of grass shoots. *Annals of Botany* 55:487–507.

Klein, R. L., and J. W. Geis, 1978. Biogenic silica in the Pinaceae. *Soil Science* 126:145–156.

Kolbe, R. W., 1957. Fresh water diatoms from Atlantic deep-sea sediments. *Science* 126:1053–1056.

Kondo, R., and T. Peason, 1981. Opal phytoliths in tree leaves (Part 2): Opal phytoliths in dicotyledon angiosperm tree leaves. *Research Bulletin of Obihiro University, Series 1* 12:217–230.

Kunoh, H., and S. Akai, 1977. Scanning electron microscopy and x-ray microanalysis of dumbbell-shaped bodies in rice lamina epidermis. *Bulletin of the Torrey Botanical Club* 104:309–313.

Kurmann, M. H., 1985. An opal phytolith and palynomorph study of extant and fossil soils in Kansas (U.S.A.) *Paleogeography, Paleoclimatology, Paleoecology* 49:217–235.

Labouriau, L., 1983. Phytolith work in Brazil, a minireview. *Phytolitharien Newsletter* 2:6–10.

Lamb, W. F., 1984. Modern pollen spectra from Labrador and their use in reconstructing Holocene vegetational history. *Journal of Ecology* 72:37–59.

Lanning, F. C., 1960. Nature and distribution of silica in strawberry plants. *Proceedings of the American Horticultural Society* 76:349–358.

Lanning, F. C., 1961. Silica and calcium in black raspberries. *Proceedings of the American Horticultural Society* 77:367–371.

Lanning, F. C., 1966. Silica and calcium deposition in the tissues of certain plants. *Advancing Frontiers of Plant Science* 13:55–66.

Lanning, F. C., 1972. Ash and silica in *Juncus*. *Bulletin of the Torrey Botanical Club* 99:196–198.

Lanning, F. C., and L. N. Eleuterius, 1981. Silica and ash in several marsh plants. *Gulf Research Reports* 7:47–52.

Lanning, F. C., and L. N. Eleuterius, 1983. Silica and ash in tissues of coastal plants. *Annals of Botany* 51:835–850.

Lanning, F. C., and L. N. Eleuterius, 1985. Silica and ash in tissues of some plants growing in the coastal area of Mississippi, U.S.A. *Annals of Botany* 56:157–172.

Lanning, F. C., B. W. X. Ponnaiya, and C. F. Crumpton, 1958. The chemical nature of silica in plants. *Plant Physiology* 33:339–343.

Lanning, F. C., T. L. Hopkins, and J. C. Loera, 1980. Silica and ash content and depositional patterns in tissues of nature *Zea mays* L. plants. *Annals of Botany* 45:549–554.

Lathrap, D. 1975. *Ancient Ecuador: Culture, Clay, and Creativity, 3000–300 B.C.* Chicago: Field Museum of Natural History.

Lee, E. G., Jr., A. S. Rand, and D. M. Windsor, eds., 1982. *The Ecology of a Tropical Forest: Seasonal Rhythms and Long-Term Changes.* Washington, D.C.: Smithsonian Institution Press.

Leeper, G. W., A. Nicholls, and S. M. Wadham, 1936. Soil and pasture studies in the Mt. Gellibrand area, Western District of Victoria. *Proceedings of the Royal Society of Victoria* 49:77–138.

Lewin, J., and B. E. F. Reimann, 1969. Silica and plant growth. *Annual Review of Plant Physiology* 20:289–304.

Lewis, R. O., 1978. Use of opal phytoliths in paleo-environmental reconstruction. *Wyoming Contributions to Anthropology* 1:127–132.

Lewis, R. O., 1981. Use of opal phytoliths in paleo-environmental reconstructions. *Journal of Ethnobiology* 1:175–181.

Leyden, B., 1984. Guatemalan forest synthesis after Pleistocene aridity. *Proceedings of the National Academy of Sciences* 81:4856–4859.

Liebowitz, H. and R. L. Folk, 1980. Archaeological geology of Tel Yin'am, Galilee, Israel. *Journal of Field Archaeology* 7:23–42.

Linares, O. F., and A. J. Ranere, eds., 1980. *Adaptive Radiations in Prehistoric Panama.* Peabody Museum Monographs, No. 5. Cambridge, Massachusetts.

Linares, O. F., Sheets, P., and E. J. Rosenthal, 1975. Prehistoric agriculture in tropical highlands. *Science* 187:137–145.

Livingston, D. A., 1982. Quaternary geography and Africa and the refuge theory. In *Biological Diversification in the Tropics* (G. T. Prance, ed.), pp. 523–536. New York: Columbia University Press.

Livingston, D. A., and W. D. Clayton, 1980. An altitudinal cline in tropical African grass floras and its paleoecological significance. *Quaternary Research* 13:392–402.

Livingston, D. A., and T. Van der Hammen, 1978. Paleogeography and paleoclimatology. In *Tropical Forest Ecosystems; A State-of-Knowledge Report.* Paris: UNESCO/UNEP/FAO.

Locker, S., and E. Martini, 1986. Phytoliths from the southwest Pacific, site 591. In *Initial Reports of the Deep Sea Drilling Project,* pp. 1079–1083. Washington, D.C.: U.S. Printing Office.

Lovering, T. S., 1959. Significance of accumulator plants in rock weathering. *Bulletin of the Geological Society of America* 70:781–800.

Lynch, T., 1976. The entry and post-glacial adaptation of man in Andean South America. *Union Internationale des Science Prehistoriques et Protohistoriques,* 9th Congress, Nice, France.

Lynch, T., 1983. The paleo-indians. In *Ancient South Americans,* (J. D. Jennings, ed.), pp. 87–137. San Francisco: Freeman.

MacNeish, R. S., 1967. A summary of the subsistence. In *The Prehistory of the Tehuacan Valley,* Vol. 1: *Environment and Subsistence,* (D. S. Byers, ed.), pp. 290–309. Austin: University of Texas Press.

Maher, L. 1972. Absolute pollen diagram of Redrock Lake, Boulder County, Colorado. *Quaternary Research* 2:531–553.

Mangelsdorf, P., 1974. *Corn: Its origin, evolution and improvement.* Cambridge, Massachusetts: Belknap Press of Harvard University Press.

Mangelsdorf, P., 1986. The origin of corn. *Scientific American* 254:80–86.

Matthews, J. V., Jr., 1982. East Beringia during late Wisconson time: A review of the biotic evidence. In *Paleoecology of Beringia* (D. M. Hopkins, J. V. Matthews, Jr., C. E. Schweger, and S. B. Young, eds.), pp. 127–150. London: Academic Press.

Maynard, N. G., 1976. Relationship between diatoms in surface sediments of the Atlantic Ocean and the biologic and physical oceanography of overlying waters. *Paleobiology* 2:99–121.

McKeague, J. A., and M. G. Cline, 1963a. Silica in soil solutions I. The form and concentration of dissolved silica in aqueous extracts of some soils. *Canadian Journal of Soil Science* 43:70–82.

McKeague, J. A., and M. G. Cline, 1963b. Silica in soil solutions II. The absorption of monosilicic acid by soil and by other substances. *Canadian Journal of Soil Science* 43:83–96.

McNaughton, S. J., and J. L. Tarrants, 1983. Grass leaf silicification: Natural selection for an inducible defense against herbivores. *Proceedings of the National Academy of Science* 80:790–791.

McNaughton, S. J., J. L. Tarrants, and R. H. Davis, 1985. Silica as a defense against herbivory and a growth promoter in African grasses. *Ecology* 66:528–535.

Mehra, P. N., and O. P. Sharma, 1965. Epidermal silica cells in the Cyperaceae. *Botanical Gazette* 126:53–58.

Mehringer, P. J., 1967. Pollen analysis of the Tula Springs area. In *Pleistocene Studies in Southern Nevada* (W. M. Wormington and D. Ellis, ed.), Nevada State Museum Anthropological Papers 13:120–200.

Melia, M. B., 1980. Distribution and provenance of palynomorphs in northeast Atlantic aerosols and bottom sediments (microfilms). Ann Arbor: University of Michigan.

Metcalfe, C. R., 1960. *Anatomy of the Monocotyledons. I. Gramineae.* London: Oxford University Press.

Metcalfe, C. R. 1971. *Anatomy of the Monocotyledons. V. Cyperaceae.* London: Oxford University Press.

Metcalfe, C. R., and L. Chalk, 1950. *Anatomy of the Dicotyledons.* London: Oxford University Press (Clarendon).

Metcalfe, C. R., and L. Chalk, 1979. *Anatomy of the Dicotyledons,* 2nd Ed. London: Oxford University Press (Clarendon).

Metcalfe, C. R., and M. Gregory, 1964. Some new descriptive terms for Cyperaceae with a discussion of variations in leaf form noted in the family. *Notes from the Jodrell Laboratory Royal Botanic Gardens, Kew* 13:1–13.

Miksicek, C., 1983. Macofloral remains of the Pulltrouser area: Settlements and fields. In *Pulltrouser Swamp* (B. L. Turner and P. Harrison, eds.), pp. 94–104. Austin: University of Texas Press.

Miller, A., 1980. *Phytoliths as indicators of farming techniques.* Paper presented at the 45th annual meeting of the Society for American Archaeology, Philadelphia.

Miller-Rosen, A., 1985. *Phytoliths of cereals from two Negev chalcolithic sites: Identification and edaphic implications.* Unpublished manuscript.

Mobius, M., 1908a. Über ein eigentümliches Vorkommen von Kieselkörpern in der Epidermis den Bau des Blattes von *Callisia repens.* (Wiesner-Festschrift: 81).

Mobius, M., 1908b. Über die Festlegung der Kalksalze und Kieselkörper in den Pflanzenzellen. *Berichte der Deutschen Botanischen Gesellschaft* 26a:29.

Moody, V., 1972. *Phytoliths as an interpretive device in archaeological sites.* Unpublished Masters thesis, Missoula: University of Montana.

Mulholland, S., 1982. Various wet-ashing procedures for phytolith extraction from plants. *Phytolitharien Newsletter* 1:5.

Mulholland, S., 1986. Classification of grass silica phytoliths. In *Plant Opal Phytolith Analysis in Archaeology and Paleoecology* (I. Rovner, ed.), pp. 41–52. Occasional Papers No. 1 of the Phytolitharien. Raleigh: North Carolina State University.

Netolitzky, F., 1900. Mikroskopische Untersuchung Gänzlich verkohlter vorgeschichtlicher Nahrungsmittel aus Tirol. *Zeitschrift Für Untersuchung der Nahrungs—Und Genussmittel* 3:401–407.

Netolitzky, F., 1914. Die Hirse aus Antiken Funden. *Sitzbuch der Keiserliche Akadamie für Wissenschaft der Mathematisch—Naturwissenschaften* 123:725–759.

Netolitzky, F., 1929. Die Kieselkörper. *Linsbauer's Handbuch der Pflanzenanatomie* 3/1a:1–19.

Neubauer, H. 1905. Mikrophotographien der für die Nahrungs—und Futtermittle-untersuchung wichtigsten Gramineenspelzen. *Landwirthschaftliche Jahrbücher* 34:793–984.

Newman, R. H., and A. L. Mackay, 1983. Silica spicules in Canary grass. *Annals of Botany* 52:927–929.

Nicholls, A., 1939. Some applications of minerology to soil studies. *Journal of the Australian Institute of Agricultural Science* 5:218–221.

Norgren, J. A., 1973. *Distribution, form and significance of plant opal in Oregon soils.* Ph.D. dissertation, Corvallis: Oregon State University.

Okuda, A., and E. Takahashi, 1961. The effect of various amounts of silicon supply on the growth of the rice plant and nutrient uptake, part 3. *Journal of the Science of Soil and Manure, Japan,* 32:533–537.

Okuda, A., and E. Takahashi, 1964. The role of silicon. In *The Mineral Nutrition of the Rice Plant.* Proceedings of the Symposium of the International Rice Research Institute, pp. 123–46. Baltimore: Johns Hopkins Press.

Oldfield, F., 1970. Some aspects of scale and complexity in pollen-analytically based palaeoecology. *Pollen Et Spores* 12:163–171.

Palmer, P. G., 1976. Grass cuticles: A new paleoecological tool for East African lake sediments. *Canadian Journal of Botany* 54:1725–1734.

Palmer, P. G., and A. Tucker, 1981. A scanning electron microscope survey of the epidermis of East African grasses, I. *Smithsonian Contributions to Botany,* No. 49. Washington, D.C.: Smithsonian Institution Press.

Palmer, P. G., and A. E. Tucker, 1983. A scanning electron microscope survey of the epidermis of East African grasses, II; *Smithsonian Contributions to Botany,* No. 53, Washington, D.C.: Smithsonian Institution Press.

Palmer, P. G., S. Gerbeth-Jones, and S. Hutchinson, 1985. A scanning electron microscope survey of the epidermis of East African grasses, III; *Smithsonian Contributions to Botany,* No. 55. Washington, D.C.: Smithsonian Institution Press.

Parmenter, C., and D. W. Folger, 1974. Eolian biogenic detritus in deep sea sediments: A possible index of equatorial ice age aridity. *Science* 184:695–698.

Parry, D. W., and M. J. Hodson, 1982. Silica distribution in the caryopsis and inflorescence bracts of foxtail millet [*Setaria italica* (L.) Beauv.] and its possible significance in carcinogenesis. *Annals of Botany* 49:531–540.

Parry, D. W., and M. Kelso, 1975. The distribution of silicon deposits in the roots of *Molinia caerulea* (L.) Moench. and *Sorghum bicolor* (L.) Moench. *Annals of Botany* 39:995–1001.

Parry, D. W., and F. Smithson, 1958a. Techniques for studying opaline silica in grass leaves. *Annals of Botany* 22:543–550.

Parry, D. W., and F. Smithson, 1958b. Silicification of bulliform cells in grasses. *Nature* 181:1549–1550.

Parry, D. W., and F. Smithson, 1963. Influence of mechanical damage on opaline silica deposition in *Molina caerulea* L. *Nature* 199:925–926.

Parry, D. W., and F. Smithson, 1964. Types of opaline silica depositions in the leaves of British grasses. *Annals of Botany* 28:169–185.

Parry, D. W., and F. Smithson, 1966. Opaline silica in the inflorescences of some British grasses and cereals. *Annals of Botany* 30:525–538.

Parry, D. W., and A. Winslow, 1977. Electron-probe microanalysis of silicon accumulation in the leaves and tendrils of *Pisum sativum* (L.) following root severance. *Annals of Botany* 41:275–278.

Parry, D. W., M. J. Hodson, and A. G. Sangster, 1984. Some recent advances in studies of silicon in higher plants. *Philosophical Transactions of the Royal Society of London Series B* 304:537–549.

Pearsall, D. M. 1977–1978. Early movements of maize between Mesoamerica and South America. *Journal of the Steward Anthropological Society* 9:41–75.

Pearsall, D. M., 1978. Phytolith analysis of archaeological soils: Evidence for maize cultivation in formative Ecuador. *Science* 199:177–178.

Pearsall, D. M., 1979. The application of ethnobotanical techniques to the problem of subsistence in the Ecuadorian formative. Ph.D. dissertation, Urbana: University of Illinois.

Pearsall, D. M., 1980. *Preliminary report on phytolith investigations at the El Balsamo site, Escuintla, Guatemala.* Unpublished manuscript.

Pearsall, D. M., 1981a. *Second report on phytolith investigations at the El Balsamo site: Surface sample analysis.* Unpublished manuscript.

Pearsall, D. M., 1981b. *Phytolith analysis at natural Trap Cave.* Unpublished manuscript.

Pearsall, D. M., 1982. Phytolith analysis: Applications of a new paleoethnobotanical technique in archaeology. *American Anthropologist* 84:862–871.

Pearsall, D. M., and D. R. Piperno, 1986. *Antiquity of maize cultivation in Ecuador: Summary and reevaluation of the evidence.* Paper presented at the 51st annual meeting of the Society for American Archaeology, New Orleans.

Pearsall, D. M., and M. Trimble, 1982. *Phytolith analysis of soil samples from the Waimea–Kawaihae project.* Unpublished manuscript.

Pearsall, D. M., and M. Trimble, 1984. Identifying past agricultural activity through soil phytolith analysis: A case study from the Hawaiian Islands. *Journal of Archaeological Science* 11:119–133.

Pease, D. S., 1967. *Opal phytoliths as indicators of paleosols.* Unpublished master's thesis, New Mexico State University, University Park.

Pease, D. S., and J. V. Anderson, 1969. Opal phytoliths in *Bouteloua Eripoda* Torr. roots and soils. *Soil Science Society of America Proceedings* 33:321–322.

Perry, C., S. Mann, R. J. P. Williams, F. Watt, G. W. Grime, and J. Takacs, 1984. A scanning electron microprobe study of microhairs from the lemma of the grass *Phalaris canariensis* L. *Proceedings of the Royal Society of London Series B* 222:439–455.

Pickersgill, B., and C. B. Heiser, 1977. Origins and distributions of plants domesticated in the New World tropics. In *The Origins of Agriculture* (C. A. Reed, ed.), pp. 803–835. The Hague: Mouton.

Piperno, D. R., 1983. The application of phytolith analysis to the reconstruction of plant subsistence and environments in prehistoric Panama (microfilms) Ann Arbor: University of Michigan.

Piperno, D. R., 1984. A comparison and differentiation of phytoliths from maize and wild grasses: Use of morphological criteria. *American Antiquity* 49:361–383.

Piperno, D. R., 1985a. Phytolith analysis and tropical paleo-ecology: Production and taxonomic significance of siliceous forms in New World plant domesticates and wild species. *Review of Paleobotany and Palynology* 45:185–228.

Piperno, D. R., 1985b. Phytolith taphonomy and distributions in archaeological sediments from Panama. *Journal of Archaeological Science* 12:247–267.

Piperno, D. R., 1985c. Phytolithic analysis of geological sediments from Panama. *Antiquity* LIX: 13–19.

Piperno, D. R., 1985d. Phytolith records from prehistoric agricultural fields in the Calima Region. *Pro Calima,* Vol. 4, pp. 37–40. Basel, Switzerland: Vereinigung Pro Calima.

Piperno, D. R., 1986a. Fitalitos, arquealogia y cambias prehistoricas en la historia vegetal de una parcela de cincuenta hectores en la Isla de Barro Colorado, Panama. In *The Ecology of a Tropical Forest: Seasonal Rhythms and Long-Term Changes* (spanish edition) (E. Leigh, A. S. Rand, and D. Windsor, eds.), Washington, D.C.: Smithsonian Institution Press.

Piperno, D. R., 1986b. The analysis of phytoliths from the Vegas site OGSE-80. Ecuador. In *The Vegas Culture: Early Prehistory of Southwestern Ecuador* (K. Stothert, ed.). Guayaquil: Museo Antropologico de Banco Central del Ecuador.

Piperno, D. R. 1986c. *Phytolith analysis of soils from the Delaware park site.* Unpublished manuscript.

Piperno, D. R., 1987a. A phytolith key for tropical grasses of panama (in preparation).

Piperno, D. R., 1987b. Phytolith evidence for cultivars, especially maize. In *Prehistoric Wetland Agriculture in Central Veracruz, Mexico.* (A. Siemens and A. Gomez-Pompa, eds.) (in preparation).

Piperno, D. R., and V. Starczak, 1985. *Numerical analysis of maize and wild grass phytoliths using multivariate techniques*. Paper presented at the 2nd Phytolith Research Workshop, Duluth, Minnesota.

Piperno, D. R., K. Clary, R. Cooke, A. J. Ranere, and D. Weiland, 1985. Preceramic maize in central Panama: Phytolith and pollen evidence. *American Anthropologist* 87:871–878.

Piperno, D. R., and K. Stothert, 1987. *Cucurbita* Phytoliths from 9,000 year old archaeological deposits in ecuador (in preparation).

Postek, M. T., 1981. The occurrence of silica in the leaves of *Magnolia Grandiflora* L. *Botanical Gazette* 142:124–134.

Prat, H., 1936. La Systematique des Graminees. *Annales des Sciences Naturelles. Botanique Series 2* 18:165–258.

Prat, H., 1948. General features of the epidermis in *Zea mays*. *Annals of the Missouri Botanical Garden* 35:341–351.

Rands, R. L., and M. M. Bargielski, 1986. Chemistry, color, and phytoliths: Mixed level ceramic research in the Palenque region, Mexico. Paper read at the 51st annual meeting of the Society for American Archaeology, New Orleans.

Ranere, A. J., 1980a. Preceramic shelters in the Talamancan Range. In *Adaptive Radiations in Prehistoric Panama* (O. F. Linares and A. J. Ranere, eds.), pp. 16–43. Peabody Museum Monographs, No. 5., Cambridge, Massachusetts.

Ranere, A. J., 1980b. Human movement in tropical America at the end of the Pleistocene. In *Anthropological Papers in Memory of Earl H. Swanson, Jr.* (L. B. Harten, C. N. Warren, and D. R. Tuohy, eds.), pp. 41–47. Pocatello: Idaho Museum of Natural History.

Ranere, A. J., and P. Hansell, 1978. Early subsistence patterns along the Pacific Coast of central Panama. In *Prehistoric Coastal Adaptations* (B. L. Stark and B. Voorhies, eds.), pp. 43–59. New York: Academic Press.

Rapp, G., 1986. Morphological classification of phytoliths. In *Plant Opal Phytolith Analysis in Archaeology and Paleoecology,* (I. Rovner, ed.), pp. 33–34. Occasional Papers No. 1 of the Phytolitharien. Raleigh: North Carolina State University.

Raven, J. A., 1983. The transport and function of silica in plants. *Biological Reviews of the Cambridge Philosophical Society* 58:179–207.

Riquier, G., 1960. Les phytolithes de certains sols tropicaux et des podzals. *Transactions of the International Congress of Soil Science* 4:425–431.

Ritchie, J. C., and L. C. Cwynar, 1982. The late Quaternary vegetation of the Yukon. In *Paleoecology of Beringia* (D. Hopkins, J. V. Matthews, Jr., C. E. Schweger, and S. B. Young, eds.), pp. 113–126. London: Academic Press.

Robinson, R., 1979. Biosilica analysis: Paleoenvironmental reconstruction of 41LL 254. In *An Intensive Archaeological Survey of Enchanted Rock State Natural Area* (C. Assad and D. Potter, eds.), pp. 125–140. Center for Archaeological Research, San Antonio: University of Texas.

Robinson, R., 1980. *Environmental chronology for central and south Texas: External correlation to the Gulf coastal plain and the southern high plains.* Paper presented at the 45th annual meeting of the Society for American Archaeology, Philadelphia.

Robinson, R., 1982. Biosilica analysis of three prehistoric archaeological sites in the Choke Canyon Reservoir, Live Oak County, Texas: Preliminary summary of climatic implications. In *Archaeological investigations at Choke Canyon Reservoir, South Texas: Phase I Findings* (G. Hall, S. Black, and C. Graves, eds.), pp. 597–610. Center for Archaeological Research. San Antonio: University of Texas.

Roosevelt, A., 1980. *Parmana: Prehistoric Maize and Manioc Subsistence along the Amazon and Orinoco.* New York: Academic Press.

Roosevelt, A., 1984. Problems interpreting the diffusion of cultivated plants. In *Pre-Colombian Plant Migration* (D. Stone, ed.), pp. 1–18. Cambridge, Massachusetts: Harvard University Press.

Rovner, I., 1971. Potential of opal phytoliths for use in paleoecological reconstruction. *Quaternary Research* 1:343–359.

Rovner, I., 1972. Note on a safer procedure for opal phytolith extraction. *Quaternary Research* 2:591.

Rovner, I., 1983. Plant opal phytolith analysis: Major advances in archaeobotanical research. In *Advances in Archaeological Method and Theory* (M. B. Schiffer, ed.), Vol. 6, pp. 225–266. New York: Academic Press.

Rovner, I., 1985. *Phytolith analysis at Alta Toquima village: Particles in the pines*. Paper read at the 2nd Phytolith Research Workshop, University of Minnesota, Duluth.

Rovner, I., 1986a. Phytolith sampling and research design in archaeology. In *Plant Opal Phytolith Analysis in Archaeology and Paleoecology* (I. Rovner, ed.), pp. 111–122. Occasional Papers No. 1 of the Phytolitharien. Raleigh: North Carolina State University.

Rovner, I., 1986b. Downward percolation of phytoliths in stable soils: A non-issue. In *Plant Opal Phytolith Analysis in Archaeology and Paleoecology* (I. Rovner, ed.) pp. 23–30. Occasional Papers No. 1 of the Phytolitharien. Raleigh: North Carolina State University.

Ruprecht, F., 1866. Geobotanical investigations on chernozem. *Acad. Sci.* (Russian)

Russ, J. C. and I. Rovner, 1987. Stereological verification of *Zea* phytolith taxonomy. *The Phytolitharien Newsletter* 4:10–18.

Sangster, A. G., 1977a. Characteristics of silica deposition in *Digitaria sanguinalis* (L.) Scop. (crabgrass). *Annals of Botany* 41:341–350.

Sangster, A. G., 1977b. Electron-probe microassay studies of silicon deposits in the roots of two species of *Andropogon*. *Canadian Journal of Botany* 55:880–887.

Sangster, A. G., 1978. Silicon in the roots of higher plants. *American Journal of Botany* 65:929–935.

Sangster, A. G., 1985. Silicon distribution and anatomy of the grass rhizome, with special reference to *Miscanthus Sacchariflorus* (Maxim.) Hackel. *Annals of Botany* 55:621–634.

Sangster, A. G., and D. W. Parry, 1969. Some factors in relation to bulliform cell silicification in the grass leaf. *Annals of Botany* 33:315–323.

Sangster, A. G., and D. W. Parry, 1971. Silica deposition in the grass leaf in relation to transpiration and the effect of Dinitrophenal. *Annals of Botany* 35:667–677.

Sangster, A. G., and D. W. Parry, 1976a. Endodermal silicon deposits and their linear distribution in developing roots of *Sorghum bicolor* L. Moench. *Annals of Botany* 40:361–371.

Sangster, A. G., and D. W. Parry, 1976b. Endodermal silicification in mature, nodal roots of *Sorghum bicolor* (L.) Moench. *Annals of Botany* 40:373–379.

Sangster, A. G., and D. W. Parry, 1976c. The ultrastructure and electron-probe microassay of silicon deposits in the endodermis of the seminal roots of *Sorghum bicolor* (L.) Moench. *Annals of Botany* **40:**373–379.

Sangster, A. G., M. J. Hodson, D. W. Parry, and J. A. Rees, 1983. A developmental study of silicification in the trichomes and associated epidermal structures of the inflorescence bracts of the grass *Phalaris canariensis* (L.) *Annals of Botany* 52:171–187.

Schellenberg, H. C., 1908. Wheat and barley from the North Kurgan, Anau. In *Exploration in Turkestan* (R. Pumpelly, ed.), Vol. 3, pp. 471–473. Washington, D.C.: Carnegie Institution.

Schreve-Brinkman, E. J., 1978. A palynological study of the upper Quaternary sequence in the El Abra Corridor and rock shelters (Colombia). *Paleogeography, Paleoclimatology, Paleoecology* 25:1–109.

Scurfield, G., C. A. Anderson, and E. R. Segnit, 1974. Silica in woody stems. *Australian Journal of Botany* 22:211–229.

Siever, R., 1957. The silica budget in the sedimentary cycle. *American Minerologist* 42:821–841.

Sih, A., and K. Milton, 1985. Optimal diet theory: Should the Kung eat mongongos? *American Anthropologist* 87:395–401.

Simpson, B. B., 1975. Glacial climates in the eastern tropical South Pacific. *Nature* 253:34–35.

Simpson, T. L., and B. Volcani, eds. 1981. *Silicon and Siliceous Structures in Biological Systems*. New York: Springer–Verlag.

Smiley, C., and L. M. Huggins, 1981. Pseudofagus Idahoensis, N. Gen. et Sp. (Fagaceae) from the Miocene Clarkia flora of Idaho. *American Journal of Botany* 68:741–761.

Smithson, F., 1956. Plant opal in soil. *Nature* 178:107.

Smithson, F., 1958. Grass opal in British soils. *Journal of Soil Science* 9:148.

Solereder, H., 1908. *Systematic Anatomy of the Dicotyledons,* (L. A. Boodle and F. E. Frisch, trans.). London and New York: Oxford University Press (Clarendon).

Stant, M. Y., 1973. Scanning electron microscopy of silica bodies and other epidermal features in *Gibasis (Tradescantia)* leaf. *Botanical Journal of the Linnaean Society* 66:233–244.

Starna, W. A., and D. A. Kane, Jr., 1983. Phytoliths, archaeology, and caveats: A case study from New York. *Man in the Northeast* 26:21–31.

Sterling, C., 1967. Crystalline silica in plants. *American Journal of Botany* 54:840–844.

Stothert, K., 1985. The preceramic Las Vegas culture of coastal Ecuador. *American Antiquity* 50:613–637.

Struve, G. A., 1835. *De Silica in Plantis Nonnullis.* Dissertation, Berlin.

Tack, M., and L. Kaplan, 1986. Phytoliths as corraborative physical findings. In *Plant Opal Phytolith Analysis in Archaeology and Paleoecology* (I. Rovner, ed.), pp. 129–132. Occasional Papers No. 1 of the Phytolitharien. Raleigh: North Carolina State University.

Tauber, H., 1965. Differential pollen dispersion and the interpretation of pollen diagrams. *Danmarks Geologiske Undersøgelse, Raekke.* II. 89:1–69.

Tauber H., 1977. Investigations of aerial pollen transport in a forested area. *Dansk Botanisk Arkiv* 32.

Ter Welle, B. J. H., 1976. Silica grains in woody plants of the neotropics, especially Surinam. *Leiden Botanical Series,* No. 3, pp. 107–142.

Thomasson, J. R., 1983. Carex Graceii SP. N., Cyperocarpus Eliasii SP.N., Cyperocarpus Terrestris SP.N., and Cyperocarpus Pulcherrima SP.N. (Cyperaceae) from the Miocene of Nebraska. *American Journal of Botany* 70:435–449.

Thomasson, J. R., 1984. Miocene grass (Gramineae: Arundinoideae) leaves showing external micromorphological and internal anatomical features. *Botanical Gazette* 145:204–209.

Thomasson, J. R., M. E. Nelson, and R. J. Zakrzewski, 1986. A fossil grass (Gramineae-Chloridoideae) from the Miocene with Krantz anatomy. *Science* 233:876–878.

Tomlinson, P. B., 1961. *Anatomy of the Monocotyledons II: Palmae.* London: Oxford University Press.

Tomlinson, P. B., 1969. *Anatomy of the Monocotyledons III: Commelinales-Zingiberales.* London: Oxford University Press.

Turner, B. L., II, and P. D. Harrison, 1981. Prehistoric raised-field agriculture in the Maya lowlands. *Science* 213:399–405.

Turner, B. L., II, and P. D. Harrison, eds., 1983. *Pulltrouser Swamp: Ancient Maya Habitat, Agriculture and Settlement in Northern Belize.* Austin: University of Texas Press.

Twiss, P. C., 1983. Dust deposition and opal phytoliths in the Great Plains. *Transactions of the Nebraska Academy of Science* XI:73–82.

Twiss, P. C., 1986a. Discussion. In *Plant Opal Phytolith Analysis in Archaeology and Paleoecology* (I. Rovner, ed.), pp. 137–140. Occasional Papers No. 1 of the Phytolitharien, Raleigh: North Carolina State University.

Twiss, P. C., 1986b. Morphology of opal phytoliths in C3 and C4 Grasses. In *Plant Opal Phytolith Analysis in Archaeology and Paleoecology* (I. Rovner, ed.), pp. 4–11. Occasional Papers No. 1 of the Phytlitharien. Raleigh: North Carolina State University.

Twiss, P. C., E. Suess, and R. M. Smith, 1969. Morphological classification of grass phytoliths. *Soil Science Society of America Proceedings* 33:109–115.

Upchurch, S. B., R. Strom, and M. J. Andreijko, 1983. A model for silicification in peat-forming environments. In *Mineral Matter in Peat: Its Occurrence, Form, and Distribution* (R. Raymond, Jr. and M. J. Andrejko, eds.), pp. 215–224. Los Alamos, New Mexico: Los Alamos National Laboratory.

Van der Hammen, T., 1974. The Pleistocene changes of vegetation and climate in tropical South America. *Journal of Biogeography* 1:3–26.

Van der Worm, P. D. J., 1980. Uptake of Si by five plant species as influenced by variations in Si-supply. *Plant and Soil* 56:153–156.

Verma, S. D., and R. H. Rust, 1969. Observations on opal phytoliths in a soil biosequence in southeastern Minnesota. *Soil Science Society of America Proceedings* 33:749–751.

Wadham, M. D., and D. W. Parry, 1981. The silicon content of *Oryza sativa* L. and its effect on the grazing behaviour of *Agriolimax reticulatus* Miller. *Annals of Botany* 48:399–402.

Walker, A., H. Haeck, and L. Perez, 1978. Microwear of mammalian teeth as an indicator of diet. *Science* 201:908–910.

Watanabe, N., 1955. Ash in archaeological sites. *Rengo–Taikai Kiji. Proceeding of the Joint Meeting of the Anthropological Society of Nippon and Japan Society of Ethnology* (Japanese) 9:169–171.

Watanabe, N., 1968. Spodographic evidence of rice from prehistoric Japan. *Journal of the Faculty of Science of the University of Tokyo,* Section V, 3:217–235.

Watanabe, N. 1970. A spodographic analysis of millet from prehistoric Japan. *Journal of the Faculty of Science of the University of Tokyo,* Section V, 3:357–359.

Weaver, F. M., and S. W. Wise, Jr., 1974. Opaline sediments of the southeastern coastal plain and horizon A: Biogenic origin. *Science* 184:899–901.

Webb, T., III, 1974. Corresponding patterns of pollen and vegetation in lower Michigan: A comparison of quantitative data. *Ecology* 55:17–28.

Webb, T., III, and R. A. Bryson, 1972. Late and post-glacial climatic change in the northern midwest, U.S.A.: Quantitative estimates derived from fossil pollen spectra by multivariate statistical analysis. *Quaternary Research* 2:70–115.

Wendorf, F., and R. Schild, 1984. The emergence of food production in the Egyptian Sahara. In *From Hunters to Farmers: The Causes and Consequences of Food Production in Africa* (J. D. Clark and S. A. Brandt, eds.), pp. 93–101. Berkeley: University of California Press.

Wendorf, F., R. Schild, N. El Hadidi, A. E. Close, M. Kobusiewicz, H. Wieckowska, B. Issawi, and H. Haas, 1979. The use of barley in the Egyptian late Paleolithic. *Science* 205:1341–1347.

Weiland, D., 1987. Preceramic settlement patterns in the Santa Maria Basin, central Pacific Panama. In *Proceedings of the 45th International Congress of the Americanists, Bogota.*

Werner, O., 1928. Blatt–Aschenbilder Heimischer Wiesengräser als Mittel ihrer Verwandtschafts—und Wertbestimmung. *Biologia Generalis* 4:125–136.

Wilding, L. P., 1967. Radiocarbon dating of biogenetic opal. *Science* 184:899–901.

Wilding, L. P., and L. R. Drees, 1968. Biogenic opal in soils as an index of vegetation history in the prairie peninsula. In *The Quaternary of Illinois* (R. E. Bergstrom, ed.), pp. 99–103. Special Publication 14. Urbana: University of Illinois College of Agriculture.

Wilding, L. P., and L. R. Drees, 1971. Biogenic opal in Ohio soils. *Soil Science Society of America Proceedings* 35:1004–1010.

Wilding, L. P., and L. R. Drees, 1973. Scanning electron microscopy of opaque opaline forms isolated from forest soils in Ohio. *Soil Science Society of America Proceedings* 37:647–650.

Wilding, L. P., and L. R. Drees, 1974. Contributions of forest opal and associated crystalline phases to fine silt and clay fractions of soils. *Clays and Clay Mineralogy* 22:295–306.

Wilding, L. P., R. E. Brown, and N. Holowaychuk, 1967. Accessibility and properties of occluded carbon in biogenetic opal. *Soil Science* 103:56–61.

Wilding, L. P., N. E. Smeck, and L. R. Drees, 1977. Silica in soils: Quartz, cristobalite, tridymite and opal. In *Minerals in Soil Environments,* pp. 471–552. Madison: Soil Science Society of America.

Wilkes, H. G., 1967. *Teosinte: The Closest Relative of Maize.* Bussey Institute, Cambridge, Massachusetts: Harvard University.

Wilson, S., 1982. Phytolith evidence from Kuk, An early agricultural site in Papua New Guinea Highlands. M. A. thesis, Camberra: Australian National University.

Wiseman, F., 1983. Analysis of pollen from the fields at Pulltrouser Swamp. In *Pulltrouser Swamp* (B. Turner and P. Harrison, eds.), pp. 105–119. Austin: University of Texas Press.

Witty, J. E., and E. G. Knox, 1964. Grass opal in some chestnut and forested soils in north central Oregon. *Soil Science Society of American Proceedings* 28:685–688.

Wright, H., 1967. The use of surface samples in quaternary pollen analysis. *Review of Paleobotany and Palynology* 2:321–330.

Zevallos Menendez, C. M., W. C. Galinat, D. Lathrap, E. R. Leng, J. Marcos, and K. M. Klumpp, 1977. The San Pablo corn kernel and its friends. *Science* 196:385–389.

Index